Introduction to Graphics Communications for Engineers

Welcome to the BEST!

McGraw-Hill's *BEST*—**B**asic **E**ngineering **S**eries and **T**ools—consists of modularized textbooks and applications appropriate for the topic covered in mostly introductory engineering courses. The goal of the series is to provide the educational community with material that is timely, affordable, of high quality, and flexible in how it is used.

McGraw-Hill's BEST!

A Beginner's Guide to Technical Communication, by A. Eisenberg, 0-07-092045-1

C Programming for Engineering and Computer Science, by A. Tan and T. D'Orazio, 0-07-913678-8

Introduction to Engineering Design, by A. Eide, et al., 0-07-018922-6

Introduction to Engineering Design and Problem Solving, by M. D. Burghardt, 0-07-012188-5

Introduction to Engineering Ethics, by R. Schinzinger and M. W. Martin, 0-07-233959-4

Introduction to Engineering Problem Solving, by A. Eide, et al., 0-07-021983-4

Introduction to Fortran 90/95, by S. Chapman, 0-07-011969-4

Introduction to Graphics Communications for Engineers, by G. Bertoline, 0-07-229144-3

Introduction to Matlab® for Engineers, by W. Palm, 0-07-047328-5

Introduction to the Internet for Engineers, by R. Greenlaw and E. Hepp, 0-07-229143-5

Mathcad: A Tool for Engineering Problem Solving, by P. Pritchard, 0-07-012189-3

Pocket Book of Technical Writing for Engineers and Scientists, by Leo Finkelstein, Jr., 0-07-237080-7

Project Management and Teamwork, by K. Smith, 0-07-012296-2

Spreadsheet Tools for Engineers: Excel '97 Version, *by* B. Gottfried, 0-07-024654-8

Spreadsheet Tools for Engineers: Excel 2000, *by* B. Gottfried, 0-07-232166-0

The Engineering Student Survival Guide, by K. Donaldson, 0-07-228647-4

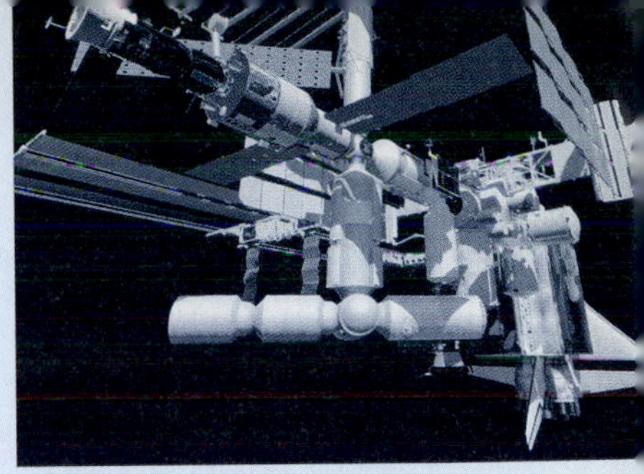

Introduction to Graphics Communications for Engineers

Third Edition

Gary R. Bertoline
Purdue University

McGraw Hill Higher Education

Boston Burr Ridge, IL Dubuque, IA Madison, WI New York San Francisco St. Louis
Bangkok Bogotá Caracas Kuala Lumpur Lisbon London Madrid Mexico City
Milan Montreal New Delhi Santiago Seoul Singapore Sydney Taipei Toronto

The McGraw·Hill Companies

Higher Education

INTRODUCTION TO GRAPHICS COMMUNICATIONS FOR ENGINEERS, THIRD EDITION

Published by McGraw-Hill, a business unit of The McGraw-Hill Companies, Inc., 1221 Avenue of the Americas, New York, NY 10020.

Some ancillaries, including electronic and print components, may not be available to customers outside the United States.

This book is printed on recycled, acid-free paper containing 10% postconsumer waste.

1 2 3 4 5 6 7 8 9 0 QPD/QPD 0 9 8 7 6 5 4

ISBN 0-07-295084-6

Managing Developmental Editor: *Emily J. Lupash*
Marketing Manager: *Dawn R. Bercier*
Senior Project Manager: *Sheila M. Frank*
Senior Production Supervisor: *Kara Kudronowicz*
Media Technology Producer: *Eric A. Weber*
Senior Designer: *David W. Hash*
(USE) Cover Images: *©PhotoDisc, Background Series Vol. 28; ©PhotoDisc, Business Today Vol. 35*
Senior Photo Research Coordinator: *John C. Leland*
Compositor: *International Typesetting and Composition*
Typeface: *10/12 Times Roman*
Printer: *Quebecor World Dubuque, IA*

Photo Credits
Introduction and Chapter Openers: © Photodisc/Vol. 85/Modern Technologies 2.
Any photos not credited on page are courtesy of the author.

Library of Congress Cataloging-in-Publication Data

Bertoline, Gary R.
Introduction to graphics communications for engineers / Gary R. Bertoline.—3rd ed.
p. cm.—(McGraw-Hill's BEST—basic engineering series and tools)
Includes bibliographical references and index.
ISBN 0-07-295084-6 (alk. paper)
1. Engineering graphics. I. Title. II. Series.

T353.B452 2006
604.2—dc22 2004023128

www.mhhe.com

For Ada, Bryan, Kevin, and Carolyn, who are my motivation and inspiration for all my books.

About the Author

Gary R. Bertoline

Gary R. Bertoline is Professor of Computer Graphics Technology at Purdue University and director of the Envision Center for Data Perceptualization. He earned his B.S. degree in Industrial Technology at Northern Michigan University in 1974, M.Ed. in Industrial Technology at Miami University in 1979, and Ph.D. at The Ohio State University in Industrial Technology in 1987. His graduate work focused on the integration of CAD into engineering graphics and visualization. He has 25 years' experience teaching graphics at all levels from elementary school to senior citizens. Prof. Bertoline taught junior high and high school graphics at St. Henry High School, St. Henry, Ohio; drafting/ design technology at Wright State University, Lake Campus, Celina, Ohio; and engineering graphics at The Ohio State University, Columbus, Ohio.

Prof. Bertoline has authored numerous publications, authored or coauthored 10 textbooks and workbooks, and made over 50 presentations throughout the world. He has won the Frank Oppenheimer Award three times for best paper at the Engineering Design Graphics Division Mid-year Meeting. He has developed many graphics courses, including CAD, solid modeling, multimedia, and virtual reality, and has integrated many modern topics into traditional engineering graphics courses, such as modeling, animation, and visualization. Prof. Bertoline has conducted research in cognitive visualization and was the co-author for a curriculum study in engineering graphics funded by SIGGRAPH. He is on the editorial board for the *Journal for Geometry and Graphics* and is the McGraw-Hill Graphics Series Editor. He was the recipient of the Orthogonal Medal for outstanding contributions to the advancement of Graphic Science by North Carolina State University in 1992, and the 1995 inaugural recipient of the Steve M. Slaby International Award for Outstanding Contributions in Graphics Education. You can contact Dr. Bertoline at bertoline@purdue.edu.

Contents

Preface

Introduction to Graphics Communications for Engineers, 3rd Edition, is part of the McGraw-Hill's BEST (Basic Engineering Series and Tools), which introduces engineering students to various topics important to their education. This workbook is an introduction to the standard practices used by engineers to communicate graphically. The primary goal of this text is to assist engineering students in learning the techniques and standards of communicating graphically so that design ideas can be clearly communicated and produced.

The text concentrates on the concepts and skills needed to sketch and create 2-D drawings and 3-D CAD models. Engineering graphics has gone through significant changes in the last decade as a result of the use of computers and CAD software. It seems as if some new hardware or software development has an impact on engineering graphics every year. Although these changes are important to the subject of technical graphics, there is much about the subject that has not changed. Engineers still find it necessary to communicate and interpret design ideas through the use of graphics methods such as sketches and CAD drawings and models. As powerful as today's computers and CAD software have become, they are of little use to engineers who do not fully understand fundamental graphics communications principles and 3-D modeling strategies, or who lack high-level visualization skills.

The workbook is divided into six chapters with multiple units of instruction. Chapter 1, "Introduction to Graphics Communications," is an introduction to graphics communications as a language for engineers and describes the tools used and some of the techniques for communicating graphically. Chapter 2, "Sketching and Text," is an introduction to sketching technique, projection theory, visualization, and the use of text on drawings. Chapter 3, "Section and Auxiliary Views," introduces the student to the use of and technique for creating sectioned drawings and models and auxiliary views. Chapter 4, "Dimensioning and Tolerancing Practices," describes how to create and read dimensional drawings. Chapter 5, "Reading and Constructing Working Drawings," describes how to read and produce working drawings. Finally, Chapter 6, "Design and 3-D Modeling," is an overview of 3-D modeling techniques and the engineering design process.

Each chapter has a list of objectives, step-by-step instructions, and a wide assortment of problems that can be assigned to reinforce the topics covered. Some chapters have a list of further readings, so that students can find more information related to the topics covered in the chapter. Sketching worksheets have been integrated into the end of each chapter. These worksheets can be used for sketching assignments to augment assignments using CAD. After completing the workbook, the student will be able to create design sketches using various projection techniques, create and read 2-D standard engineering drawings, and create and visualize 3-D computer models.

Special thanks to James Mohler and Amy Fleck for their work on the illustrations and Jim Leach for some of the drawing problems added in the second edition.

Gary R. Bertoline, PhD
Professor Computer Graphics Technology
Purdue University
West Lafayette, IN

Chapter

Introduction to Graphics Communications 1

OBJECTIVES

After completing this chapter, you will be able to:

1. Describe why technical drawings are an effective communications system for technical ideas about designs and products.
2. Identify important parts of a CAD system.
3. Identify important traditional tools.
4. Identify standard metric and U.S. drawing sheet sizes.
5. Identify the types and thicknesses of the various lines in the alphabet of lines.

1.1 INTRODUCTION

Graphics communications using engineering drawings and models is a language—a clear, precise language—with definite rules that must be mastered if you are to be successful in engineering design. Once you know the language of graphics communications, it will influence the way you think, the way you approach problems. Why? Because humans tend to think using the languages they know. Thinking in the language of technical graphics, you will visualize problems more clearly and will use graphic images to find solutions with greater ease.

In engineering, 92 percent of the design process is graphically based. The other 8 percent is divided between mathematics and written and verbal communications. Why? Because graphics serves as the primary means of communication for the design process. Figure 1.1 shows a breakdown of how engineers spend their time. 3-D modeling and documentation, along with design modeling, comprise more than 50 percent of the engineer's time and are purely visual and graphical activities. Engineering analysis depends largely on reading technical graphics, and manufacturing engineering and functional design also require the production and reading of graphics.

Why do graphics come into every phase of the engineer's job? To illustrate, look at the jet aircraft in Figure 1.2. Like

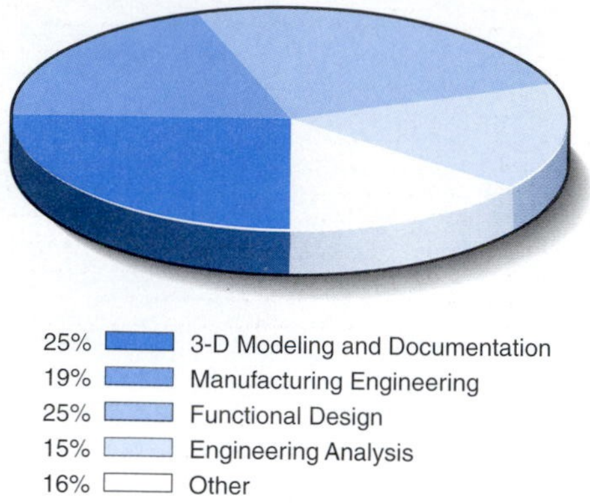

Figure 1.1 A Total View of Engineering Divided into Its Major Activities

Graphics plays a very important role in all areas of engineering; for documentation, communications, design, analysis, and modeling. Each of the activities listed is so heavily slanted toward graphics communications that engineering is 92 percent graphically based. Information from Dataquest, Inc. CAD/CAM/CAE/GIS Industry Service.

Figure 1.2

This jet aircraft would be impossible to create without computer graphics models and drawings. Drawings are the road maps that show how to manufacture or build products and structures.
(© The Boeing Company.)

any new product, it was designed for a specific task and within specified parameters; however, before it could be manufactured, a 3-D model and engineering drawings like that shown in Figure 1.3 had to be produced. Just imagine trying to communicate all the necessary details verbally or in writing. It would be impossible!

A designer has to think about the many features of an object that cannot be communicated with verbal descriptions (Figure 1.4). These thoughts are dealt with in the mind of the designer using a visual, nonverbal process. This "visual image in the mind" can be reviewed and modified to test different solutions before it is ever communicated to someone else. As the designer draws a line on paper or creates a solid cylinder image with a computer, he or she is translating the mental picture into a drawing or model that will produce a similar picture in the mind of anyone who sees the drawing. This drawing or graphic representation is the medium through which visual images in the mind of the designer are converted into the real object.

Technical graphics can also communicate solutions to technical problems. Such technical graphics are produced according to certain standards and conventions so they can be read and accurately interpreted by anyone who has learned those standards and conventions.

The precision of technical graphics is aided by tools; some are thousands of years old and still in use today, and others are as new and rapidly changing as computer-aided design/drafting (CAD). This book will introduce you to the standards, conventions, techniques, and tools of technical graphics and will help you develop your technical skills so that your design ideas become a reality.

Engineers are creative people who use technical means to solve problems. They design products, systems, devices, and structures to improve our living conditions. Although problem solutions begin with thoughts or images in the mind of the designer, presentation devices and computer graphics hardware and software are powerful tools for communicating those images to others. They can also aid the visualization process in the mind of the designer. As computer graphics have a greater impact in the field of engineering, engineers will need an ever-growing understanding of and facility in graphics communications.

Practice Exercise 1.1

1. Try to describe the part shown in Figure 1.15 using written instructions. The instructions must be of such detail that another person can make a sketch of the part.
2. Now try verbally describing the part to another person. Have the person make a sketch from your instructions.

These two examples will help you appreciate the difficulty in trying to use written or verbal means to describe even simple mechanical parts. Refer to Figure 1.3 and others in this text to get an idea of how complicated some parts are compared with this example. It is also important to note that air and water craft have thousands of parts. For example, the nuclear powered *Sea Wolf* class submarine has more than two million parts. Try using verbal or written instructions to describe that!

GROUP	QUAN-TITY	PART NUMBER		PIECE OF GROUP	NAME OF PART	MATERIAL	SYM-BOL
		DRAWING NO.					○
	1	126257		20	U - BOLT	✓	A
	1	2 3Y104		K	FRAME	BABBITED	B
	2				HEX NUT	✓	C
A	2				LOCK WASHER	✓	D
	1				PIPE NIPPLE	✓	E
	1	1041Y33			FRAME		F
	1				PIPE COUPLING	✓	G

342
300
.25—18 NPT
⌀12
44
26
24
152
38
50

R	W	R	L	R	N	R	C			
DIMENSION TOLERANCES EXCEPT AS SPECIFIED		R		J				◇	BRONZE CAP NOTE ADDED	E.F.C.
		R		D				◇	PART No. 283Y112-C ADDED	R.C.
TITLE No. 198 HANGAR ASSEMBLY		R		T				◇	GROUP B, NOTES & DIMENSIONS FOR GROUP C, REMOVED - FRAME WAS	
		I		M				◇	1041 Y 33 - B FOR GROUP C ONLY. C.W.	
		O		R				REV	DATE DESCRIPTION OF REVISION	
DRAWING	CHECKED	APPROVED		FORM	REFERENCE					
DATE										
SCALE	LINK - BELT COMPANY							**162Y259**		

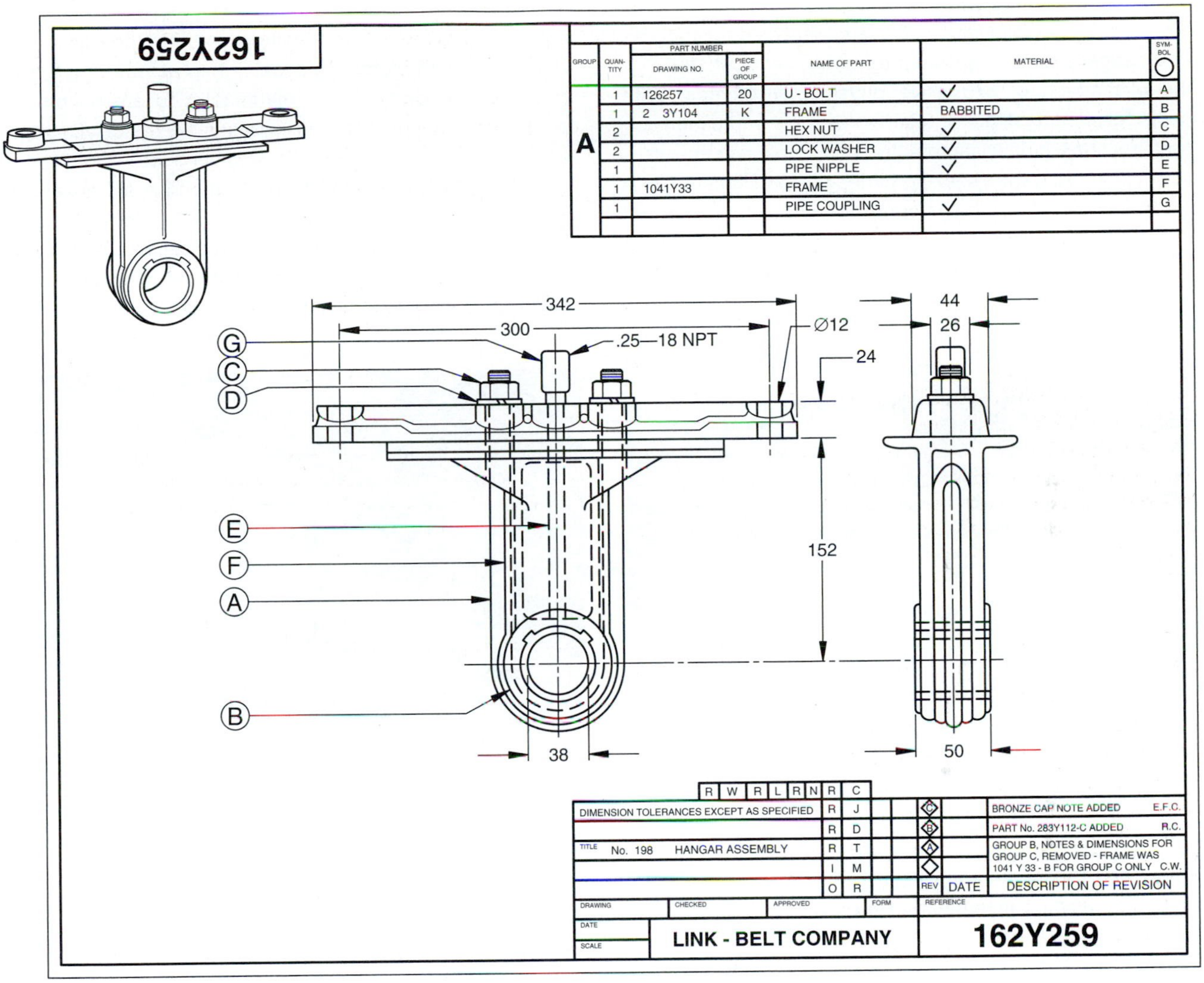

Figure 1.3 **Engineering Drawing**

Engineering drawings and computer models such as these were needed to produce the hanger assembly shown. The 3-D model is used to design and visualize the hanger. The engineering drawings are used to communicate and document the design process.

1.2 | TECHNICAL DRAWING TOOLS

Just as the graphics language has evolved over the years into a sophisticated set of standards and conventions, so have the tools used to graphically communicate technical ideas. Tools are used to produce three basic types of drawings: freehand sketches, instrument drawings, and computer drawings and models. The tools have evolved from pencils, triangles, scales, and compasses to **computer-aided**

design/drafting (CAD) systems. **CAD** is computer software and related computer hardware that supplements or replaces traditional hand tools for creating models and technical drawings (Figure 1.5).

Since many industries have not fully integrated CAD into their design offices, it is necessary to learn both traditional and computer design methods. Also, traditional tools are used for sketching, which is one of the most effective methods available to represent design ideas quickly.

Design in Industry

Going Virtual

Virtual reality and simulation software tools hold the promise of drastically slashing product development costs through the elimination of expensive physical prototypes. With costs for the latest virtual reality (VR) tools and simulation systems coming down, automotive and aerospace manufacturers increasingly are seeking to deploy sophisticated, collaborative visualization systems throughout their product development planning organizations, as well as using virtual simulations for designing overall plant layouts and within manufacturing cells.

Although VR tools historically have been the domain of researchers, commercial applications in automotive, aerospace, and medical device manufacturing are becoming much more common. Using VR systems like the CAVE (Computer Automated Visualization Environment), developed in the early 1990s by the Electronic Visualization Laboratory at the University of Illinois at Chicago (EVL, UIC), automakers and aircraft manufacturers can review realistic virtual model prototypes, avoiding the expense of $200,000 for a fiberglass auto prototype to upwards of $3 million for an aircraft prototype.

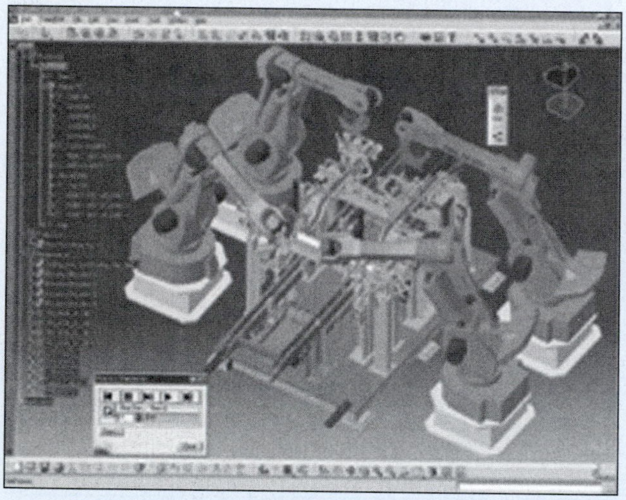

Factory Floor Layout and Cell Designs Like This GM Powertrain Engine Assembly Line Can Be Generated with Delmia's Factory Simulation Software.

(Image courtesy of DELMIA Corp. All rights reserved.)

With Fakespace Systems' WorkWall, Teams Can View Realistic Stereoscopic Images During Product Development Team Design Reviews.

(Image courtesy of Fakespace Systems Inc.)

1.3 COMPUTER-AIDED DRAWING TOOLS

Traditional tools will continue to be useful for sketching and rough layout work; however, good CAD software can create virtually any type of technical drawing. Circle commands replace the compass, line commands replace the T-square and triangles, and editing commands replace the dividers and erasing shield.

A CAD system consists of hardware devices used in combination with specific software. The **hardware** for a CAD system consists of the physical devices used to support the CAD software. There are many different hardware manufacturers and types of hardware devices, all of which are used to create, store, or output technical drawings and models. It is not uncommon in industry to have multiple input, storage, and output devices for a CAD system.

Over the past few years, the addition of more realistic visualization software has also furthered VR's acceptance, with efforts like the partnership between software developer Engineering Animation Inc. (Ames, IA), workstation supplier Silicon Graphics Inc. (Mountain View, CA), and General Motors Corp. (Detroit) offering EAI's VisConcept, a software suite providing a true 1:1, or human-scale, immersive visualization environment. In addition, projection and display technologies have improved to the point where it's possible to easily create high-resolution stereoscopic images—seeing an image in each eye with depth and volume just as in the real world.

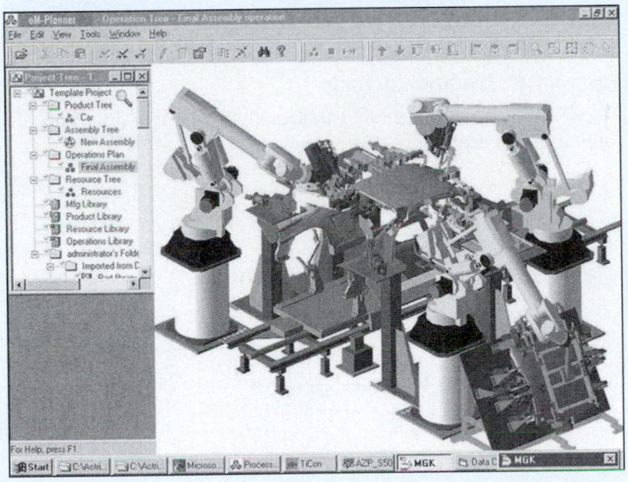

Simulation of a Manufacturing Cell.

(Image courtesy of Tecnomatix Technologies Ltd.)

Collaborative visualization may represent a new opportunity to manufacturers, particularly in the automotive industry where many major auto manufacturers are trying to persuade their top suppliers to adopt visualization technology. Large-scale displays like the WorkWall enable manufacturing teams to collaborate in much the same way they used to work around drafting tables, but with realistic, full-scale 3-D models.

This article (or exerpts thereof) appears with permission from *Manufacturing Engineering*, the official publication of the Society of Manufacturing Engineers (SME).

Users of the Fakespace Wall Can Review Styling and Component Changes on Virtual Models before Committing to Final Product Designs.

(Image courtesy of Fakespace Systems, Inc.)

1.4 TRADITIONAL TOOLS

The traditional tools used to create technical drawings have evolved over time. Many tools were originally used in ancient Greece to study and develop geometry. Although computers may someday replace the need for some traditional tools, they are still useful today for drawing, and more importantly, for sketching. **Traditional tools** are devices used to assist the human hand in making technical drawings. The assistance includes drawing lines straighter, making circles more circular, and increasing the speed with which drawings are made. The tools typically used to create mechanical drawings or sketches (Figure 1.6) consist of the following:

1. Wood and mechanical pencils
2. Instrument set, including compass and dividers
3. 45- and 30/60-degree triangles

Figure 1.4 **Engineering Drawings Used for Communications**

Engineering drawings are a nonverbal method of communicating information. Descriptions of complex products or structures must be communicated with drawings. A designer uses a visual, nonverbal process. A visual image is formed in the mind, reviewed, modified, and is ultimately communicated to someone else, all using visual and graphics processes.

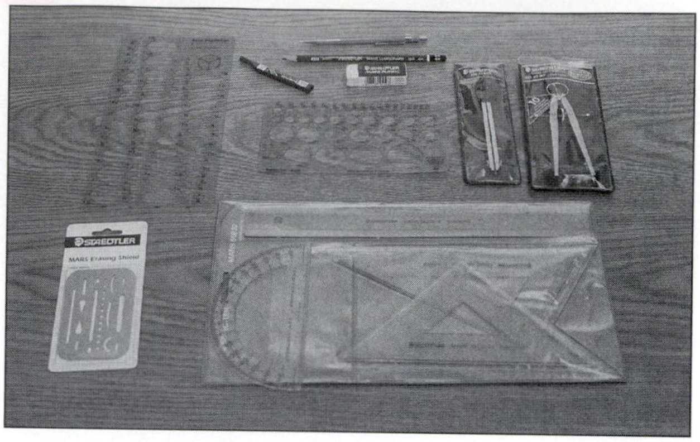

Figure 1.6 **Traditional Tools**

These are some of the many traditional mechanical drawing tools used for engineering drawings.

4. Scales
5. Irregular curves
6. Protractors
7. Erasers and erasing shields
8. Drawing paper
9. Circle templates
10. Isometric templates

Figure 1.5 **CAD Workstations**

Typical CAD workstations used in industry have large color monitors.

1.5 MEDIA

Media are the surfaces upon which an engineer or technologist communicates graphical information. The media used for technical drawings are different types or grades of paper, such as tracing paper, vellum, and polyester film. Tracing paper is a thin, translucent paper used for detail drawings. Vellum is a tracing paper chemically treated to improve translucency. Polyester film, or its trade name Mylar, is transparent, waterproof, and difficult to tear. Mylar can be used for lead pencil, plastic-lead pencil, or ink drawings. Mylar is an excellent drawing surface that leaves no trace of erasure.

Special papers have also been developed for CAD plotters. For example, plotter paper used for fiber-tipped pens has a smooth or glossy surface to enhance line definition and minimize skipping. Often, the paper comes with a preprinted border, title block, and parts list (Figure 1.7).

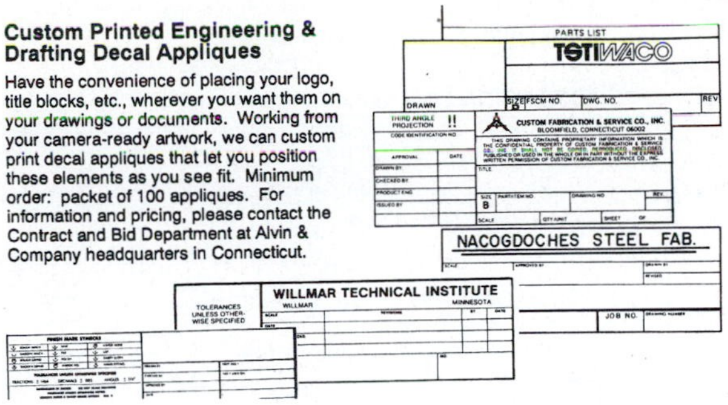

Figure 1.7 **Preprinted Title Blocks**

Preprinted standard borders and title blocks on drafting paper are commonly used in industry. Courtesy of Alvin & Company.

Table 1.1 **ANSI Standard Sheet Sizes**

Metric (mm)	U.S. Standard	Architectural
A4 210 × 297	A-Size 8.5″ × 11″	9″ × 12″
A3 297 × 420	B-Size 11″ × 17″	12″ × 18″
A2 420 × 594	C-Size 17″ × 22″	18″ × 24″
A1 594 × 841	D-Size 22″ × 34″	24″ × 36″
A0 841 × 1189	E-Size 34″ × 44″	36″ × 48″

The American National Standards Institute (ANSI) has established standard sheet sizes and title blocks for the media used for technical drawings. Each paper size is designated by a letter, as shown in Table 1.1, and title block sizes are shown in Figure 1.16 at the end of the chapter.

1.6 ALPHABET OF LINES

The **alphabet of lines** is a set of standard linetypes established by the American Society of Mechanical Engineers (ASME) for technical drawing. Figure 1.8 shows the alphabet of lines and the approximate dimensions used to create different linetypes, which are referred to as **linestyles** when used with CAD. ASME Y14.2M-1992 has established these linetypes as the standard for technical drawings. Two line weights are sufficient to follow the standards, a 0.6 mm and a 0.3 mm. These approximate widths are intended to differentiate between thin and thick lines and are not for control of acceptance or rejection of drawings. Thick lines are drawn using soft lead, such as F or HB. Thin lines are drawn using a harder lead, such as H or 2H. Construction lines are very light and are drawn using 4H or 6H lead. A good rule of thumb for creating construction lines is to draw them so that they are difficult to see if your drawing is held at arm's length.

Following are the standard linetypes and their applications in technical drawings:

Center lines are used to represent symmetry and paths of motion and to mark the centers of circles and the axes of symmetrical parts, such as cylinders and bolts.

Break lines come in two forms: a freehand thick line and a long, ruled thin line with zigzags. Break lines are used to show where an object is broken to save drawing space or reveal interior features.

Dimension and extension lines are used to indicate the sizes of features on a drawing.

Section lines are used in section views to represent surfaces of an object cut by a cutting plane.

Cutting plane lines are used in section drawings to show the locations of cutting planes.

Visible lines are used to represent features that can be seen in the current view.

Hidden lines are used to represent features that cannot be seen in the current view.

Phantom lines are used to represent a movable feature in its different positions.

Stitch lines are used to indicate a sewing or stitching process.

Chain lines are used to indicate that a surface is to receive additional treatment.

Symmetry lines are used as an axis of symmetry for a particular view.

It is important that you understand and remember these different linetypes and their definitions and uses, because they are referred to routinely throughout the rest of this book.

CAD software provides different linestyles for creating standard technical drawings. Figure 1.9 shows the linestyle menu for a typical CAD system. The thicknesses of lines on a CAD drawing are controlled by two different means: (1) controlling the thickness of the lines drawn on the display screen and (2) controlling the plotted output of lines on pen plotters by using different

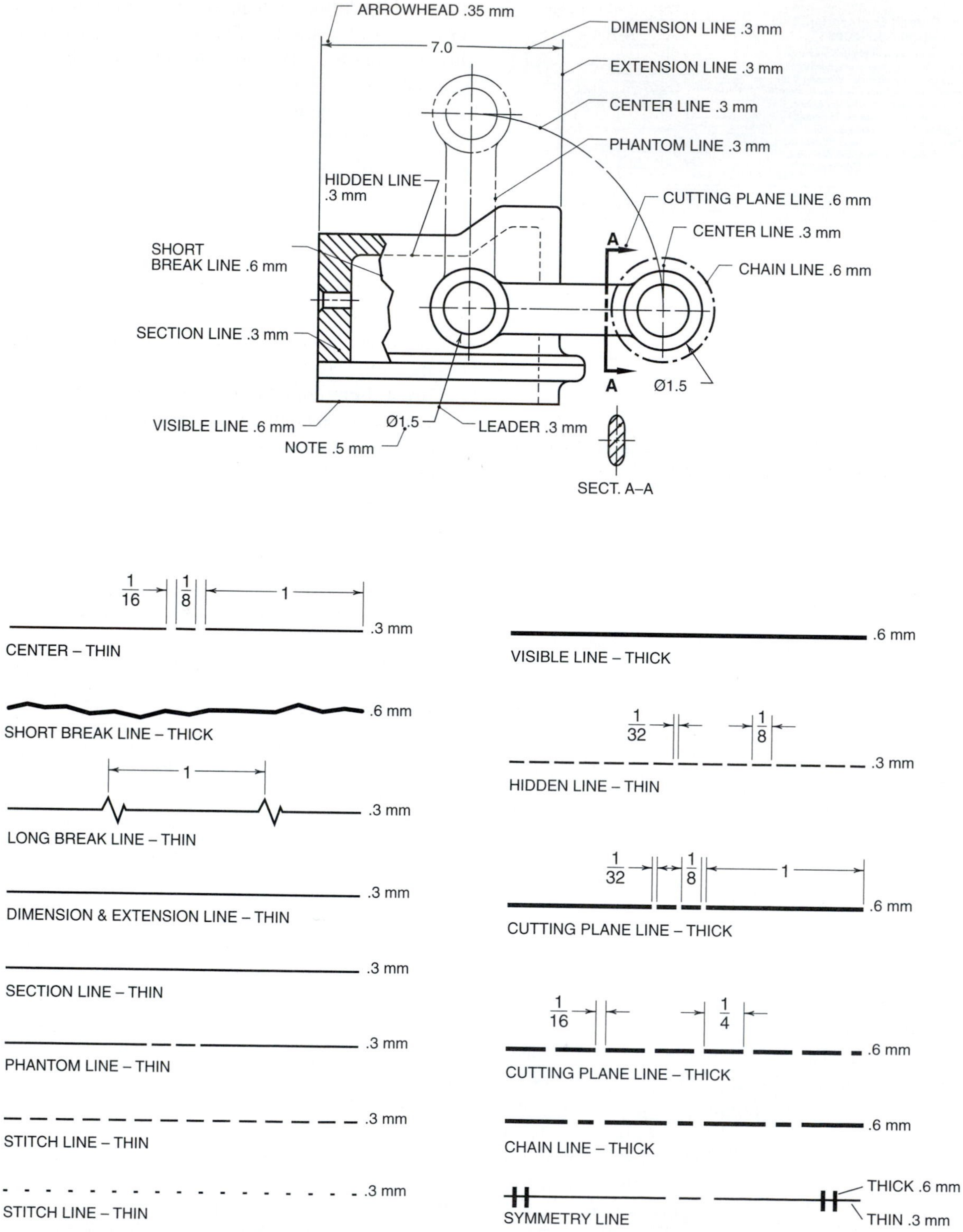

Figure 1.8 **The Alphabet of Lines**

The alphabet of lines is a set of ASME standard linetypes used on technical drawings. The approximate dimensions shown on some linetypes are used as guides for drawing them with traditional tools. The technical drawing at the top shows how different linetypes are used in a drawing.

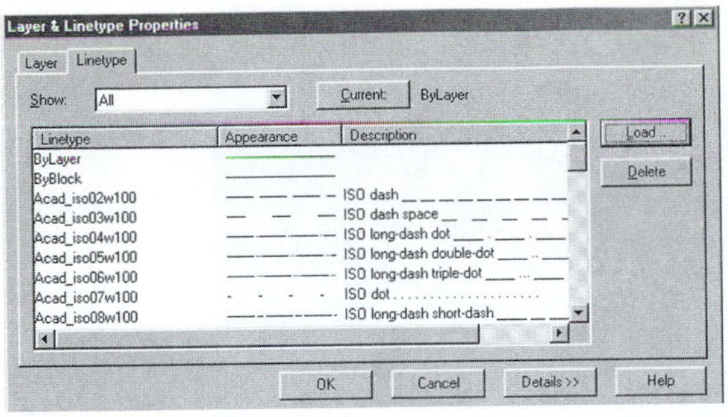

Figure 1.9 AutoCAD's Linestyle Menu Showing Some of the Linetypes Available

pen numbers for different linestyles, where different pen numbers have different thicknesses, such as a 0.7 mm and 0.3 mm.

1.7 WHAT YOU WILL LEARN

In this text, you will learn the six important areas in technical graphics:

Visualization—the ability to mentally control visual information.
Graphics theory—geometry and projection techniques.
Standards—sets of rules that govern how parts are made and technical drawings are represented.
Conventions—commonly accepted practices and methods used for technical drawings.
Tools—devices used to create engineering drawings and models, including both handheld and computer tools.
Applications—the various uses for technical graphics in engineering design, such as mechanical, electrical, and architectural.

Each chapter in the text will explain the graphics theory important for a topic, integrate the visualization practices, explain the relevant standards and conventions, demonstrate the tools used to create drawings, and apply the topic to engineering design.

Learning to communicate with drawings is very similar to learning to write or speak in a language. For someone inexperienced in technical drawing, the learning process is very similar to learning a new language. There is a set of rules that must be learned in order to communicate graphically or when learning a new language. You will soon find out that graphics is a very effective method of supporting the design process.

1.8 FUTURE TRENDS

The convergence of technology, knowledge, and computer hardware and software is resulting in a number of nontraditional processes that can be used in the engineering design process. These new processes and technologies can extend the circle of people in an organization who are involved in design. Many of these future trends are beginning to combine the design and manufacturing processes into a whole. A few trends even take a more global approach of attempting to control the entire enterprise.

1.8.1 Visualization Tools

The sharing of design ideas has always been important for the engineer. Today, however, the importance of sharing design ideas with others is even more important. One technique that is becoming popular is the sharing of design ideas through various computer graphics techniques. The following is a list in order of realism and interactivity that can be produced with computer graphics tools.

- High-resolution rendered images are a static means of showing initial design ideas (Figure 1.10).
- Computer animations or simulation can also be very effectively used to share design ideas with others.
- 3-D stereo graphics can be used to enhance the viewing of static and animated 3-D computer images on the computer screen (Figure 1.11).
- Holographic and volumetric displays of 3-D computer images are being developed that allow designers to literally walk around and through the design as it is being created (Figure 1.12).
- Virtual reality tools can be employed to get an even higher level of realism by immersing the user in a 3-D world (Figure 1.13).
- Rapid prototyping systems are used to create real prototype models directly from CAD models (Figure 1.14).

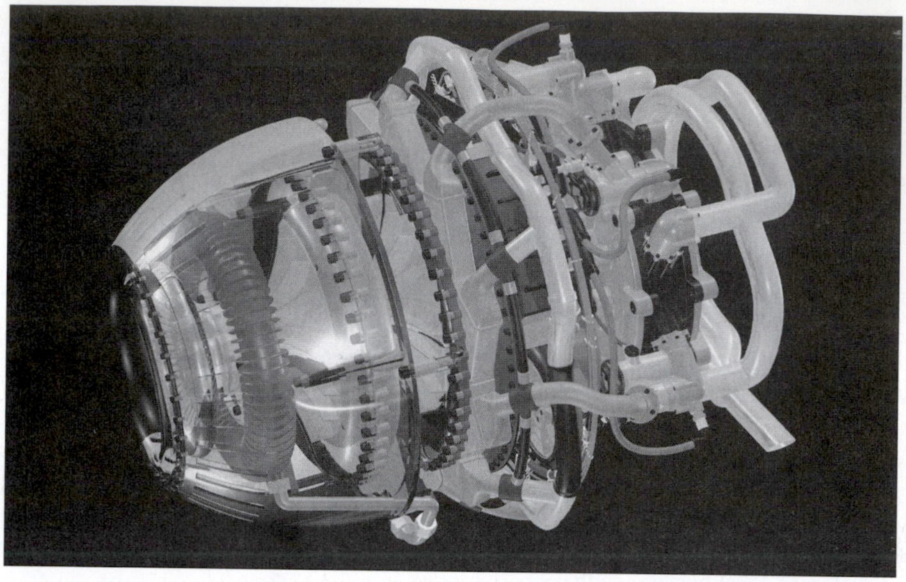

Figure 1.10 High-Resolution Rendered Image of a CAD Model
(Courtesy of Simon Floyd Design Group.)

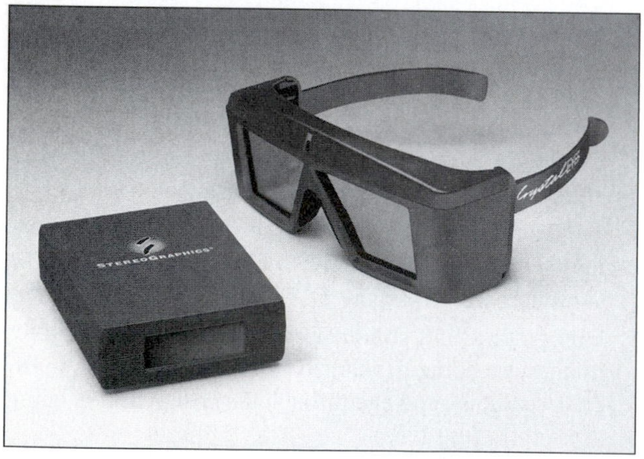

Figure 1.11 3-D Stereo Glasses Used to Enhance the Viewing of CAD Models
(Courtesy of StereoGraphics Corporation.)

Figure 1.12 Volumetric Display Device
(© Lou Jones.)

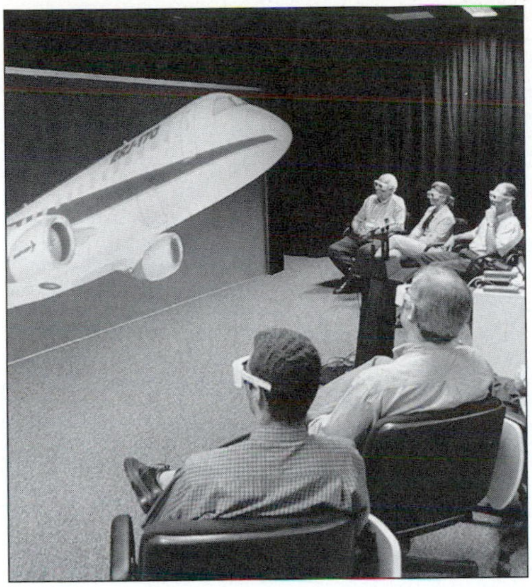

Figure 1.13 Stereoscopic Imagery Displayed on a Large Display

(Courtesy of Fakespace Systems, Inc.)

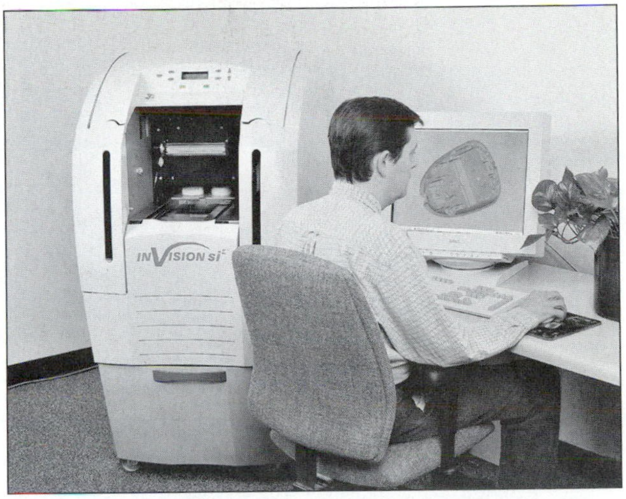

Figure 1.14 Rapid Prototyping System

(Courtesy of 3D Systems.)

Problems

Use the worksheets provided at the end of this section to complete the problems.

1.1 Research and report on an important historical figure in engineering design, such as Henry Ford, Thomas Edison, the Wright brothers, or Alexander Graham Bell.

1.2 Identify at least five other individuals who worked as engineers and had an impact on society.

1.3 Research and report on an important historical engineering achievement, such as airplanes, space flight, computers, or television.

1.4 Identify three new products that have appeared on the market in the last five years.

1.5 Research and report on an important historical figure in graphics, such as Gaspard Monge, M. C. Escher, Thomas Edison, Leonardo da Vinci, Albrecht Durer, or Frank Lloyd Wright.

1.6 To demonstrate the effectiveness of graphics communications, write a description of the object shown in Figure 1.15. Test your written description by having someone attempt to make a sketch from your description.

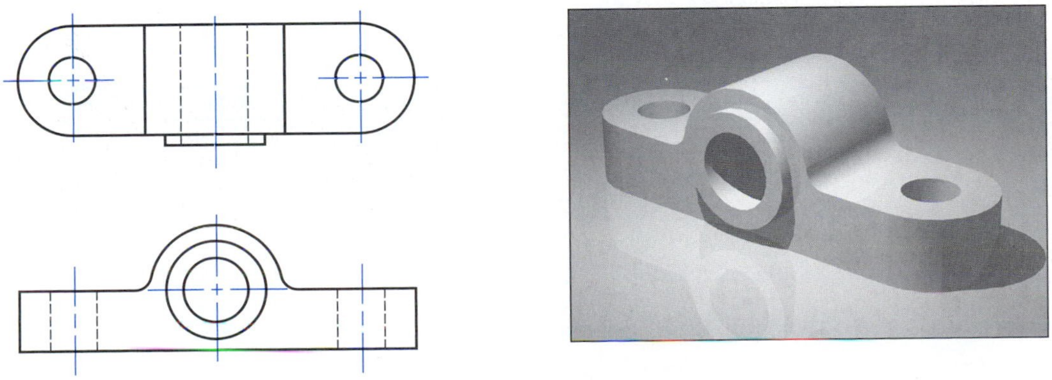

Figure 1.15 Problem 1.6 Bearing Block to Be Described Verbally

Title Block for A, B, C.

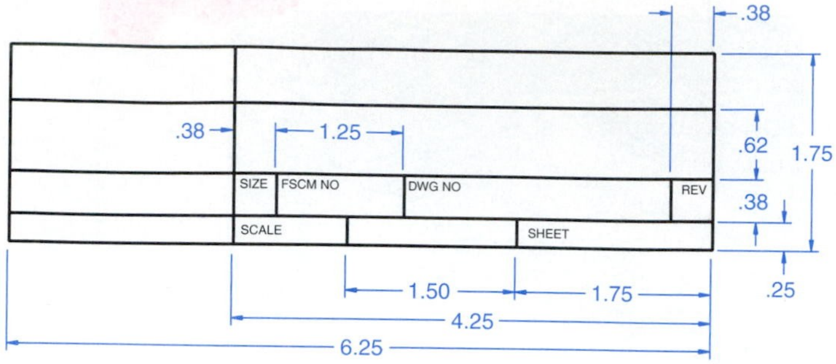

Title Block for D and E.

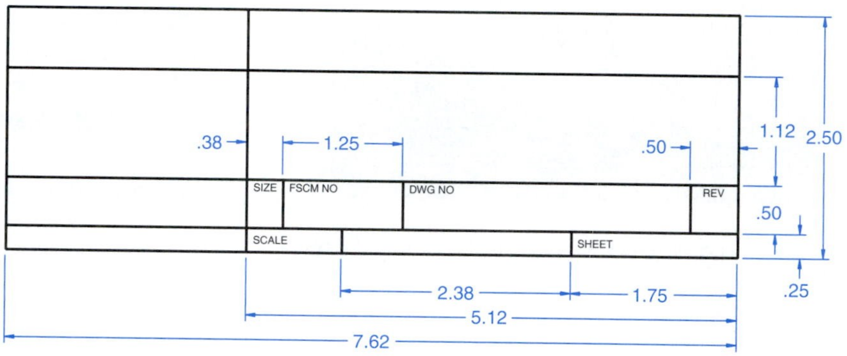

Continuation Sheet Title Block for A, B, C.

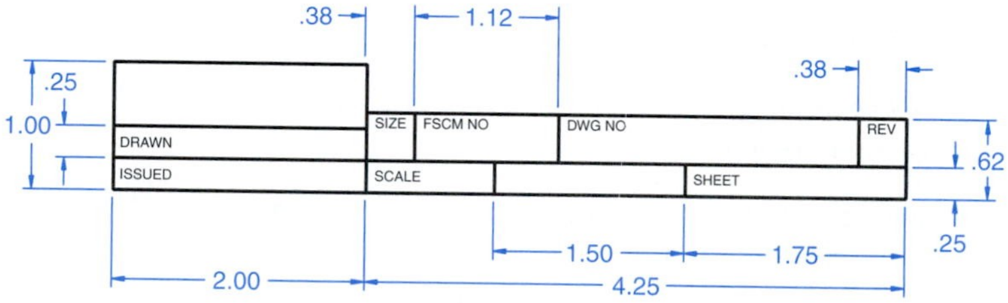

Continuation Sheet Title Block for D and E.

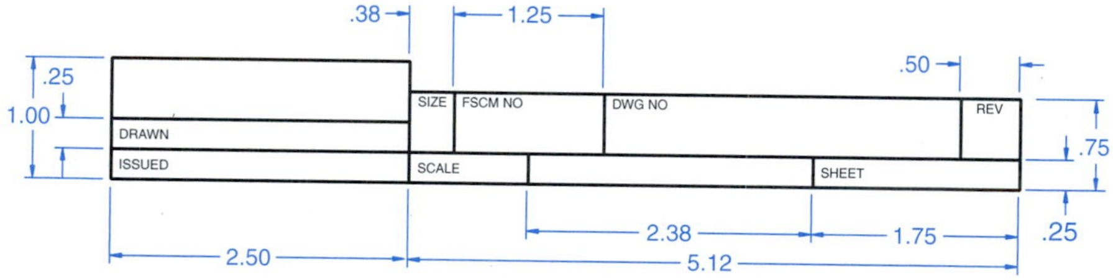

Size Designation	Width (Vertical)	Length (Horizontal)	Margin		International Designation	Width		Length	
			Vertical	Horizontal		mm	in.	mm	in.
A (Horiz)	8.5	11.0	0.38	0.25	A4	210	8.27	297	11.69
A (Vert)	11.0	8.5	0.25	0.38	—	—	—	—	—
B	11.0	17.0	0.38	0.62	A3	297	11.69	420	16.54
C	17.0	22.0	0.75	0.50	A2	420	16.54	594	23.39
D	22.0	34.0	0.50	1.00	A1	594	23.39	841	33.11
E	34.0	44.0	1.00	0.50	A0	841	33.11	1189	46.11

Figure 1.16 **Problem 1.12 ANSI Standard Title Blocks and Border Lines**

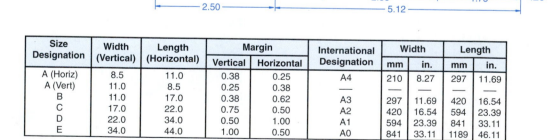

1.7 Make a sketch of a common device, such as a telephone, automobile, computer mouse, or coffee cup.

1.8 Get a clear mental picture of a television, then sketch what you see in your mind. Is this mental image 2-D or 3-D? Try to put words to each feature of the TV you are drawing. In this problem you will experience the difficulty in trying to verbally describe an object with enough detail for it to be manufactured.

1.9 Interview a practicing engineer or technologist and ask how graphics are used in his or her daily work.

1.10 Ask the practicing engineer or technologist what changes are taking place in his or her profession.

1.11 Research and report on an important historical figure in computer graphics, such as Ivan Sutherland, Steve Coons, R. E. Bezier, or George Lucas.

1.12 Draw the border lines and title blocks for the ANSI and ISO drawing sheets, using the dimensions shown. Add text as shown, using ⅛″ (3 mm) all-caps text (Figure 1.16).

1.13 See Figure 1.17. Using a scale of ⅛″=1′–0″, draw the truss shown in the figure. The rise (R) is one-fourth the span of the truss.

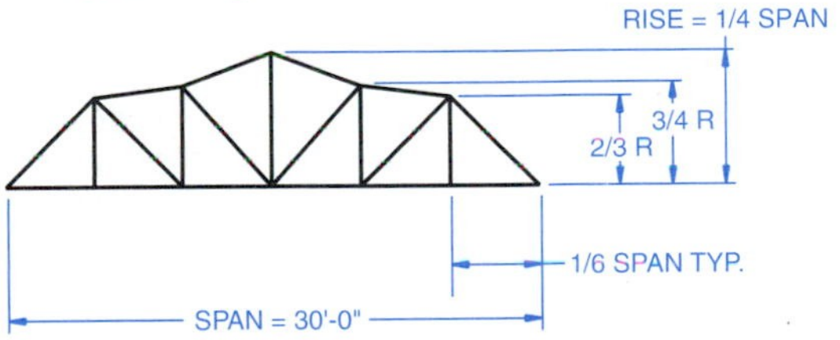

Figure 1.17 **Problem 1.13 Truss Bridge**

1.14 See Figure 1.18. Construct the irregular polygon shown in the figure, using the given dimensions, on an A- or A4-size sheet. Do not dimension.

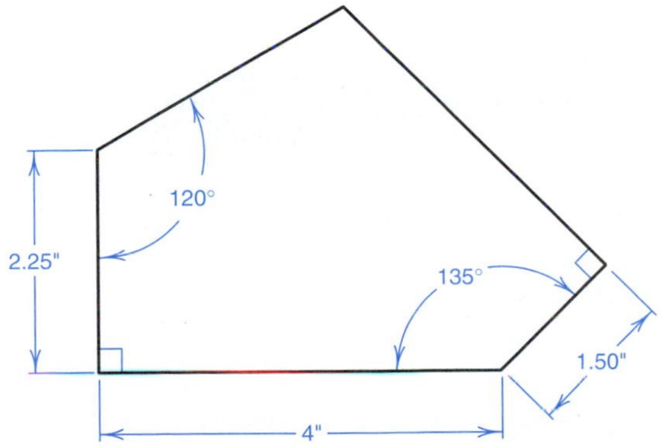

Figure 1.18 **Problem 1.14 Angle Polygon**

1.15 See Figure 1.19. Construct the irregular polygon shown in the figure, using the given dimensions, on an A- or A4-size sheet. Do not dimension.

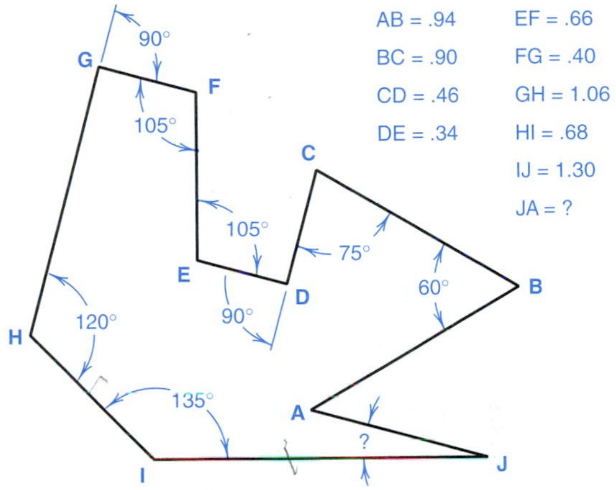

Figure 1.19 **Problem 1.15 Irregular Polygon**

1.16 See Figure 1.20. Construct the centering plate, using the given dimensions. All of the angles are proportional to angle A. Place the drawing on an A- or A4-size sheet. Do not dimension.

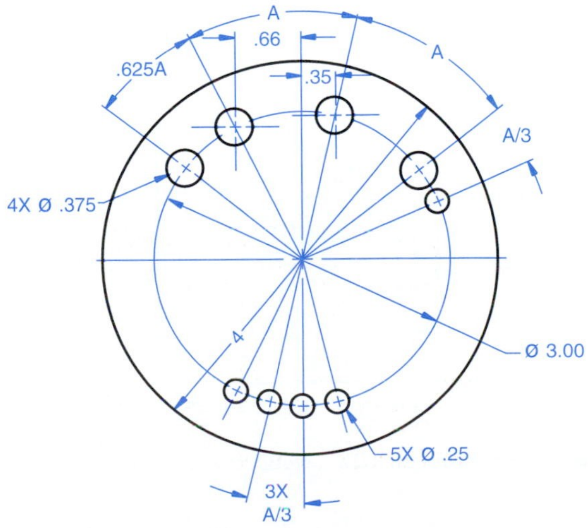

Figure 1.20 **Problem 1.16 Centering Plate**

1.17 See Figure 1.21. Construct the retaining ring shown in the figure. Use an A-size sheet and triple the size of all radii.

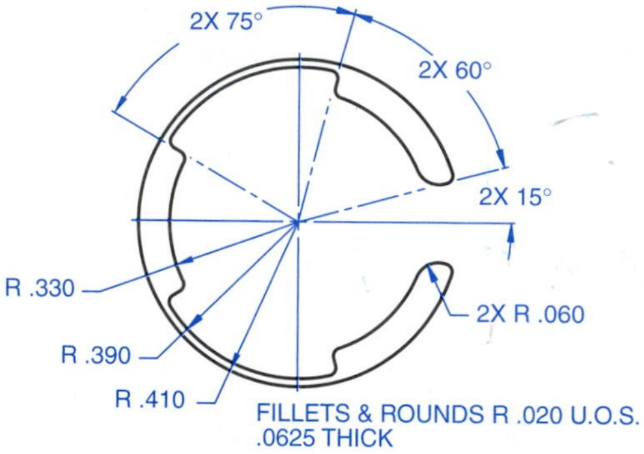

Figure 1.21 Problem 1.17 Retaining Ring

1.18 See Figure 1.22. A laser beam directed from source A is reflected at a 45-degree angle from mirror B to mirror C, then onto the horizontal machine surface. Draw the mirrors, machine surface, and light path. Determine angle X for mirror C. [Hint: Angle Y must equal angle Z (angle of incidence equals angle of reflection).] Use a scale of ¼″ equals 1′–0″ and draw on an A-size sheet.

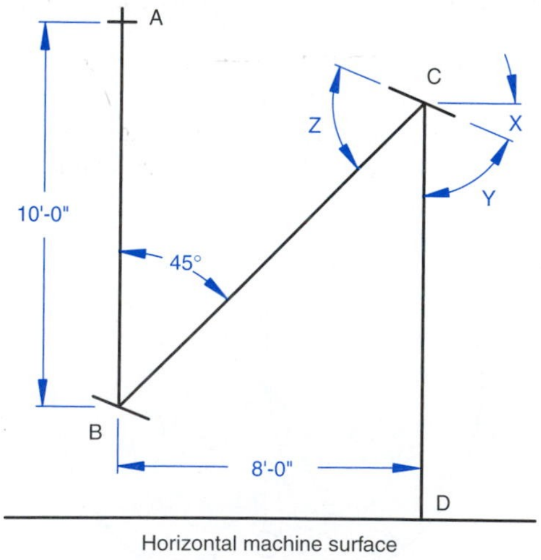

Figure 1.22 Problem 1.18 Reflector

1.19 See Figure 1.23. Construct the pump gasket shown in the figure, using a B-size sheet.

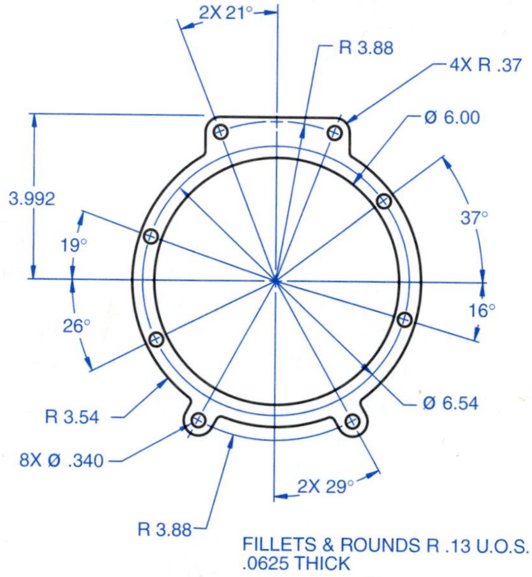

Figure 1.23 Problem 1.19 Pump Gasket

1.20 See Figure 1.24. Construct the chamber clip shown in the figure, using a B-size sheet.

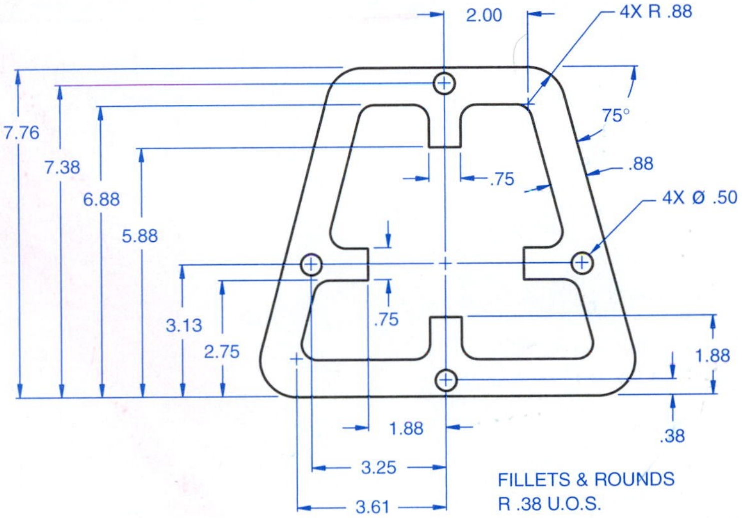

Figure 1.24 **Problem 1.20 Chamber Clip**

Further Reading

Booker, P. *A History of Engineering Drawing*. London: Chatto & Windus, 1962.

Ferguson, E. S. "The Mind's Eye: Nonverbal Thought in Technology." *Science* 197, no. 4306 (August 26, 1977), pp. 827–36.

Grown, J. R., et al. *Visualization: Using Computer Graphics to Explore Data and Present Information.* New York: John Wiley and Sons.

Higbee, F. G. "The Development of Graphical Representations." In *Proceedings of the Summer School for Drawing Teachers,* edited by R. P. Hoelscher and J. Rising. New York: McGraw Hill, 1949, pp. 9–26.

LaCourse, D. E. *Handbook of Solid Modeling.* New York: McGraw-Hill, 1995.

Land, M. H. "Historical Developments of Graphics." *Engineering Design Graphics Journal 40*, no. 2 (Spring 1976), pp. 28–33.

Reynolds, T. S. "Gaspard Monge and the Origins of Descriptive Geometry." *Engineering Design Graphics Journal 40*, no. 2 (Spring 1976), pp. 14–19.

Orthographic Sketch Paper

Sketch Number: ____________________

Name: ____________________

Div/Sec: ____________________

Date: ____________________

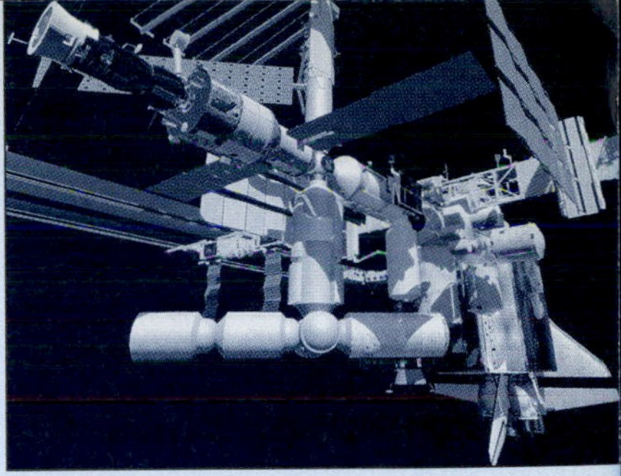

Chapter

Sketching and Text 2

OBJECTIVES

After completing this chapter, you will be able to:

1. Define technical sketching.
2. Understand how sketching integrates into the design process.
3. Identify and define two types of sketches.
4. Create a design sketch using pencil or computer.
5. Identify and use sketching tools.
6. Use grid paper to create sketches.
7. Lay out a sketch using proportions.
8. Understand the difference between pictorial and multiview projection.
9. Create a perspective sketch.
10. Create an isometric sketch.
11. Create an oblique sketch.
12. Create a multiview sketch.
13. Identify the types and precedence of lines.
14. Follow good hand-lettering practice.
15. Identify important practices when using CAD for lettering.

2.1 TECHNICAL SKETCHING

Technical sketching is the process of producing a rough, preliminary drawing representing the main features of a product or structure. Such sketches have traditionally been done freehand; today, CAD systems can also be used. A technical sketch is generally less finished, less structured or restricted, and it takes less time than other types of freehand illustrations. Also, a technical sketch may communicate only selected details of an object, using lines; whole parts of an object may be ignored, or shown with less emphasis, while other features may be shown in great detail.

Technical sketches can take many different forms, depending on the clarity needed and the purpose of the sketch, both of which depend on the audience for which the

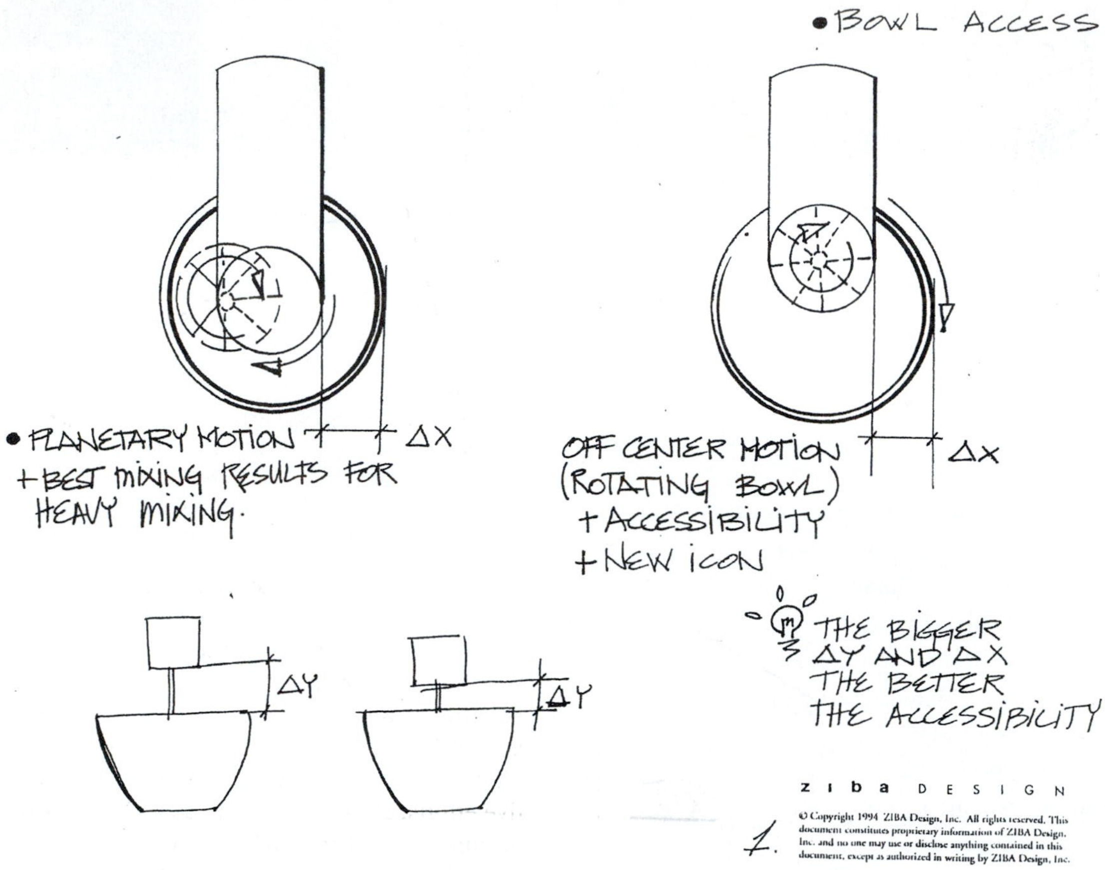

Figure 2.1 Technical Sketch

A rough technical sketch can be made to capture a design idea quickly.

sketch is intended. For example, a sketch made quickly to record a fleeting design idea may be very rough (Figure 2.1). This type of sketch is for personal use and is not meant to be understood by anyone but the individual who produced it. A sketch may also use the format of a more formal, multiview drawing intended to be used by someone who understands technical drawings (Figure 2.2). However, this type of sketch would not be appropriate for a nontechnical person. Pictorial sketches would be used to further clarify the design idea and to communicate that idea to nontechnical individuals (Figure 2.3). Shading can be used to further enhance and clarify a technical sketch (Figure 2.4).

Technical sketches are used extensively in the first (ideation) stage of the design process and are an informal tool used by everyone involved in the design and

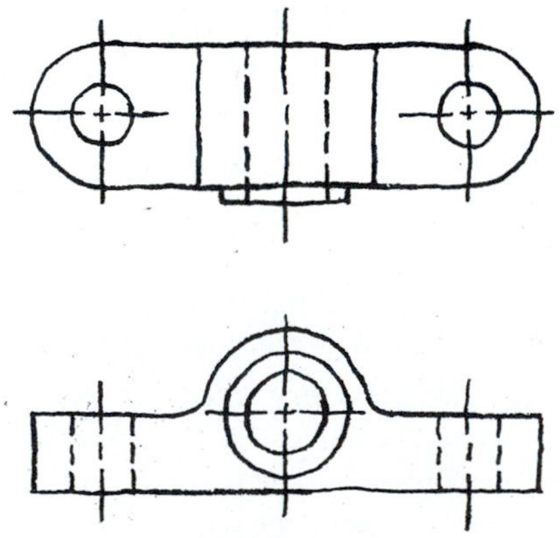

Figure 2.2 Multiview Sketch of a Mechanical Part, Used by the Engineer to Communicate Technical Information about the Design to Others

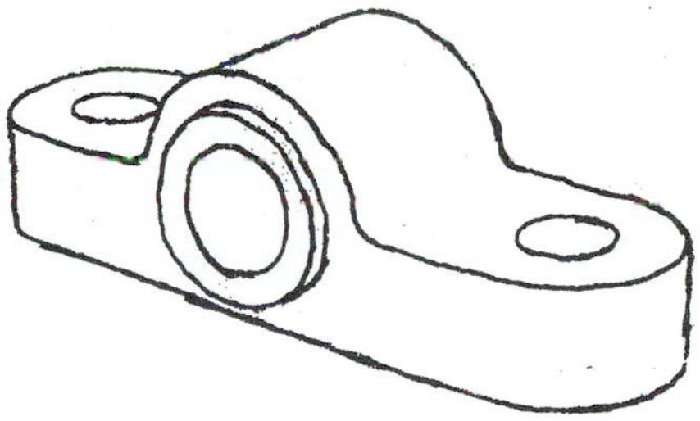

Figure 2.3 Pictorial Sketch

Pictorial sketches are used to communicate technical information in a form that is easy to visualize.

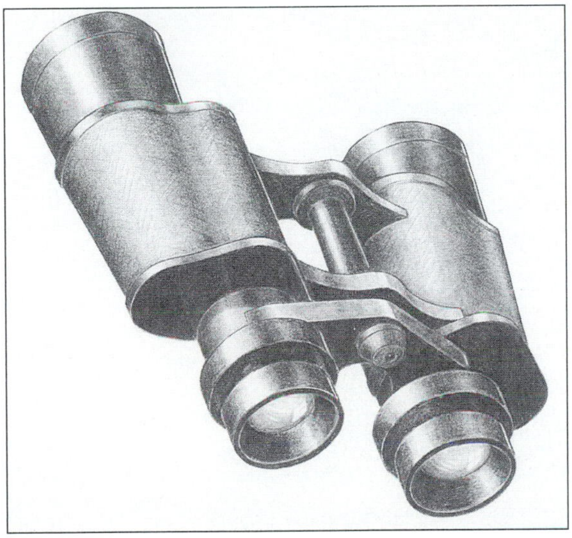

Figure 2.4 Shaded Sketch

This rendered sketch is an example of the amount of detail that can be used when creating sketches. This type of sketch is more appropriate for technical illustrations than for design communications. Irwin drawing contest winner Tim Brummett, Purdue University.

manufacture of a product (Figure 2.5). For example, an industrial engineer might make several sketches of a layout for a factory floor.

Many designers find that sketching becomes part of their creative thinking process. Through the process of *ideation,* sketching can be used to explore and solidify design ideas that form in the *mind's eye,* ideas that are often graphic in nature. Sketching helps capture these mental images in a

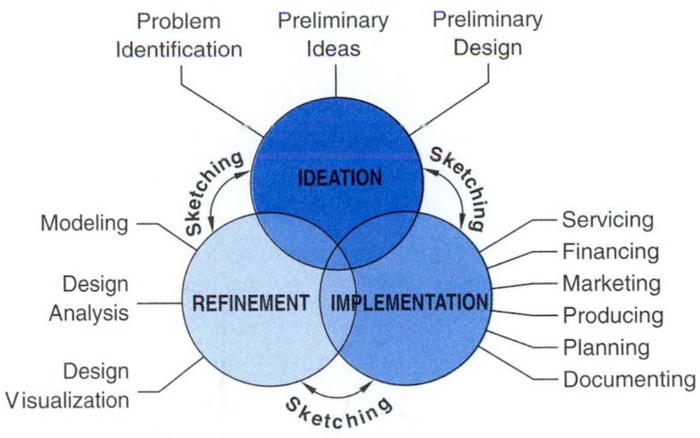

Figure 2.5

Sketching is used throughout the design process to communicate information.

permanent form. Each sketch is used as a stepping stone to the next sketch or drawing, where ideas are refined, detail is added, and new ideas are formed.

On a large project, hundreds of sketches are created, detailing both the successful and the unsuccessful approaches considered for solving the design problem. Since all but the smallest of design projects are collaborative efforts, sketches become important tools for communicating with other members of the design team.

At the early stages of the design process, highly refined, detailed drawings can actually impede the exploration of alternative ideas. What is needed are informal, nonrestrictive sketches that can communicate both geometric and nongeometric information and can be produced quickly and changed easily. Technical sketching, being fast and less restrictive, can convey ideas at a level of detail that communicates the design intent and, at the same time, can allow the viewers to imagine for themselves how different solutions might further the design. Sketches as communications tools encourage collaborative contributions from other members of the design team.

2.1.1 Freehand Sketching Tools

Normally, tools used for sketching should be readily available and usable anywhere: pencil, paper, and eraser. Although variations on these tools are numerous and sophisticated, the goal of technical sketching is simplification. Just a couple of pencils, an eraser, and a few sheets of paper should be all that is needed. Many a great design idea was born on the back of a napkin with a No. 2 wooden

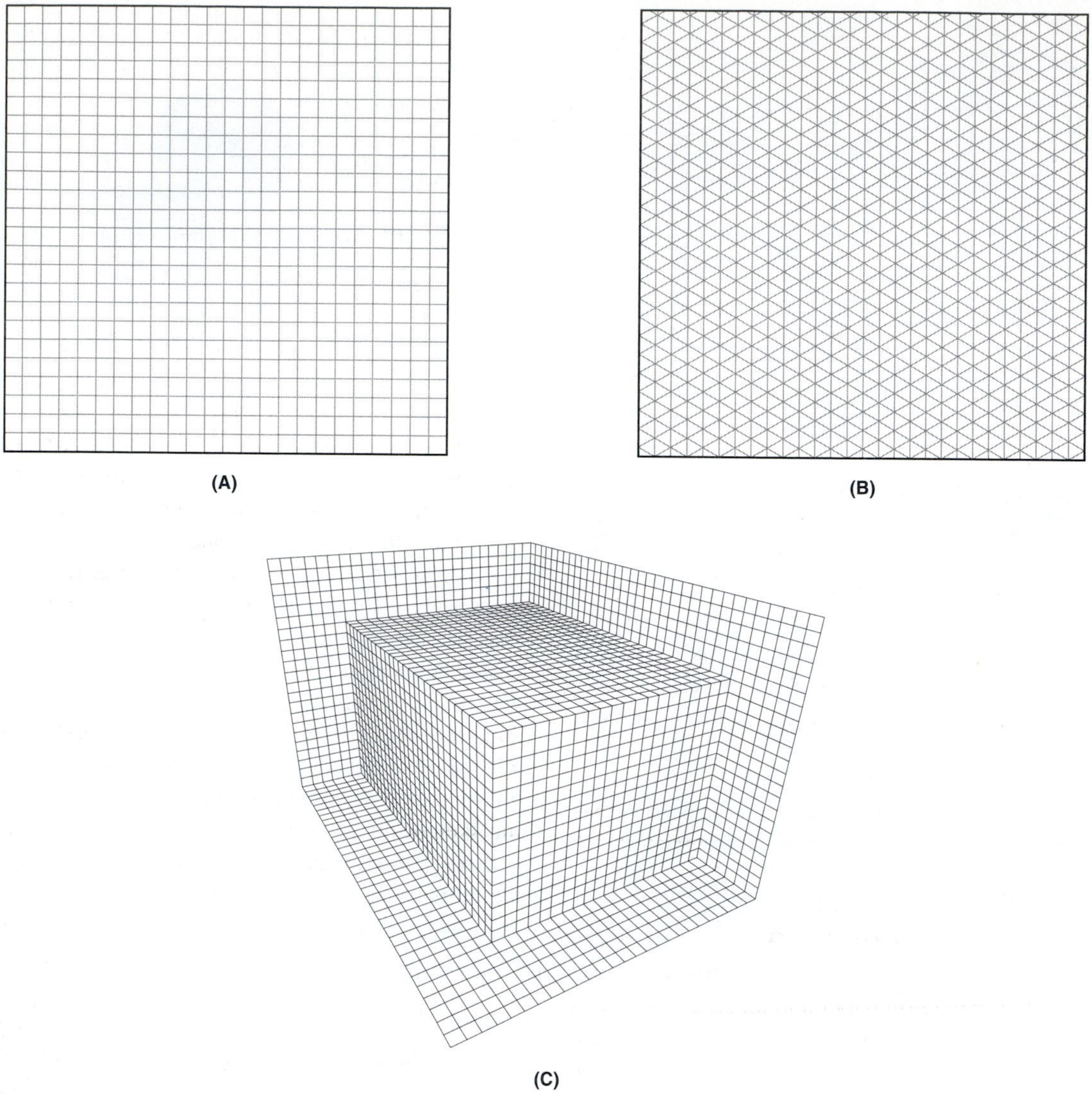

Figure 2.6 **Square (A), Isometric (B), and Perspective (C) Grids Used for Sketching**
The grid lines are used as an aid in proportioning the drawing and sketching straight lines freehand.

pencil! Although there may be a temptation to use straightedges, such as T-squares and triangles, a minimum amount of practice should allow you to draw lines good enough for sketches without these aids. Mechanically drawn lines can slow you down, add a level of accuracy not needed in the early stages of a design, and restrict the types of forms explored.

Pencils The lead used in pencils comes in many different hardnesses; the harder the lead, the lighter and crisper the line. For general-purpose sketching, leads in the H and HB range will give you acceptable lines. If the lead is much harder, the lines will be too light and hard to see. In addition, hard lead has a tendency to puncture and tear some of the lighter-weight papers used in sketching. On the other hand, if the lead is too soft, too much graphite is deposited on the paper and can be smudged easily. Leads in the middle range allow for a dark, relatively crisp line.

Eraser Erasing should only be used to correct mistakes in a line, not to make changes in a design. Such changes should be recorded on a separate sketch, and the original sketch should be preserved. Still, most people find that a small amount of erasing is helpful. Usually, the eraser on the end of the pencil is sufficient. However, if you are going to do a lot of sketching, you may need a separate eraser, and one of any size or shape will do. You might consider a gum eraser, since they leave less residue when used.

Paper There is a wide range of paper choices for sketching (including a napkin you could draw on during lunch). The most accessible and easiest to use is notebook-size (8-½″ × 11″) paper. Because of the difficulty of drawing long lines freehand, paper much larger than that is normally not useful for a single sketch. On the other hand, larger paper is useful for drawing multiple sketches that should be visually grouped together.

Plain bond paper with no lines offers the highest degree of flexibility; lined paper tends to lock you in visually to drawing along the lines. However, when you want the guidance of existing lines on the paper, it is most useful to have the lines running along both dimensions, forming a grid. Two of the most common **grid papers** used in sketching are **square grid** (Figure 2.6A) and **isometric grid** (Figure 2.6B) for use in certain types of pictorial sketches. Common grid densities run from 4 to 10 lines per inch. A less common type of grid paper is perspective, which is used to create another type of pictorial sketch (Figure 2.6C).

Often, it would be useful to have grid lines for the sketch, but not for the final drawing. One way this can be achieved is to sketch on thin, plain, semitransparent **tracing paper** laid over the grid paper and taped down so that the grid lines show through. When the sketch is done, it is untaped from the grid paper and viewed without the grid lines behind it. This technique is also a money saver because grid paper is more expensive than tracing paper (often called trash paper), which can be bought in bulk on rolls. The other advantage to tracing paper is that it can be laid over other sketches, photos, or finished technical drawings. A light table can be used to improve the tracing process. Tracing is a fast, accurate method for refining a design idea in progress or for using an existing design as the starting point for a new one.

2.2 SKETCHING TECHNIQUE

It takes practice and patience to produce sketches that are both legible and quickly made. The following sections describe common techniques used to produce good sketches quickly. The discussions cover the tools and the techniques for creating straight lines, curves (such as circles and arcs), and proportioned views. With patience and practice, it is possible for you to become good at making quick, clear sketches, regardless of your experience and natural drawing ability.

2.2.1 Seeing, Imaging, Representing

There are certain fundamental skills that must be learned in order for sketching to be used as a tool for design. Over a period of time and with practice you will be able to acquire the skills and knowledge necessary to create design sketches. Sketching is based on seeing (perception) and visual thinking through a process of seeing, imaging, and representing (Figure 2.7). Seeing is our primary sensory channel because so much information can be gathered through our eyes. It is our best-developed sense and one we take for granted every day as we easily move through our environment. Seeing empowers us to sketch.

Imaging is the process that our minds use to take the visual data received by our eyes to form some structure and meaning. The mind's eye initially creates the images whether real or imagined, and these are the images used to create sketches. **Representing** is the process of creating sketches of what our minds see.

Seeing and imaging is a pattern-seeking process in which the mind's eye actively seeks those features that fit within

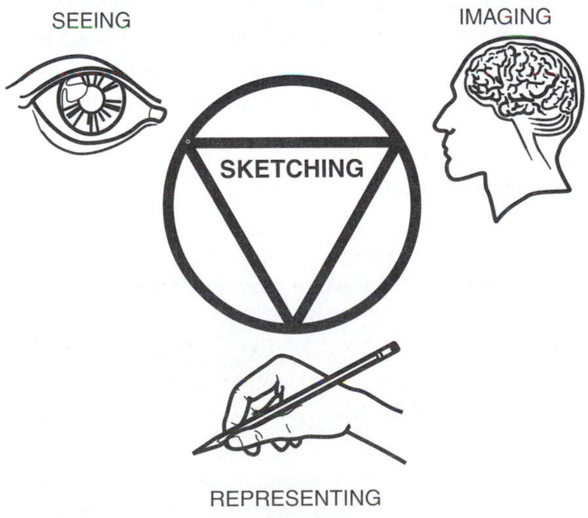

Figure 2.7 The Sketching Process
Sketching is based on the interactive process of seeing, imaging, and representing.

Figure 2.8 Pattern-Seeking Process of the Human Mind
In the illusion on the left, created by psychologist E. G. Boring in 1930, you can see either the head of an older woman or the profile of a younger woman. The illustration on the right can be viewed as either a vase or two profiles of the same person looking at each other.

our interests, knowledge, and experiences. Figures 2.8 and 2.9 are examples of sketches that can be interpreted in more than one way. It is also possible to make sketches of objects that cannot exist in the real world. M.C. Escher was a genius at creating sketches and drawings of objects or environments that could not exist in reality (Figure 2.10).

Practice Exercise 2.1

Our perception is not limited to what we can see. Images often appear spontaneously in response to a memory recall. In this exercise, read the words and see if visual images are created in your mind's eye.

1. Your bedroom where you grew up as a child, or the street you lived on.
2. A close relative, a famous actor, or a close friend from high school.
3. A basketball sitting at center court on your high school gym floor. Try sketching the basketball on the floor.

Your response to these written prompts is an example of your visual memory. You are thinking visually, which is a very powerful way of thinking when designing.

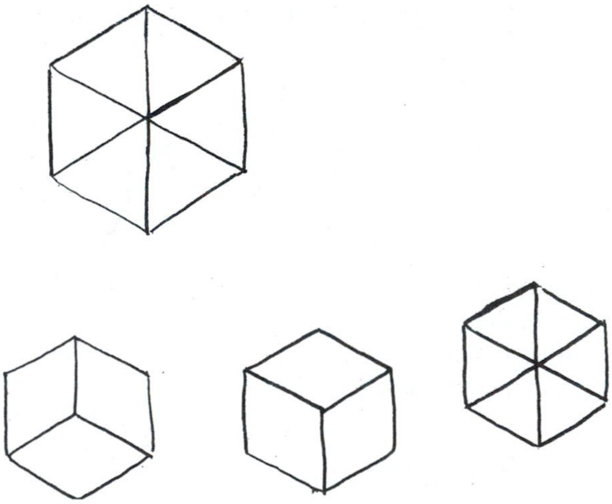

Figure 2.9 Different Interpretations of the Same Object
In this example a cubelike illustration can be interpreted as a hexagonlike figure or as a cube viewed from below, above, or with transparent sides.

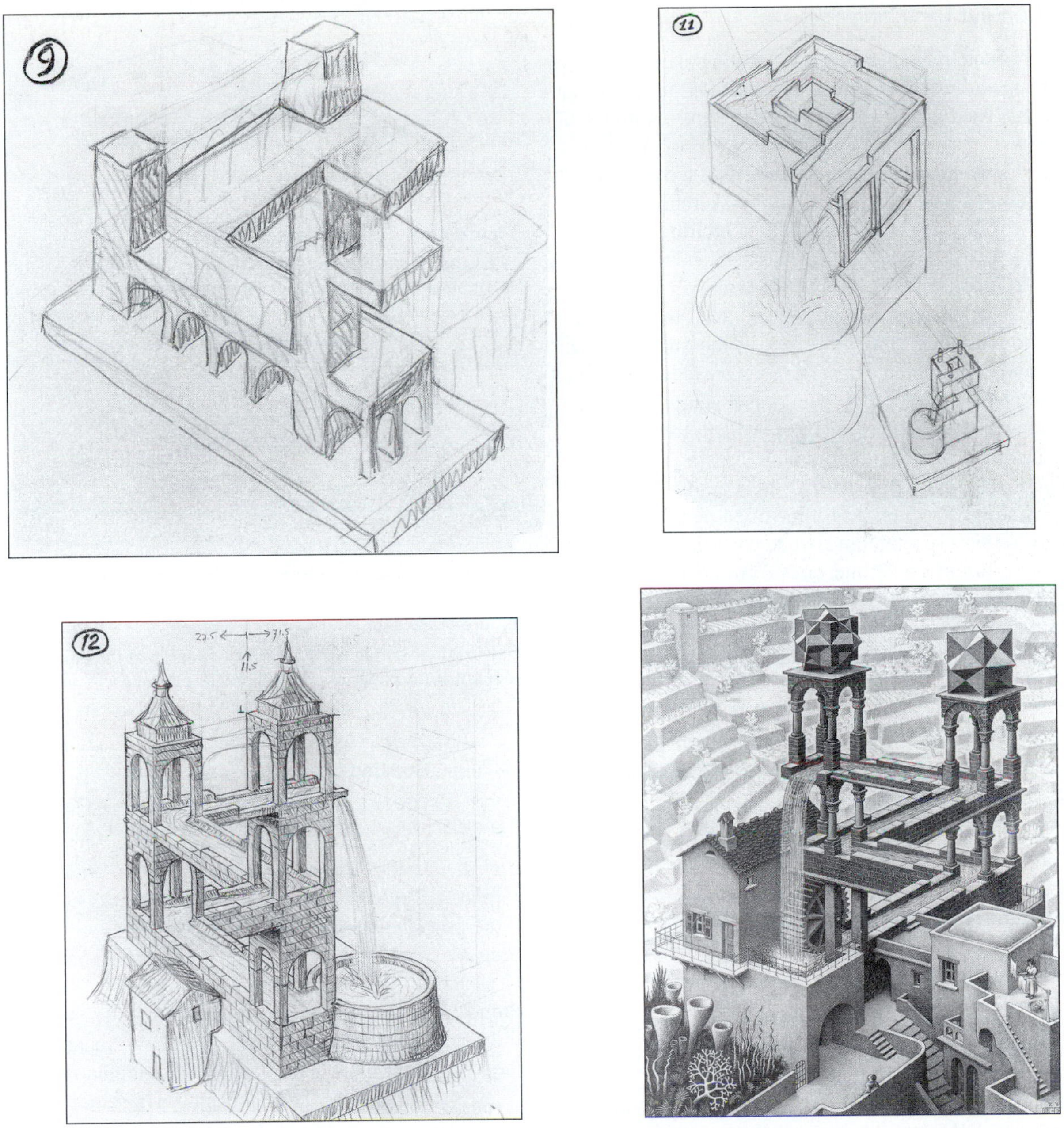

Figure 2.10 **Optical Illusion**

Follow the path of the water in this illustration to see the optical illusion created by M. C. Escher. You can also see Escher's early design sketches of the waterfall. (M. C. Escher's "Waterfall" © 2000 Cordon Art B.V. -Baarn-Holland. All rights reserved.)

2.2.2 Contour Sketching

The most fundamental element to creating sketches is the line or the outline of an object. The lines or outlines of an object are used to represent the edges and contours of objects we see in the world. If we sketch the boundaries, an object slowly takes shape and we begin to recognize it as a familiar object. This technique of sketching the outline of an object is called **contour sketching** and is an important technique used by novice sketchers to gain confidence in their sketching ability. Contours on objects can take the form of edges of an object, lines that separate contrasting light or color, changes in the surface of an object, and overlapping parts. The primary reason for contour sketching is to develop your visual acuity and sensitivity to important object features, which are needed to create accurate sketched representations.

When you first learn how to use contour sketching, begin by slowly tracing the outline of an object with your eyes while slowly sketching what you see. At first the sketch may seem crude and out of proportion, but with practice your sketches will be quite good. Figure 2.11 shows an example of a contour sketch created by carefully looking at the outline of the object and sketching what you see without looking at the paper. Figure 2.12 is a sketch created by carefully looking at the outline of the object and looking at the paper as you sketch. Both techniques are useful when learning how to observe and create sketches of what you see.

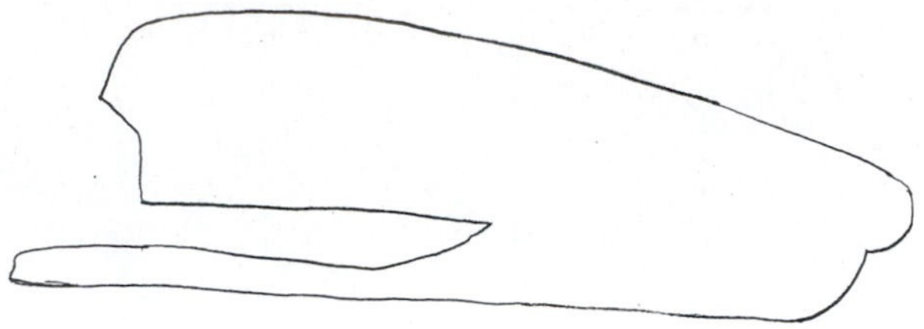

Figure 2.11 Contour Sketch

A contour sketch is created by carefully observing the outline of an object while sketching. This technique is used to improve your sketching ability. In this example, the contour sketch was created without looking at the paper.

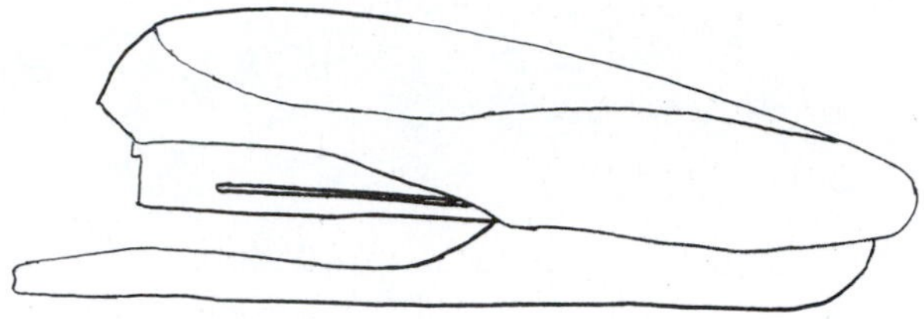

Figure 2.12 Modified Contour Sketch

This contour sketch was created by looking at the object, then looking at the paper as the sketch was produced.

Making a Contour Sketch

In this exercise, you are to create a sketch of the stapler shown in Figure 2.11 using the contour sketching technique.

Step 1. Using a plain piece of white paper and a soft lead pencil, place your drawing hand with the pencil near the center of the paper.

Step 2. Orient the paper in a comfortable position for sketching.

Step 3. Comfortably and in a relaxed manner, very slowly begin to trace the outline of the object with your eyes.

Step 4. Slowly move your pencil across the paper as your eyes scan the outline of the object. Do not erase or sketch over lines and do not look at your sketch. Sketch very slowly and deliberately.

Step 5. Continue to draw each edge as you view it at a slow and deliberate pace.

Step 6. Look at your sketch after you have finished viewing the contours of the object.

Making a Modified Contour Sketch

In this exercise, you are to create a contour sketch, but you will be able to look at your sketch as you are working (Figure 2.12).

Step 1. Using a plain piece of white paper and a soft lead pencil, place your drawing hand with the pencil near the center of the paper.

Step 2. Orient the paper in a comfortable position for sketching.

Step 3. Comfortably and in a relaxed manner, very slowly begin to trace the outline of the object with your eyes.

Step 4. Slowly move your pencil across the paper as your eyes scan the outline of the object. Do not erase or sketch over lines. Sketch very slowly and deliberately.

Step 5. Occasionally look at your sketch to match it with the object being drawn.

Step 6. Continue to draw each edge and interior edges at a slow and deliberate pace as you view the object.

2.2.3 Negative Space Sketching

Another useful technique novice sketchers can try to improve their sketching technique is called **negative space sketching.** In this technique you concentrate on the spaces between the objects and not on the objects themselves. In other words, you concentrate on the geometry of the objects, such as lines, curves, angles, and tangencies, and not on the names of the objects, such as handle, hole, base, cube. An example of a negative space sketch is shown in Figure 2.13. Notice that the object itself is not shaded and lacks details, but the space surrounding the object is shaded.

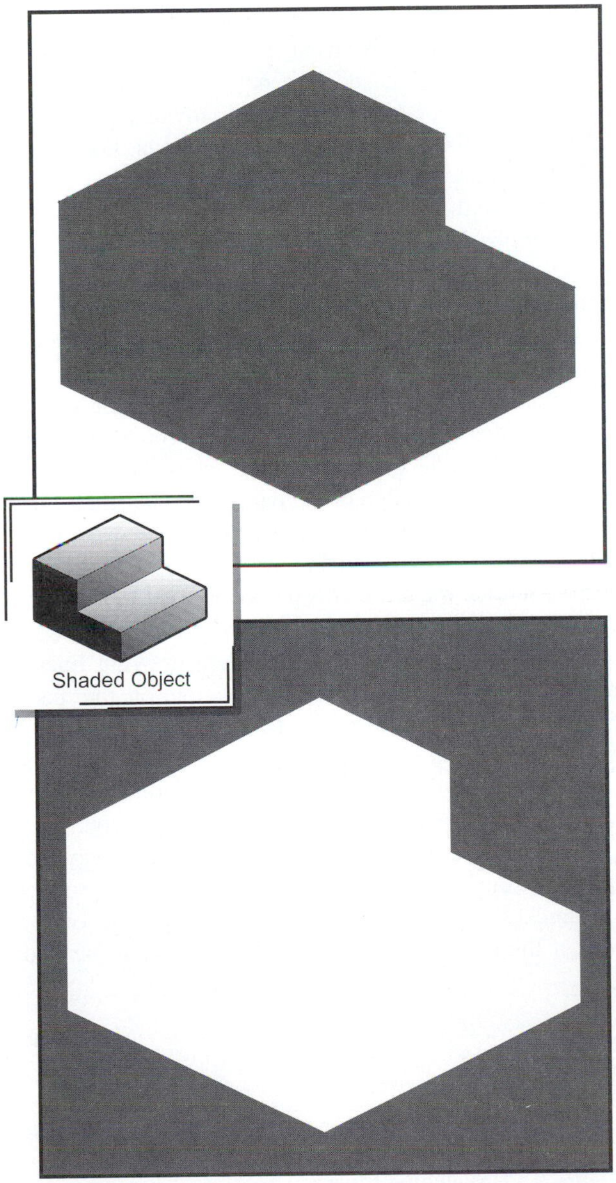

Figure 2.13 **Negative Space Sketching**
Negative space sketching produces a sketch having only the spaces between the object and not the object itself.

Making a Negative Space Sketch

For this exercise, you are to create a negative space sketch of the object shown in Figure 2.13.

Step 1. Use a plain sheet of white paper and begin by sketching the box surrounding the object.

Step 2. Sketch over the top of the negative spaces in the figure to reinforce that you are going to be sketching the negative spaces and not the object itself.

Step 3. Focus on one of the outlined negative spaces just created in step 2 until you can visualize the negative space.

Step 4. Now begin sketching the negative space form on your sheet of paper. Concentrate on drawing lines and curves by determining the angles, lengths, tangencies, and other geometric characteristics.

Step 5. Repeat steps 3 and 4 until all the negative space has been created.

2.2.4 Upside-Down Sketching

Upside-down sketching is another method that you can use to improve your sketching ability. In this technique you take a photograph of a recognizable object, such as a chair, and turn it upside-down before sketching it. By turning it upside-down you can concentrate on the shape and form of the object, allowing you to create a better sketch. Figure 2.14 is a photograph of a table that is

Figure 2.14 **Upside-Down Sketching**
Sketch the outline of the object by concentrating on the geometric forms. (Courtesy of Lunar Design Incorporated.)

upside-down. Carefully sketch the outline of the object by concentrating on the geometry or form and not the names of the part, such as legs or feet. By doing so you will be able to create a more accurate sketch of the object.

2.2.5 Straight Lines

All sketches are made up of series of lines. Lines created for sketches differ from mechanically produced lines in that they are not constrained or guided by instruments, such as a T-square, template, or compass. Instead, the lines are guided strictly by the eye and hand. Such lines have a different aesthetic quality than mechanical lines (Figure 2.15). At a micro level, sketched straight lines are uneven; at a macro level, they should appear to follow a straight path without any interruptions (Figure 2.16).

One of the easiest guides to use for sketched lines is grid paper. Lines drawn right on the grid are the easiest to produce, and even those lines that are offset but parallel to a grid line are fairly easy to produce. The idea is to keep your sketched line a uniform (but not necessarily equal) distance between two existing grid lines.

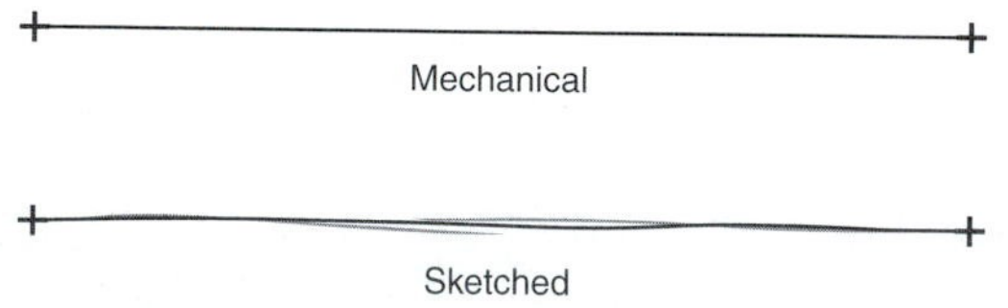

Figure 2.15 A Comparison of Mechanically Drawn and Sketched Lines

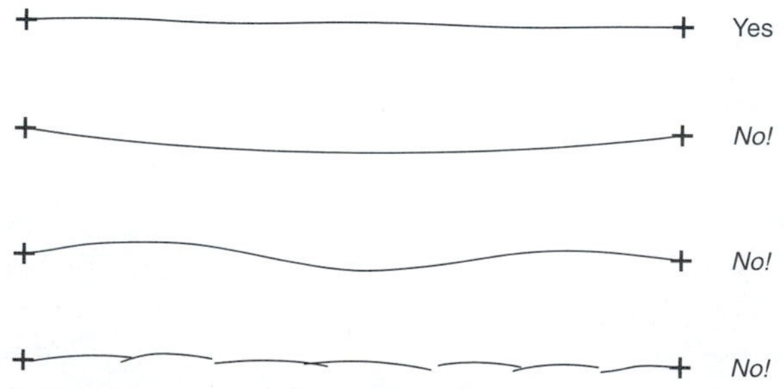

Figure 2.16 Examples of Good and Bad Straight Line Technique

Sketched lines should be straight and dark and should have a consistent thickness.

Curved lines, straight lines not parallel to a grid line, and lines drawn without the aid of a grid are more difficult. In all of these cases, the lines are drawn as interpolations between two or more points. The points are typically marked on an engineering drawing as two intersecting lines, one horizontal and one vertical, and each approximately 3⁄16″ long. Your eye should take a "global" view of all the points to be connected and should guide your hand as it goes from point to point.

Quite often, the sketched line is built up from a sequence of two or three passes with the pencil (Figure 2.17). The first pass is drawn light, using a hard lead, such as a 4H, sharpened to a point, and may not be as straight as your final line will be; however, it should provide a path on top of which the final, even, darker line is drawn. For particularly long lines, the initial line may be drawn in segments, coming from the two endpoints and meeting in the middle; however, the final line should be drawn in one single pass to avoid choppiness. If necessary, another pass can be used to darken or thicken the line.

Long lines are difficult to control, even for someone with a lot of experience. If you cannot choose a drawing scale that reduces the size of the sketch, use grid paper as a guide, drawing either directly on the grid paper or on tracing paper placed on top of the grid paper. If the line is parallel and relatively close to the edge of the paper, you can rest a finger or a portion of your palm along the edge of the paper to stabilize your drawing hand (Figure 2.18). If necessary, you can use a ruler or a scrap of paper to mark a series of points on the sketch, but this will slow you down a bit.

Another technique that helps when drawing lines of any length is changing the orientation of the paper. Sketching paper should not be fixed to your drawing surface. Instead, you should be able to rotate the paper freely, orienting it in the direction that is most comfortable. Practice will determine which orientation is best for you. Many people find that drawing the lines by moving away from or toward the body, rather than from left to right, produces the quickest, straightest lines; others find it most comfortable if the paper is angled slightly away from the body.

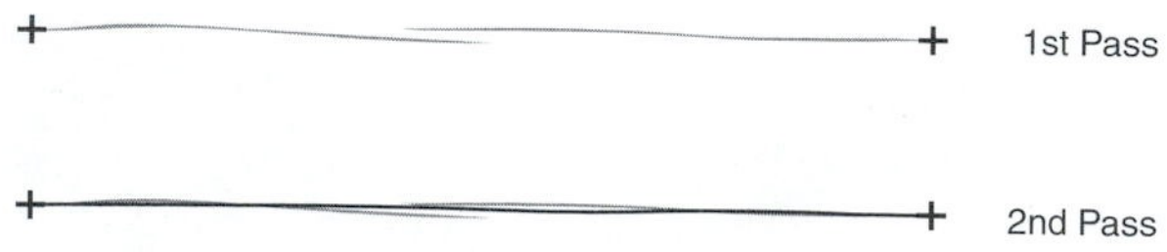

Figure 2.17 Sketching Lines

The sequential drawing of a straight line is done by first drawing a very light line, using short strokes. The light line is then drawn over and darkened.

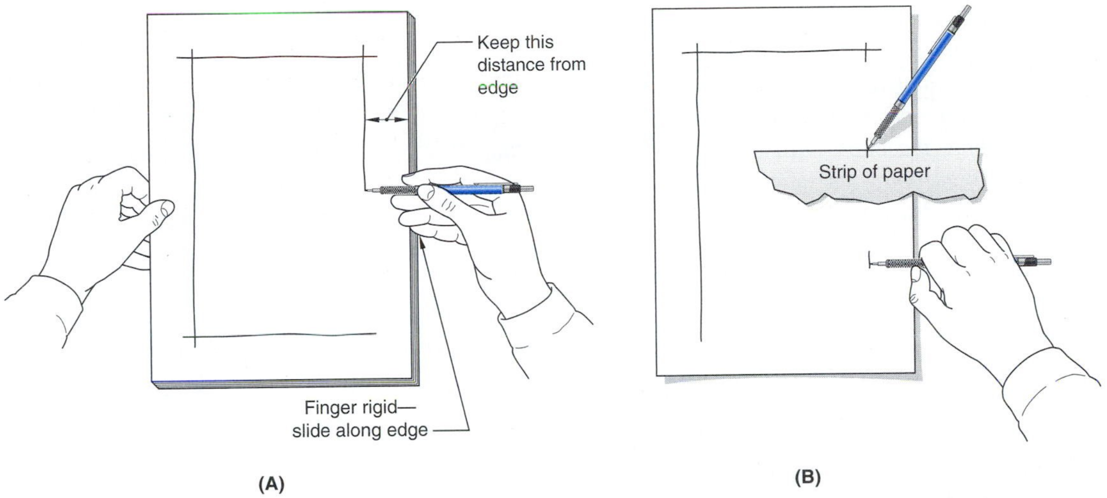

Figure 2.18 **Sketching Long Lines**
Very long lines can sometimes be more difficult to sketch. One technique is to use the edge of the paper as a guide for your hand (A). Another technique is to mark equal distances from the edge of the paper using a marked scrap of paper as a guide (B). The marks are then used as a guide to produce the line.

Again, the longer the line, the more important it is that the paper be positioned comfortably for you.

The following summarizes the techniques used to sketch straight lines:

- Orient the paper to a comfortable position. Do not fix the paper to the surface.
- Mark the endpoints of the lines to be sketched.
- Determine the most comfortable method of creating lines, such as drawing from left to right, or drawing either away from or toward your body.
- Relax your hand and the rest of your body.
- Use the edge of the paper as a guide for making straight lines.
- Draw long lines by sketching a series of connected short lines.
- If necessary, draw on grid paper or on tracing paper that is overlaid on grid paper.

Sketching Straight Lines

In this exercise, you are to create a series of 5″ long parallel lines equally spaced at 0.5″. Refer to Figures 2.16 and 2.17.

Step 1. Lightly mark the endpoints of the lines to be sketched on 8-½″ × 11″ paper.

Step 2. Orient the paper in a comfortable position for sketching.

Step 3. Comfortably and in a relaxed manner, position your hand so that the pencil is close to one of the marked endpoints of the first line to be sketched. Sketch the top line first, to avoid smearing newly sketched lines with your hand.

Step 4. Quickly scan the two endpoints of the first line to determine the general direction in which you will be sketching.

Step 5. Lightly sketch a short line, approximately 1″ long, by moving your hand and the pencil in the general direction of the other end of the line.

Step 6. Repeat steps 4 and 5 until the other end of the line is reached.

Step 7. Return to the starting point of the line and overdraw the line segments with a slightly longer, heavier stroke, to produce a thick, dark, more continuous straight line.

Step 8. Repeat steps 3 through 7 to sketch the remaining straight lines.

2.2.6 Curved Lines

Curved lines need multiple guide points. The most common curve is a circle or circular arc. Although very small circles and arcs can be drawn in one or two strokes and with no guide points, larger circles need some preliminary points. The minimum number of points for a circle is four, marked on the perimeter at equal 90-degree intervals. For an arc, use at least one guide point for every 90 degrees and one at each end.

There are a number of ways to lay out the guide points for circular curves quickly. One way is to draw a square box whose sides are equal to the diameter of the circle (Figure 2.19A). The midpoints on each side of the square mark the points where the circle will touch the square. These points are called points of tangency. More guide points can be added by drawing the two diagonals across the square. The center of the circle being sketched is the point where the diagonals cross (Figure 2.19B). Mark the guide points on each diagonal approximately two-thirds the distance from the center of the circle to the corner of the square. This distance is the approximate radius of the circle (Figure 2.19C).

As with longer straight lines, large arcs and circles are harder to draw and may need guide points marked at more frequent intervals. To do this, it is handy to use a scrap of paper with the radius marked on it (Figure 2.19D).

Circular arcs are drawn the same way as circles, adjusting the number of points to suit the degree of curvature (i.e., the length) of the arc. Noncircular arcs, however, can be more difficult. Since these lines are only to be sketched, calculating the points that the curve should pass through is too involved and is not recommended. Simply use the eye to estimate guide points and then gradually draw a curve to pass through those points. (Ellipses and curves in multiview drawings are two special cases treated later in this chapter.)

As with straight lines, positioning the paper and using a relaxed grip are important for helping you create good curves. Unlike straight lines, curves are usually best drawn in a series of arcs of not more than 90 degrees. After each arc is drawn, rotate the paper for the next segment of arc. With practice you may be able to eliminate rotating the paper for smaller arcs, but you will probably still have to do so for larger ones.

Sketching a Circle or Arc

The following steps demonstrate how to sketch a circle or arc. Refer to Figure 2.19 as a guide.

Step 1. Orient the paper in a comfortable position and relax your grip on the pencil. Lightly mark the corners of a square with sides equal in length to the diameter of the circle or arc to be sketched.

Step 2. Lightly sketch the square, using short strokes to create the straight lines.

Step 3. Mark the midpoints of the four sides of the square. This gives you four marks on the perimeter of the circle.

Step 4. Sketch diagonals across the corners of the square. Where the diagonals cross is the center of the circle.

Step 5. Mark the diagonals at two-thirds the distance from the center of the circle to the corner of the square. This gives you four more marks on the circle's perimeter.

Step 6. Sketch the circle by creating eight short arcs, each between two adjacent marks on the perimeter. Start at any mark and sketch an arc to the next mark (on either side of the first one, whichever is most comfortable for you).

Step 7. Rotate the paper and sketch the next arc from the last mark you touched to the next mark on the perimeter. Repeat this step until all eight arc segments have been sketched. For smoother sketches, rotate the paper in the opposite direction from the one you used to draw the arc.

Step 8. Overdraw the arcs with a thick, black, more continuous line to complete the sketched circle.

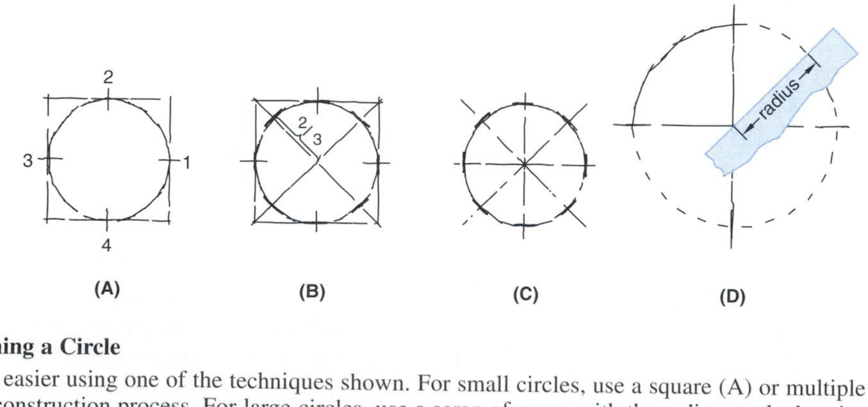

Figure 2.19 Sketching a Circle

Sketching a circle is easier using one of the techniques shown. For small circles, use a square (A) or multiple center lines (B)(C) to guide the construction process. For large circles, use a scrap of paper with the radius marked on it as a guide (D).

2.3 | PROPORTIONS AND CONSTRUCTION LINES

Frequently, in the sketch of an object, the relative proportions of its primary dimensions—width, height, and depth—are more important than their actual physical sizes. A **proportion** is the ratio between any two dimensions of an object. These proportions are represented in the sketch by a series of preliminary lines, which are drawn light and fast, and which may or may not represent the locations of the final lines in the sketch. Their purpose is to form a backbone, a structure inside which the final linework can be drawn.

The first step in a sketch involves drawing the construction lines, which guide a sketch's overall shape and proportion. **Construction lines** are very light, thin lines used to roughly lay out some of the details of sketches or drawings. Do not try to draw the construction lines to exact lengths since lengths are marked later, either by intersecting lines or short tick marks.

Construction lines have two primary features: the lines themselves and the intersections created where two lines cross. For example, the construction lines become the paths for the final straight lines. Points marked by the intersections of construction lines guide the drawing of circles. Usually, both of these features are used in creating sketches. Since all the dimensions of a sketch are estimated, groups of construction lines forming boxes and other shapes are an important tool for preserving the shape and proportion of the object and its features as the sketch is developed.

Grid paper can be used as a guide in creating construction lines but should not be thought of as a substitute, since the grid does not directly represent the proportions of the object, and there are many more grid lines than there are features on the object. The goal is to draw construction lines on top of the grid to reveal the form of the object. With experience, you may be able to make do with fewer construction lines, but while you are learning how to create properly proportioned sketches, you should use more, rather than fewer, construction lines to guide you.

Each feature has a proportion that can be represented by a series of construction lines. The following steps describe how to proportion a drawing by breaking it down into its component features.

Creating a Proportioned Sketch

Step 1. Refer to Figure 2.20. Gage the proportion of the overall size of the object. For the first sketch, use two overall dimensions of the object: width and height. Lightly sketch a box that represents the ratio of these two dimensions (Figure 2.20, step 1). This box is called a **bounding box** because it represents the outer dimensional limits of the feature being drawn. If the object is rectangular in shape, the final linework will follow the perimeter of the bounding box. In most cases, however, the final linework will only touch on a portion of the box's edges.

Step 2. Inside the first bounding box, draw other boxes to represent the larger features of the object, and within those boxes draw still others to represent the smaller features of the object. Often, a construction line can be used for more than one box. The final boxes each show the proportions of one feature of the object.

Step 3. Continue to draw bounding boxes until the smallest features of the object have been represented. As you gain experience, you may find that some of these smaller features need not be boxed; instead, their final lines can be sketched directly.

Step 4. When all of the features of the object have been boxed, begin sketching the final linework, which is done significantly darker than the construction lines.

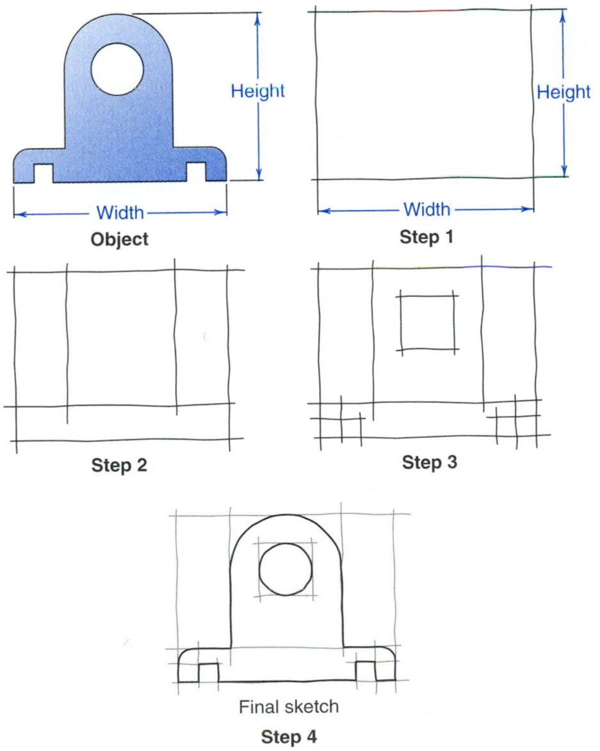

Figure 2.20 Creating a Proportioned Sketch
To create a well-proportioned sketch, use multiple steps to create lightly sketched rectangles and squares that are then used as guides for the final sketch.

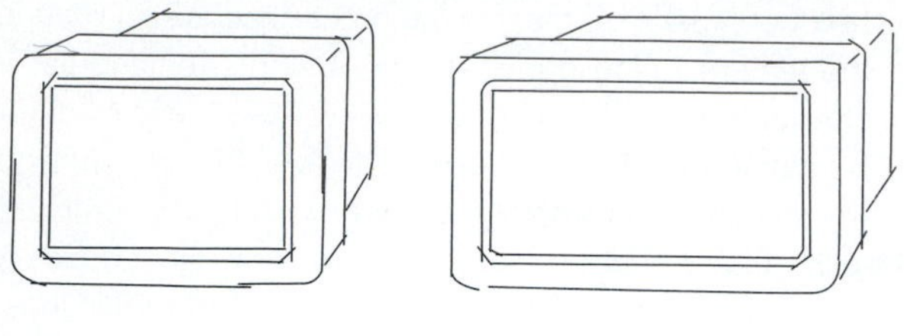

Figure 2.21 **Good and Poor Proportions**
One well and one poorly proportioned sketch of a computer monitor. The poorly proportioned monitor looks too wide.

The goal is, if you hold the drawing at arm's length, the construction lines are hard to see, and the final linework is clear and sharp. If there is not enough contrast between the construction lines and the final linework, then the construction lines become a distraction. Make the final lines darker, or the construction lines lighter, or both; however, do not erase your construction lines.

One of the most difficult sketching techniques to learn is making a sketch look well proportioned. For example, Figure 2.21 shows a well-proportioned and a poorly proportioned sketch of a computer monitor. Proportioning skills will improve with practice. A good rule of thumb is, if the drawing does not look or feel right, it probably is not. In the poorly proportioned monitor in Figure 2.21, the ratio of the height to the width is incorrect.

Sketching Objects

Step 1. Collect magazine photographs or clippings that show 2-D images or patterns. These can range from pictures of faces, to company logos, to fronts of buildings, etc. Stick with images that look flat; that is, images that don't show a depth dimension.

Step 2. Lay tracing paper over an image and tape the paper down.

Step 3. Lightly sketch an overall bounding box of the object. Look at the image contained in the bounding box. Mentally identify as many features on the object as you can. The features may be small and self-contained or a collection of several smaller features.

Step 4. Refine the drawing by sketching a series of progressively smaller bounding boxes. Start with the larger features and work downward.

Step 5. If desired, you can then darken some of the lines representing the image, to highlight the most important lines of a feature. What are the most important lines of a feature? Experiment with different lines to see which are more critical than others in representing the form of the image.

Hint: Buy a roll of tracing paper from your local blueprint or art supply store. It's cheaper than individual sheets, and you won't run out as often.

2.4 INTRODUCTION TO PROJECTIONS

Both ideation and document sketches can represent the objects being designed in a number of different ways. We live in a three-dimensional (3-D) world, and representing that world for artistic or technical purposes is largely done on two-dimensional (2-D) media. Although a sheet of paper is technically three-dimensional, the thickness of the paper (the third dimension) is useless to us. It should be noted that the computer screen is a form of two-dimensional medium, and images projected on it are governed by the same limitations as projections on paper. Modern techniques, such as holograms, stereograms, and virtual reality devices, are attempts to communicate three-dimensional ideas as three-dimensional forms. However, drawings are still the primary tool used for representing 3-D objects.

Most projection methods were developed to address the problem of trying to represent 3-D images on 2-D media (Figure 2.22). Projection theory and methods have taken hundreds of years to evolve, and engineering and technical graphics is heavily dependent on projection theory.

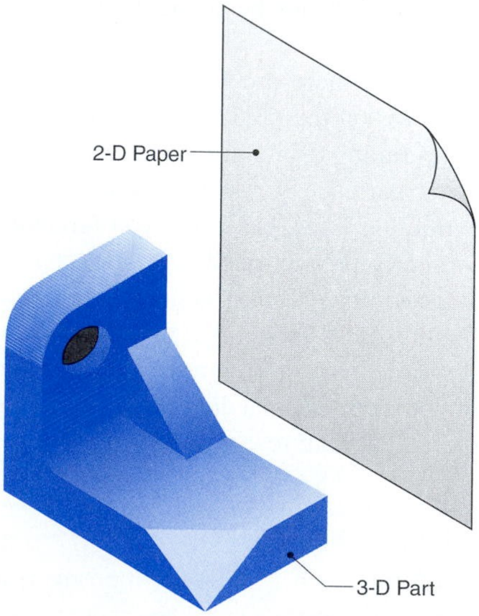

Figure 2.22 **3-D Object on 2-D Medium**
For centuries, graphicians have struggled with representing 3-D objects on 2-D paper. Various projection techniques have evolved to solve this problem.

▼ Practice Problem 2.1

Sketch the object on the grid.

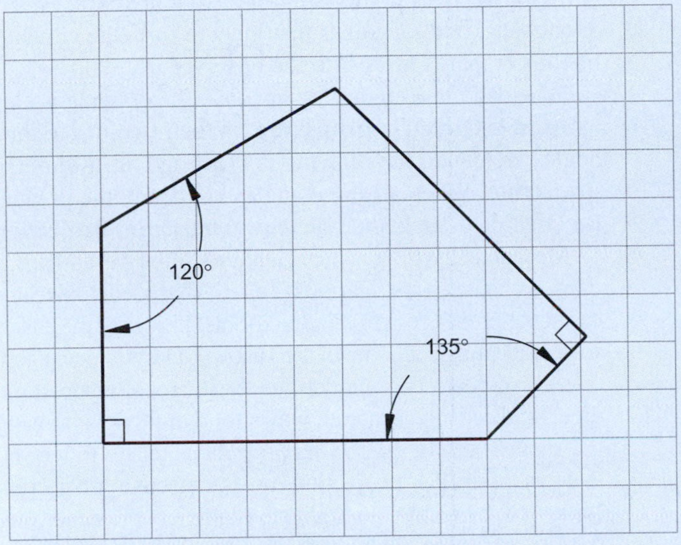

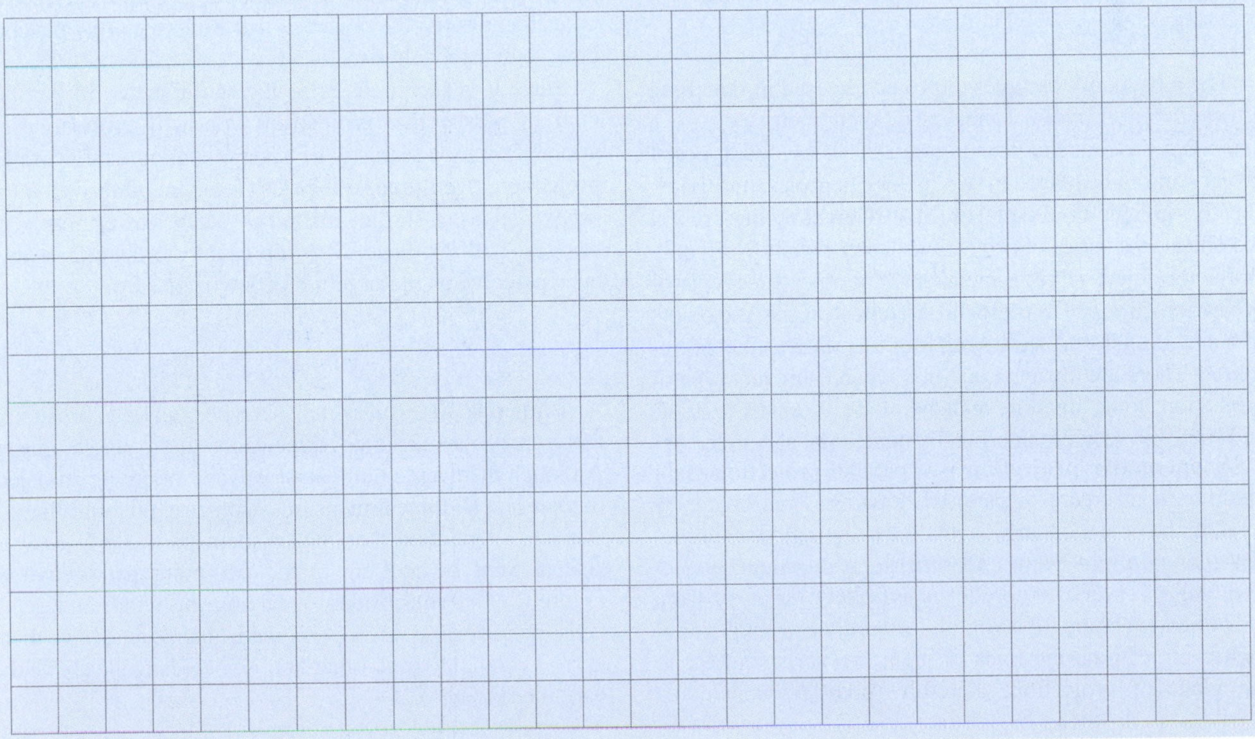

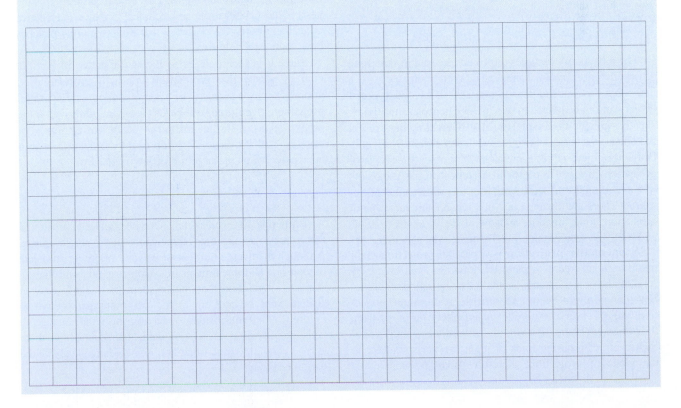

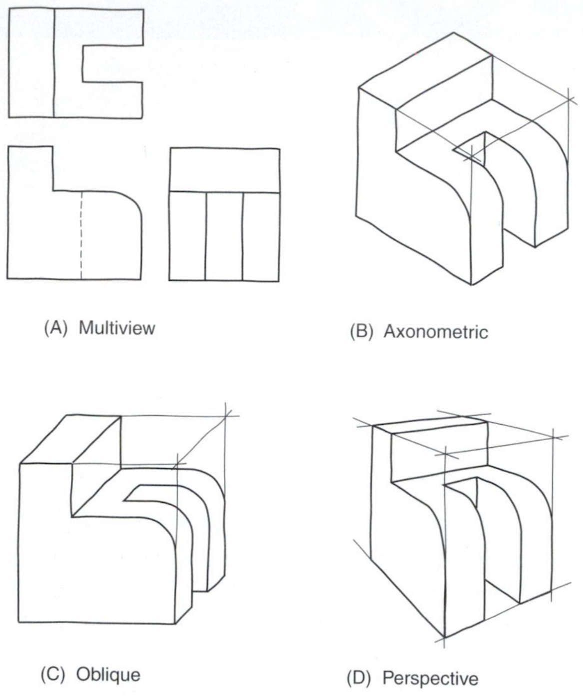

Figure 2.23 **Classification of Sketches**

Various projection techniques are used to create four basic types of sketches: multiview, axonometric, oblique, and perspective. The sketches shown in B, C, and D are called pictorial because they represent the object as a 3-D form. The multiview sketch uses multiple flat views of the 3-D object to accurately represent its form on 2-D paper.

The most common types of projection used in sketching are *multiview, isometric* (one type of axonometric), *oblique,* and *perspective,* as shown in Figure 2.23. These four types of projection can be placed in two major categories: multiview sketches and pictorial sketches. **Multiview sketches** present the object in a series of projections, each one showing only two of the object's three dimensions. The other three types of projection, grouped as **pictorial sketches,** present the object in a single, pictorial view, with all three dimensions represented. There are always trade-offs when using any type of projection; some are more realistic, some are easier to draw, and some are easier to interpret by nontechnical people.

Axonometric projection is a parallel projection technique used to create a pictorial drawing of an object by rotating the object on an axis relative to a *projection,* or *picture plane*. In multiview, axonometric, and oblique projection, the observer is theoretically infinitely far away from the projection plane. In addition, for multiviews and axonometric projections the lines of sight are perpendicular to the plane of projection; therefore, both are considered *orthographic projections*. The differences between multiview drawing and an axonometric drawing are that, in a multiview, only two dimensions of an object are visible on each view and more than one view is required to define the object, whereas in an axonometric drawing, the object is rotated about an axis to display all three dimensions, and only one view is required.

Axonometric drawings are classified by the angles between the lines comprising the axonometric axes. The axonometric axes are axes that meet to form the corner of the object that is nearest to the observer.

When all three angles are unequal, the drawing is classified as a **trimetric projection.** When two of the three angles are equal, the drawing is classified as a **dimetric projection.** When all three angles are equal, the drawing is classified as an **isometric (equal measure) projection.**

Mechanically drawn pictorials can often be as hard to draw as multiviews. Various 2-D CAD-based tools have eased the process of creating pictorials. Probably the easiest way of creating such views is to use a 3-D CAD package to create a model. This model can easily represent pictorial views and can also generate views for a multiview drawing.

Another way of classifying projections relates to whether they use **parallel projection** or **perspective projection.** Multiview, isometric, and oblique multiview projections use parallel projection, which preserves the true relationships of an object's features and edges. This type of projection is the basis of most engineering and technical graphics. Perspective projection distorts the object so that it more closely matches how you perceive it visually.

Since it is much easier to lay out a sketch in parallel than in perspective projection, you will probably find yourself doing a majority of your sketching using parallel projection, even though it is less realistic. Only when the object spans a large distance—such as a house or bridge—will it be useful to represent the distortion your eyes perceive as the object recedes from view.

2.4.1 Isometric Pictorials

An **isometric pictorial** sketch is a type of parallel projection that represents all three dimensions in a single image. Although there are a number of ways of orienting an object to represent all three dimensions, isometric pictorials have a standard orientation that makes them particularly easy to sketch. Start by looking at the two-point perspective in Figure 2.24. Then, instead of having the width and depth construction lines converge on vanishing points, have them project parallel to each other at a 30-degree angle above the baseline (Figure 2.25).

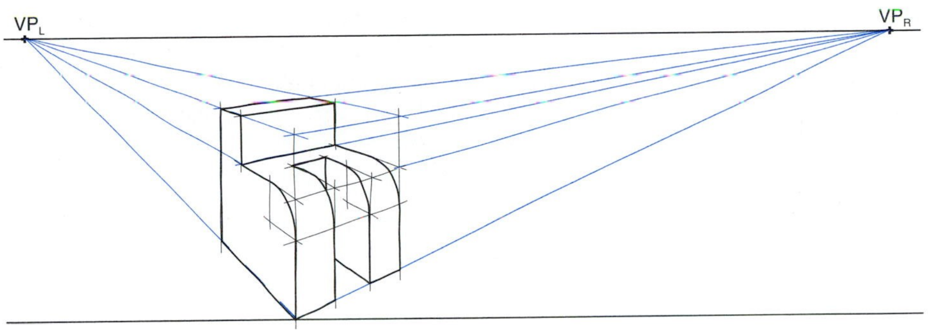

Figure 2.24 **Perspective Sketch**
For perspective projection, the width and depth dimensions converge on vanishing points.

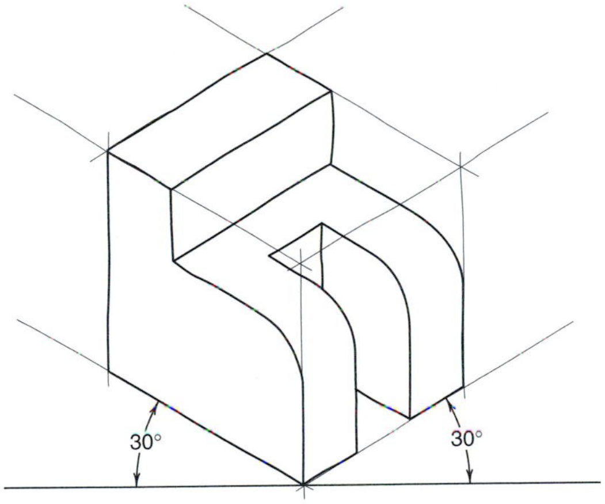

Figure 2.25 **Isometric Sketch**
For this isometric sketch, the width and depth dimensions are sketched 30 degrees above the horizontal.

Many CAD systems will automatically produce an isometric view of a 3-D model when the viewing angle is specified. Some CAD systems have predefined views, such as isometric, which are automatically created after selection.

Making an Isometric Sketch

Make an isometric sketch of the object shown in Figure 2.26.

Sketching the isometric axis

Step 1. Isometric sketches begin with defining an **isometric axis,** which is made of three lines, one vertical and two drawn at 30 degrees from the horizontal. These three lines of the isometric axis represent the three primary dimensions of the object: width, height, and depth. Although they are sketched at an angle of only 60 degrees to each other, they represent mutually perpendicular lines in 3-D space.

Step 2. Begin the sketch by extending the isometric axes shown in Step 1, Figure 2.26. Sketch a horizontal construction line through the bottom of the vertical line. Sketch a line from the base of the vertical line to the right, at an approximate angle of 30 degrees above the horizontal construction line. Sketch a line from the base of the vertical line to the left, at an approximate angle of 30 degrees above the horizontal construction line.

The corner of the axis is labeled point 1; the end of the width line is labeled point 2; the end of the depth line is labeled point 4; and the top of the height line is labeled point 3. The lengths of these lines are not important, since they will be treated as construction lines, but they should be more than long enough to represent the overall dimensions of the object. Estimate the overall width, height, and depth of the object using the estimating techniques described earlier in this chapter. Use these dimensions to sketch a block that would completely enclose the object.

Blocking in the object

Step 3. Sketch in the front face of the object by sketching a line parallel to and equal in length to the width dimension, passing the new line through point 3. Sketch a line parallel to and equal in length to the vertical line (1–3), through points 5–2. The front face of the object is complete.

Step 4. From point 3, block in the top face of the object by sketching a line parallel to and equal in length to line 1–4. This line is labeled 3–6. Sketch a line parallel to and equal in length to line 3–5, from point 6. This line is labeled 6–7. Sketch a line from point 5 to point 7. This line should be parallel to and equal in length to line 3–6. Block in the right side face by sketching a line from point 6 to point 4, which is parallel to line 1–3. The bounding box of the object, sketched as construction lines, is now finished. The box serves the same purpose as the one drawn in Figure 2.20, but it represents all three dimensions of the object instead of just two.

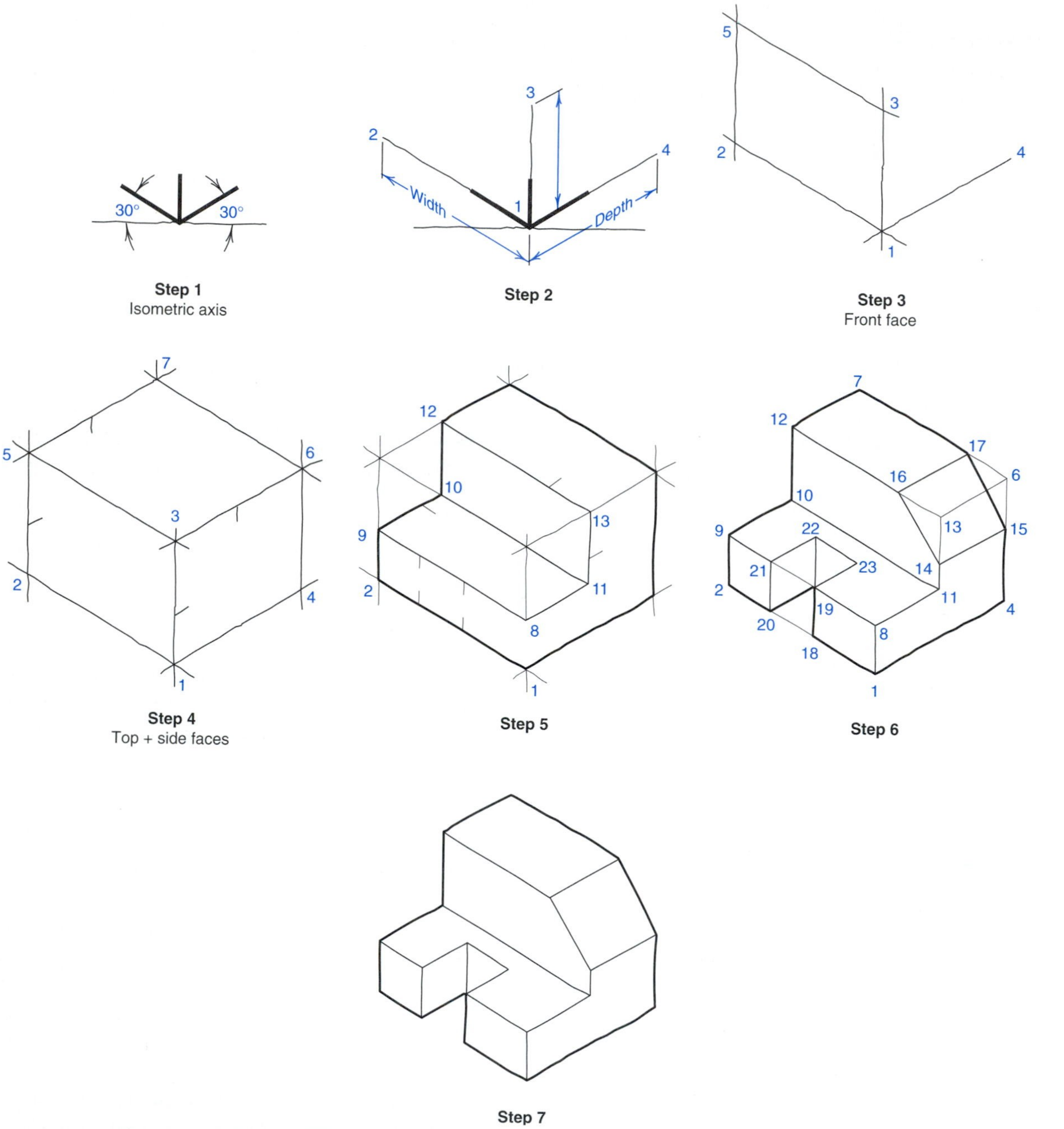

Figure 2.26 **The Basic Steps Used to Create an Isometric Sketch of an Object**

Adding details to the isometric block

Step 5. Begin by estimating the dimensions to cut out the upper front corner of the block, and mark these points as shown in Step 4. Sketch the height along the front face by creating a line parallel to line 1–2; label it 8–9. Sketch 30-degree lines from points 8 and 9 and label these lines 9–10 and 8–11. Now sketch a line from point 10 to point 11. Sketch vertical lines from points 10 and 11 and label the new lines 10–12 and 11–13. Sketch a line from point 12 to point 13 to complete the front cutout of the block.

With a simple sketch, you can often lay out all of your construction lines before having to darken in your final linework.

With more complicated sketches, the sheer number of construction lines can often cause confusion as to which line belongs to which feature. The confusion can be worse in an isometric sketch, where the lines represent three dimensions rather than two. Therefore, after the marks are made for the last two features in Step 5, you can begin darkening in some of the lines representing the final form.

Step 6. Estimate the distances to create the angled surface of the block, and mark these points, as shown in Step 5. From the marked point on line 11–13, sketch a 30-degree line to the rear of the block on line 4–6. Label this new line 14–15. From the marked point on line 12–13, sketch a 30-degree line to the rear of the block on line 6–7. Label this new line 16–17. Sketch a line from point 14 to point 16 and from point 15 to point 17 to complete the sketching of the angled surface. Lines 14–16 and 15–17 are referred to as **nonisometric lines** because they are not parallel to the isometric axis.

Estimate the distances for the notch taken out of the front of the block, and mark these points, as shown in Step 5. Draw vertical lines from the marked points on line 1–2 and line 8–9. Label these lines 18–19 and 20–21, as shown in Step 6. Sketch 30-degree lines from points 19, 20, and 21 to the estimated depth of the notch. Along the top surface of the notch, connect the endpoints of the 30-degree lines, and label this new line 22–23. The 30-degree line extending back from point 20 is stopped when it intersects line 18–19, as shown in Step 6. To complete the back portion of the notch, drop a vertical line from point 22, as shown in Step 6. Stop this new line at the intersection point of line 19–23. The rough isometric sketch of the block is complete.

Note that we have not mentioned hidden features representing details behind the visible surfaces. The drawing convention for isometric sketches calls for disregarding hidden features unless they are absolutely necessary to describe the object.

Step 7. Darken all visible lines to complete the isometric sketch. Since the construction lines are drawn light, there is no need to lighten them in the completed sketch.

2.4.2 Creating an Isometric Sketch from a Three-View Drawing

The following steps will describe how to create an isometric sketch if you are given a three-view drawing. The basic procedures are to determine the desired view, create the isometric axes, box in the object using the overall length, width, and depth dimensions from the three-view drawing (Figure 2.27), then sketch in the details to create the final isometric sketch.

Step by Step: Constructing an Isometric Sketch from a Three-View Drawing

Step 1. Determine the desired view of the object, then sketch the isometric axes.

Step 2. Construct the front isometric plane using the W and H dimensions. Construct the top isometric plane using the W and D dimensions. Construct the right side isometric plane using the D and H dimensions.

Step 3. Determine dimensions X and Y from the front view and transfer them to the front face of the isometric drawing. Project distance X along an isometric line parallel to the W line. Project distance Y along an isometric line parallel to the H line. Point Z will be located where the projectors for X and Y intersect.

Step 4. Sketch lines from point Z to the upper corners of the front face. Project point Z to the back plane of the box on an isometric line parallel and equal in length to the D line. Sketch lines to the upper corner of the back plane to complete the isometric sketch of the object.

Notice that the 45-degree angles do not measure 45 degrees in the isometric view. This is an example of why no angular measures are taken from a multiview to construct an isometric sketch.

2.4.3 Isometric Ellipses

An isometric ellipse is a special type of ellipse used to represent holes and ends of cylinders in isometric drawings. In an isometric drawing, the object is viewed at an angle, which makes circles appear as ellipses. When sketching an isometric ellipse, it is very important to place the major and minor axes in the proper positions. Figure 2.28 is an isometric cube with ellipses drawn on the three visible surfaces: top, profile, and front. Remember Figure 2.28A, because those are the three positions of isometric ellipses found on most isometric sketches and drawings. The following are the key features of the isometric ellipse on each plane:

- The major and minor axes are always perpendicular to each other.
- On the top plane, the major axis is horizontal, and the minor axis is vertical.
- On the front and profile planes, the major axes are measured 60 degrees to the horizontal.
- The major axis is always perpendicular to the axis running through the center of the hole or cylinder.

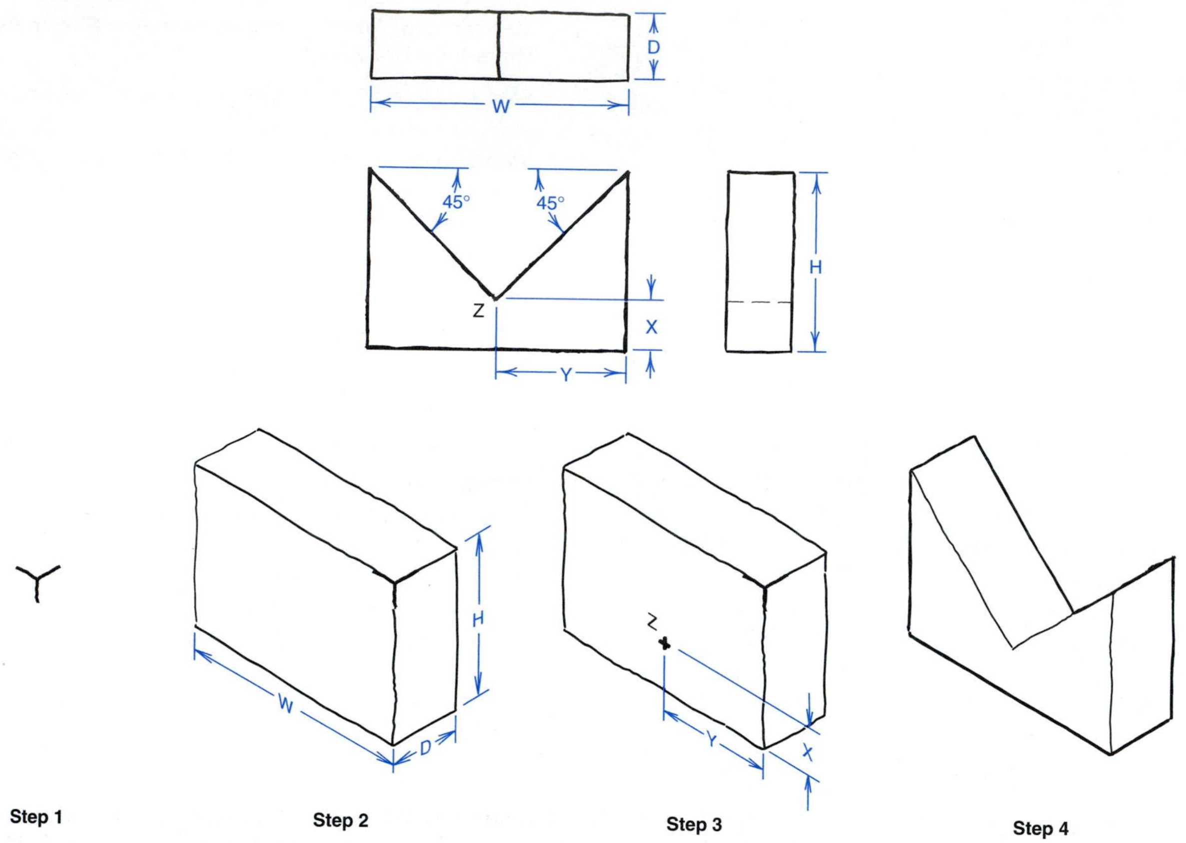

Figure 2.27 **Constructing an Isometric Sketch from a Three-View Drawing**

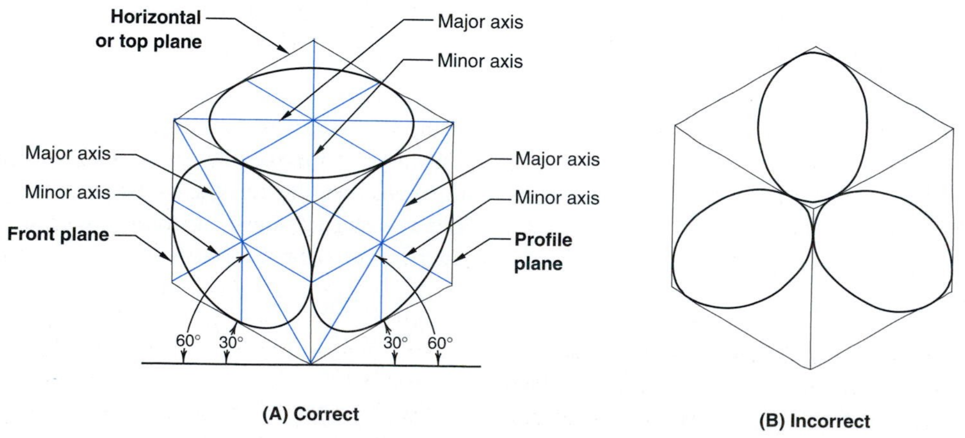

Figure 2.28 **Isometric Representation of Circles**

Circular features appear as ellipses in isometric sketches. The orientation of the ellipse is set according to the face on which the circle lies. The correct orientation is shown in (A), and examples of incorrect orientations are shown in (B).

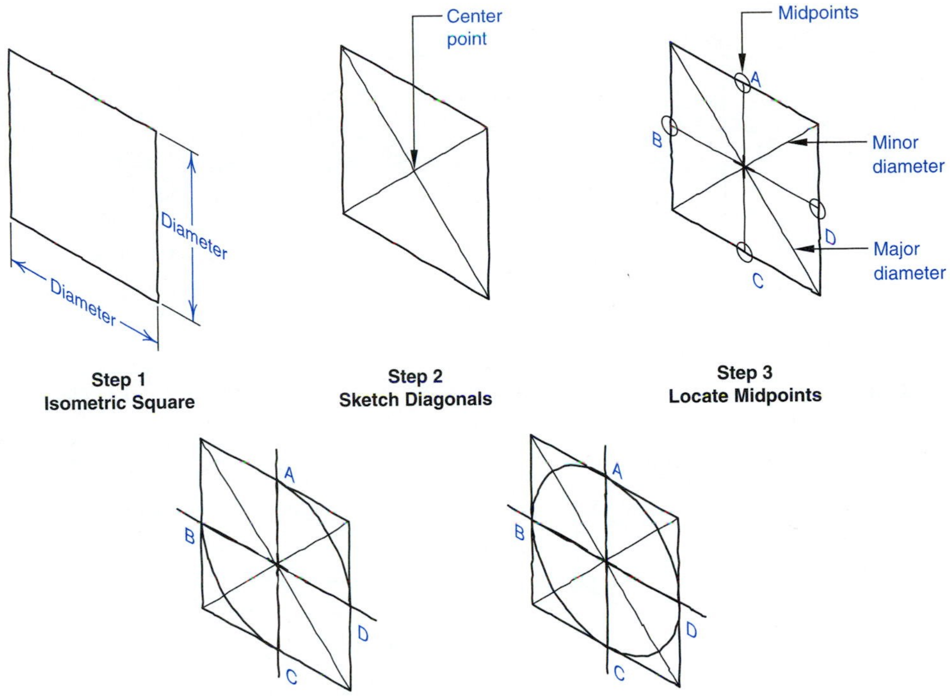

Figure 2.29 Sketching an Isometric Ellipse

The steps used to create a sketch of an isometric ellipse begin with constructing an isometric box whose sides are equal to the diameter of the circle. The center of the box and the midpoints of the sides are found, and arcs are then drawn to create the ellipse.

Sketching an Isometric Ellipse

Figure 2.29 shows the steps for creating an isometric ellipse. Notice that the steps are almost identical to those for sketching a circle as explained earlier in this chapter. The difference is in the orientation and proportion of the primary axes.

Step 1. This isometric ellipse will be drawn on the front plane. Begin by sketching an isometric square whose sides are equal to the diameter of the circle.

Step 2. Add construction lines across the diagonals of the square. The long diagonal is the **major axis,** and the short diagonal is the **minor axis** of the ellipse. The two diagonals intersect at the center of the square, which is also the center of the isometric ellipse.

Step 3. Sketch construction lines from the midpoints of the sides of the square through the center point. These lines represent the center lines for the isometric ellipse. The midpoints of the sides of the isometric square will be tangent points for the ellipse and are labeled points A, B, C, and D.

Step 4. Sketch short, elliptical arcs between points B and C and points D and A.

Step 5. Finish the sketch by drawing the elliptical arcs between points C and D and points A and B, completing the ellipse.

Sketching an Isometric Cylinder

Figure 2.30 shows the steps for creating an isometric view of a cylinder.

Step 1. Sketch the isometric axis. To sketch the bounding box for the cylinder, begin on one 30-degree axis line and sketch an isometric square with sides equal to the diameter of the cylinder. This square will become the end of the cylinder. Next, mark the length of the cylinder on the other 30-degree axis line, and sketch the profile and top rectangles of the bounding box. For the profile rectangle, the length represents the length of the cylinder, and the height represents the diameter of the cylinder. For the top rectangle, again the length represents the length of the cylinder, but the width represents the diameter of the cylinder. Note that only three long edges of the bounding box are drawn (the hidden one is not), and only two lines for the back end

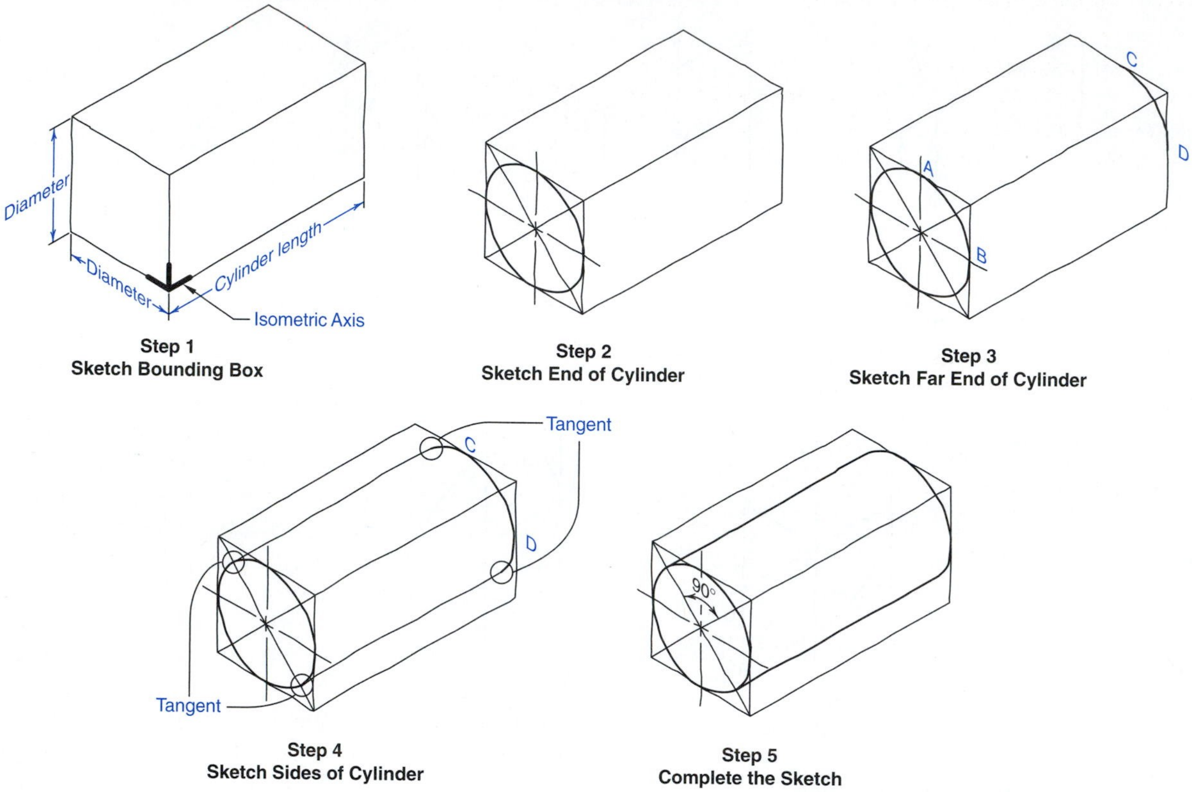

Figure 2.30 Steps Used to Construct a Sketch of an Isometric Cylinder

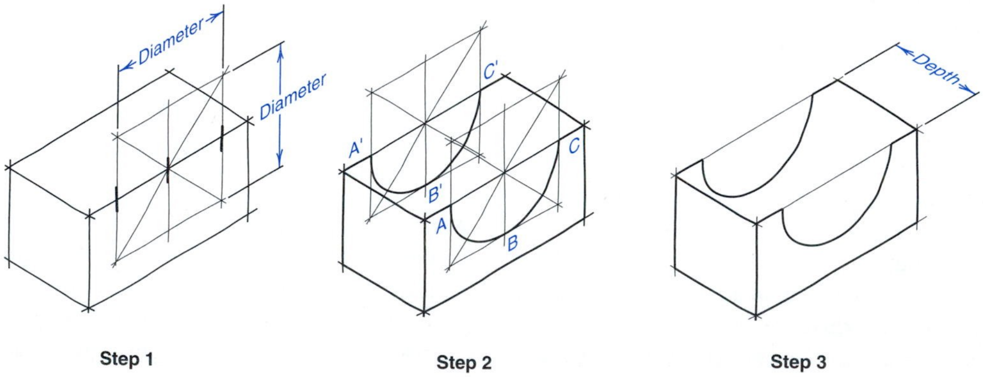

Figure 2.31 Steps Used to Construct a Sketch of a Semi-Ellipse

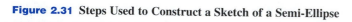

of the bounding box are drawn (the two hidden ones are not).

Step 2. Draw diagonals and center lines on the isometric square, and sketch in the ellipse to create the end of the cylinder, as described in "Sketching an Isometric Ellipse."

Step 3. Where the center lines intersect with the top and front sides of the isometric square, mark points A and B. Sketch construction lines from points A and B to the back end of the bounding box and mark points C and D. Sketch an arc between points C and D.

Step 4. On the isometric square, locate the two points where the long diagonal intersects with the ellipse. From those two points, sketch two 30-degree lines to the back of the bounding box. (These 30-degree lines are tangent to the ellipse on the front of the cylinder.) Then, sketch short elliptical arcs from points C and D to the tangent lines, as shown in the figure. The cylinder should now be visible in the bounding box.

Step 5. Darken all visible lines to complete the cylinder. Note that the major axis of the ellipse is perpendicular to the axis running through the center of the cylinder, and the minor axis is coincident to it.

Sketching Semi-Ellipses

Figure 2.31 shows how to sketch a semi-ellipse.

Step 1. This isometric ellipse will be drawn on the profile plane. Begin by sketching an isometric square whose sides are equal to the diameter of the arc. Add construction lines across the diagonals of the square. The two diagonals intersect at the center of the square, which is also the center of the isometric ellipse. Sketch construction lines from the midpoints of the sides of the square through the center point. These lines represent the center lines for the isometric ellipse.

Step 2. The midpoints of the sides of the isometric square will be tangent points for the ellipse and are labeled points A, B, and C. The long diagonal is the major axis, and the short diagonal is the minor axis. Sketch short, elliptical arcs between points B and C and points B and A, which creates the elliptical arc on the near side of the object. The back part of the semi-ellipse can be sketched by constructing 30-degree parallel lines that are equal in length to the depth of the part, from points A, B, and C. This locates points A′, B′, and C′ on the back side of the object. Add the back ellipse by sketching an arc between points B′ and C′ and points B′ and A′.

Step 3. Finish by darkening the final lines and lightening the construction lines.

2.4.4 Isometric Grid Paper

The use of isometric grid paper can improve your technique and decrease the time necessary to create an isometric sketch. Isometric grid paper is made of vertical and 30-degree grid lines, as shown in Figure 2.6B. There are two primary advantages to using isometric grid paper. First, there is the advantage obtained by using any kind of grid paper. Proportions of the object's features can be translated into certain numbers of blocks on the grid. This can be especially useful when transferring the dimensions of a feature from one end of the object to the other. Unlike square grid paper, each intersection on an isometric grid has three lines passing through it, one for each of the primary axis lines. This can create some confusion when counting out grid blocks for proportions. Just remember which axis line you are using and count every intersection as a grid block.

The second advantage of the isometric grid is the assistance it provides in drawing along the primary axis lines. Although estimating a vertical line is not difficult, estimating a 30-degree line and keeping it consistent throughout the sketch is more challenging. Remember that the only dimensions that can be transferred directly to an isometric sketch are the three primary dimensions. These dimensions will follow the lines of the grid paper. When blocking in an isometric sketch, lay out the dimensions on construction lines that run parallel to the grid lines. Angled surfaces are created using construction lines that are nonisometric; that is, they do not run parallel to any of the grid lines and are drawn indirectly by connecting points marked on isometric construction lines.

Practice Exercise 2.2

Using isometric grid paper, sketch common, everyday objects. Some examples are given in Figure 2.32. Sketch objects with a variety of features. Some should require sketching isometric ellipses, while others should have angled surfaces that require nonisometric lines. Start with simpler forms that only contain isometric lines and work toward more complex forms. Another approach is simply to leave out some of the details. You can capture the essence of the form by representing just its primary features. This is a common approach in creating ideation sketches.

The cost and availability of isometric grid paper can be a discouraging factor in using it to create lots of sketches. You can minimize the expense by using roll tracing paper over a sheet of grid paper. The two sheets can be held together with low-tack tape or put in a clipboard. With practice, you will find that grid paper is not needed and you can create sketches on the tracing paper alone.

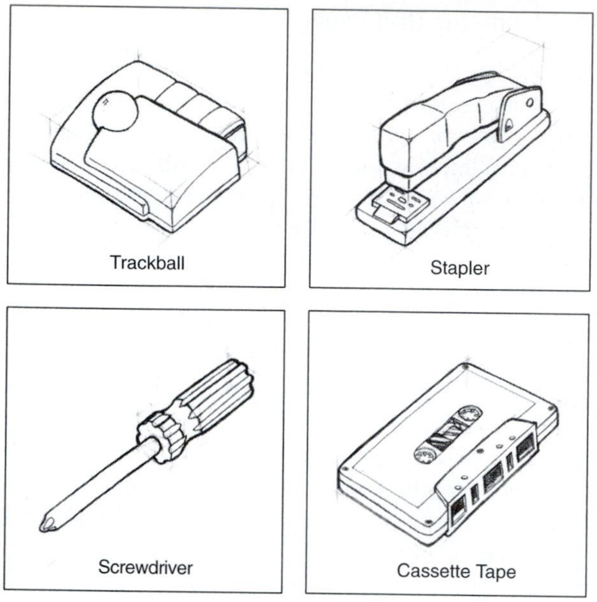

Figure 2.32 **Isometric Sketches of Common Objects**

2.4.5 Multiview Projections

Multiview drawings are based on parallel projection techniques and are used when there is a need to represent the features of an object more accurately than is possible with a single (pictorial) view. A multiview drawing is a collection of *flat* 2-D drawings that work together to give you an accurate representation of the overall object. With a pictorial drawing, all three dimensions of the object are represented in a single view. The disadvantage of this approach is that not all the features in all three dimensions can be shown with optimal clarity. In a multiview projection, however, each view concentrates on only two dimensions of the object, so particular features can be shown with a minimum of distortion (Figure 2.33). Enough views are generated to capture all the important features of the object.

Given their advantages, why are multiview drawings not always used? For one thing, there are the multiple views, rather than a single view, to be drawn. These views must be coordinated with one another to represent the object properly. You have to carefully visualize the views as you sketch them, and so does the person viewing them. Without training and experience, you might find it hard to

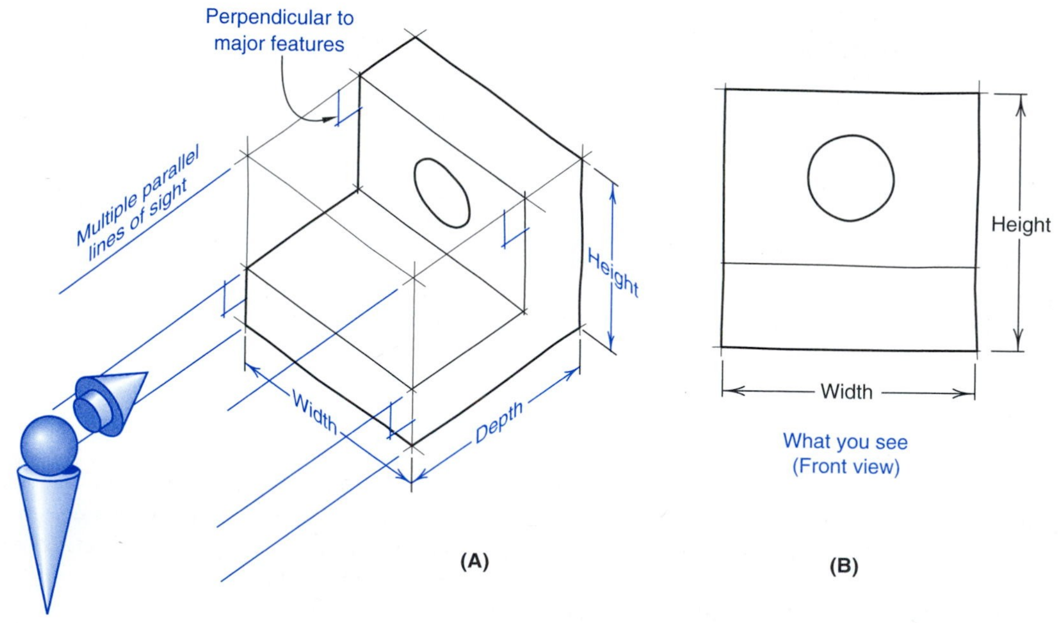

Figure 2.33 **Multiview Drawings**

Multiview drawings are classified as a parallel projection, because the lines of sight used to view the object are parallel. This method of viewing an object results in a single view, with only two of the three dimensions represented. Therefore, it takes multiple views to show all three dimensions.

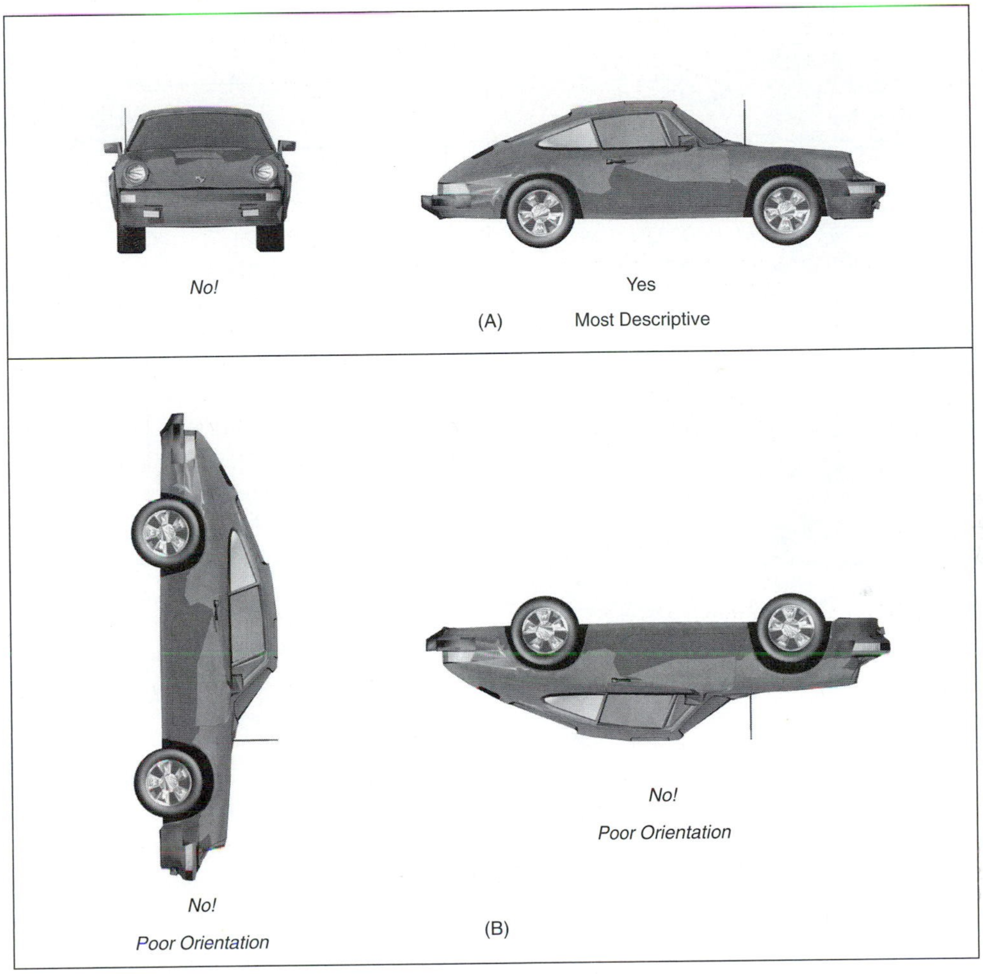

Figure 2.34 **Most Descriptive View**
Proper orientation and the most descriptive features of an object are important when establishing the front view for a multiview sketch. Objects should be positioned in their natural orientation (A); for this car, that position is on its wheels.

interpret multiview drawings. The choice between multiviews and pictorials is often one of exact representation of features versus ease of sketching and viewing.

Orienting and Selecting the Front View When creating a multiview drawing of a design, the selection and orientation of the front view is an important first step. The front view is chosen as the most descriptive of the object; for example, what would normally be thought of as the side of the car is chosen as the front view because it is the most descriptive (Figure 2.34A). In addition, the object must be properly oriented in the view. The orientation of the object is based on its function. For example, for an automobile, the normal or operating position is on its wheels, rather than on its roof or bumpers (Figure 2.34B).

Choosing the Views for a Multiview Drawing Another way to understand the views of an object is to pick it up and turn it around. This may be hard to imagine with something like a car, but many of the objects you will be sketching are considerably smaller and can be held in the hand. Imagine picking up the object shown in Figure 2.35 and holding it so that you are looking at the front. Now, rotate the object

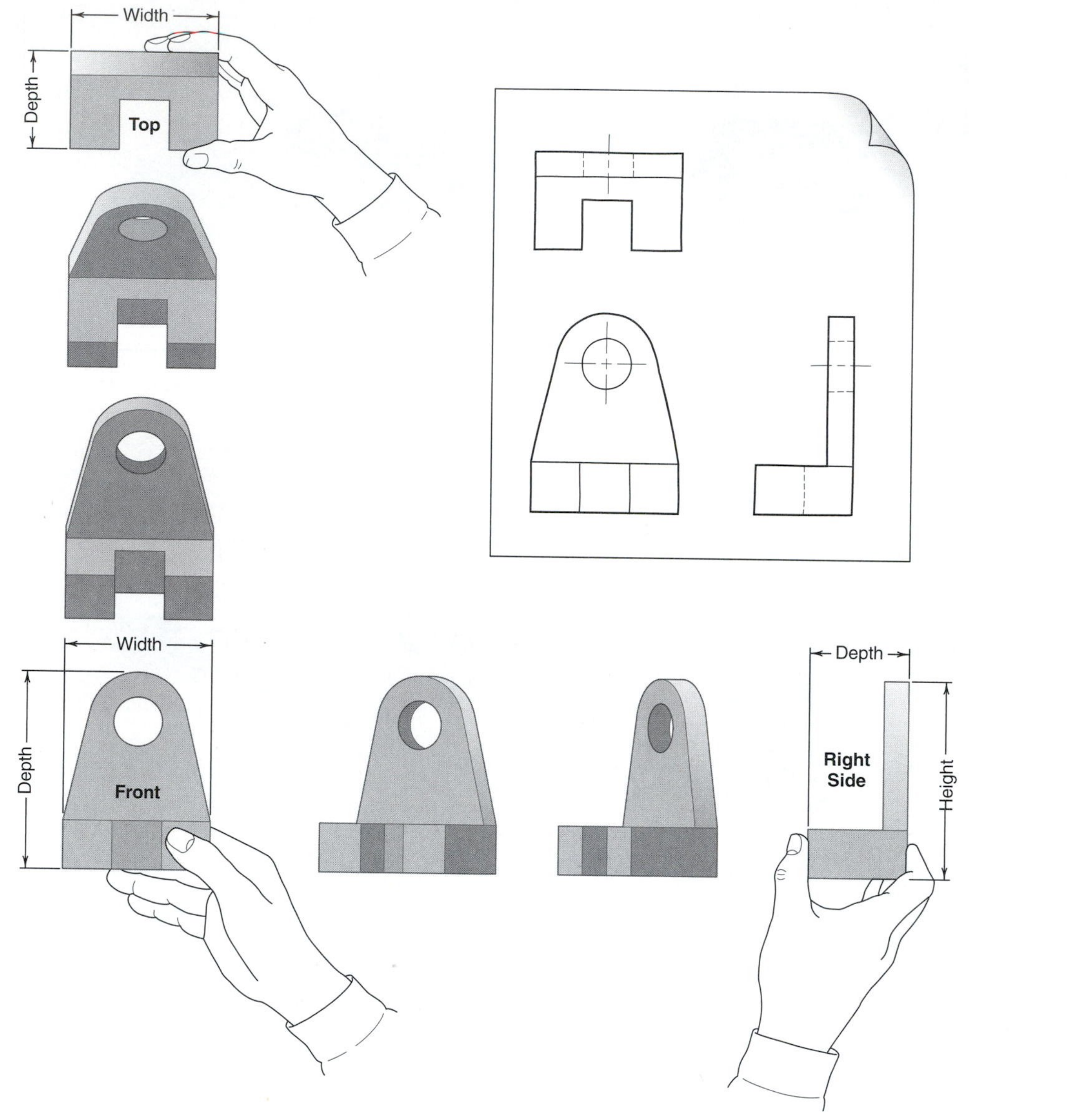

Figure 2.35 Visualizing a Multiview Drawing

By rotating a real object in your hand, you can simulate how a multiview drawing is created. A different principal view of the object is produced for every 90 degrees of rotation.

so that you are looking at its top. Rotate it back to where you started and then rotate it so you are looking at its right side. There are an infinite number of intermediate views of the object between the points of rotation at which you stopped; for right now, however, consider only those views that are rotated 90 degrees from each other. These are considered *regular* or *principal views,* and each represents two primary dimensions of the object. If you continue rotating the object in 90-degree increments, you will see as many as six regular views (Figure 2.36).

A multiview drawing should have the minimum number of views necessary to describe an object completely. Normally, three views are all that are needed; however, the three views chosen must be the most descriptive ones.

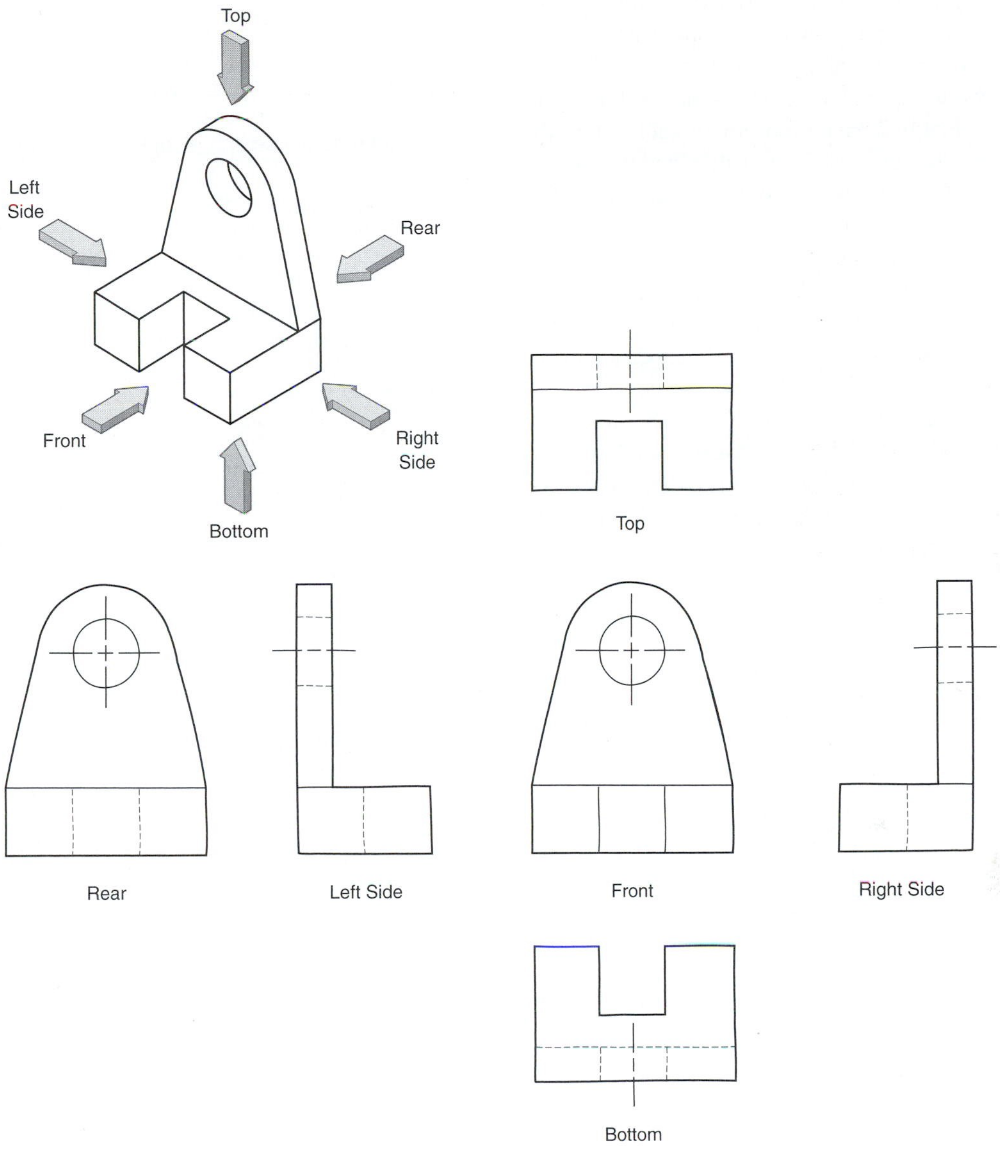

Figure 2.36 Six Principal Views

A multiview drawing of an object will produce six views, called regular or principal views. The object is viewed from six mutually perpendicular viewpoints. The six views are called front, top, bottom, rear, right side, and left side.

The most descriptive views are those that reveal the most information about the design, with the fewest features hidden from view.

For example, Figure 2.36 shows one isometric view and six orthographic views of a simple object. Of the six orthographic views, which are the most descriptive? Although all the views given reveal much about the size and shape of the object, some are more descriptive than others. The choice of views is determined as follows:

1. Identify the most descriptive or important features of the object.
2. Determine the views that best represent those features.

After deciding on the most descriptive features of the part, choose the views that show these features. Part of this selection will involve determining which views do the best job of neither obscuring nor hiding other features. For example, the object in Figure 2.36 has four important features: the hole, the rounded top, the L-shaped profile, and the slot cut out of the base in front. There are only two views that show the hole and rounded top: the front and rear. Although both views show these features equally well, the front view is chosen over the rear view because it does not hide the slot cut in the base. The L-shaped profile is shown equally well in both the right and left side views, and they have an equal number of hidden features. Although either view can be used, convention favors choosing the right side view. The slot in the base is shown in both the top and bottom views. However, the top view has fewer hidden lines, so it is preferred over the bottom view for the sketch. For this object then, the preferred views are the front, top, and right side views.

Practice Exercise 2.3

Find a number of small objects that can be picked up in your hand and rotated. Look at them carefully and identify their important features. Orient each object so that the most important features are seen. Make this view the front view. Before rotating them, try to imagine what the top and right side views will look like. Now rotate them and see if that is how they look. Do all of the important features show up in these three views? If not, what other front view can you select that would result in seeing all of the important features in the three views?

Next, look at some larger items you cannot pick up. For each object, move around it and look at it from different viewpoints. Try to imagine picking it up and rotating it. Is there any difference in how the object looks if you pick it up and rotate it rather than walk around it?

Figure 2.37 shows some common objects represented in multiview drawings.

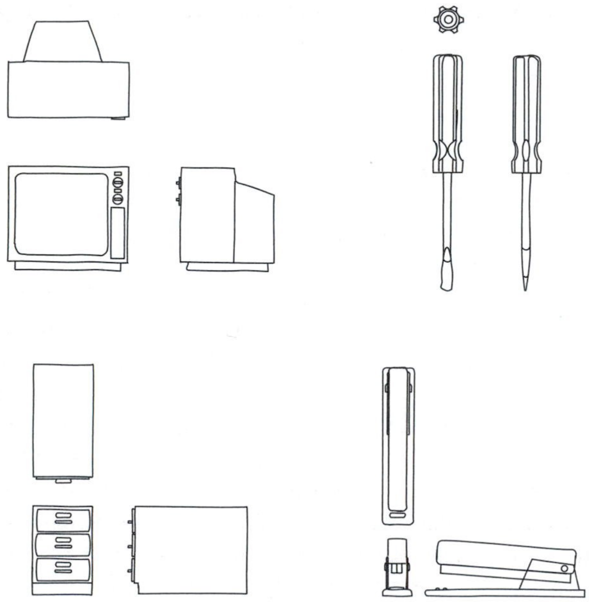

Figure 2.37 Multiviews of Common Objects

These multiview drawings of common objects show the front, top, and right side views. Hidden lines have been omitted for clarity.

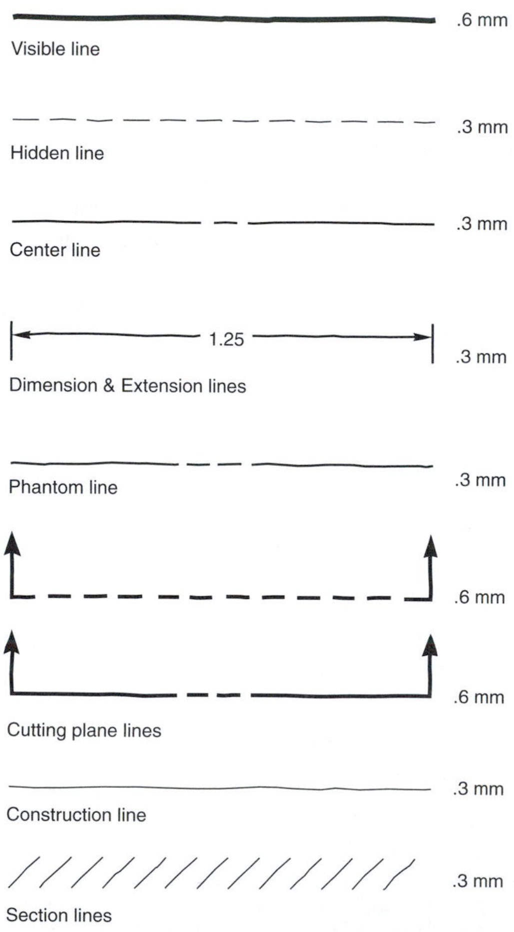

Figure 2.38 Sketched Alphabet of Lines

Standard engineering drawing practice requires the use of standard linetypes, which are called the alphabet of lines. The sizes show the recommended line thicknesses.

2.5 | MULTIVIEW SKETCHING TECHNIQUE

As with other types of drawings, there is a certain technique, or standard set of methods, followed when creating multiview sketches. The technique includes line conventions, proportioning, and methods for creating circles, arcs, and ellipses. A knowledge of the proper and effective technique will assist the beginner in creating legible multiview sketches.

2.5.1 Line Conventions

As in all engineering and technical drawings, multiview drawings and sketches require adherence to the proper use of the alphabet of lines. Figure 2.38 shows the alphabet of lines, sketched rather than drawn with drawing instruments, and includes the recommended pencil thicknesses. Figure 2.39 shows the application of each linetype on an engineering sketch.

In engineering and technical drawing, it is important that hidden features be represented so that the reader of the drawing can clearly understand the object. Many conventions related to hidden lines have been established over the years. Figure 2.40 shows the **hidden line conventions** that must be followed when creating technical sketches or instrument drawings unless a CAD system has limitations that make it difficult to comply with these requirements. These conventions are as follows:

- There should be no gap when a hidden line intersects a visible line (Figure 2.40A).
- Corners on hidden lines should be joined (Figure 2.40B).
- There should be a gap when a hidden line intersects either a visible corner or a visible arc (Figure 2.40C).
- Three (hidden) intersecting corners, found in holes that are drilled and that end inside the object (i.e., do not go all the way through the object), should be joined as shown in Figure 2.40D.
- At the bottom of the drilled hole, the lines indicating the tip (created by the drill, which has a pointed tip) are joined (Figure 2.40E).
- Hidden arcs are started on the center line or the point of tangency (Figure 2.40F).
- When a hidden line passes behind a visible line (i.e., does not intersect the visible line), do not put a hidden-line dash on the visible line (Figure 2.40G).
- At the point where one hidden line crosses in front of another hidden line (indicating two hidden features, one closer to the visible view than the other), use a dash for the hidden line in front; that is, if the front hidden line is horizontal, use a horizontal dash at the point of crossing (Figure 2.40H).

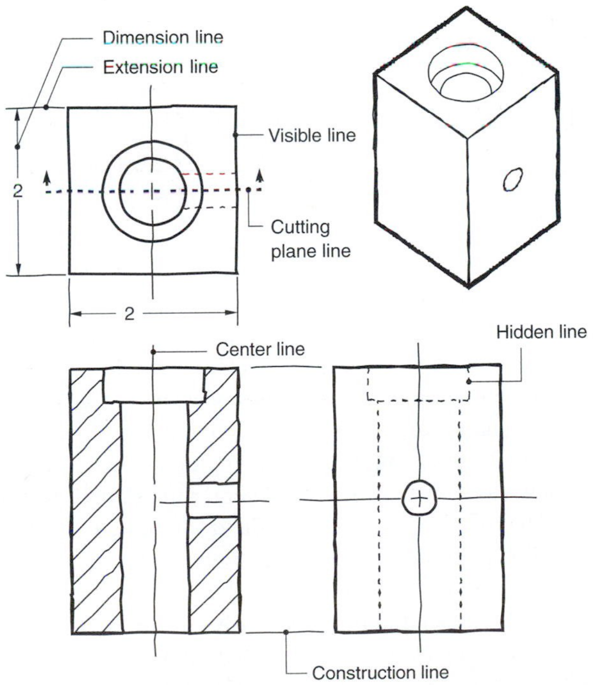

Figure 2.39
This engineering sketch has labels to identify some of the alphabet of lines.

2.5.2 Precedence of Lines

It is fairly common in technical drawings to have two lines in a view coincide. When this happens, the conventional practice called the **precedence of lines** dictates the linetype to draw when two or more lines in a view overlap (Figure 2.41).

For example, in Figure 2.42A a visible line in the top view coincides with the hidden lines for the hole. The precedence of lines requires that the visible lines be drawn and the hidden lines not be shown in the top view. Figure 2.42B shows an example of a hidden line that has precedence over a center line. Figure 2.42C is an example of a visible line having precedence over a center line. Notice that whenever a hidden or visible line has precedence over a center line, the center line is still drawn in the view by leaving a space and then extending it beyond the edge (Figure 2.42D).

2.5.3 Conventional Practices for Circles and Arcs

Circles are used to represent holes and the bases of cones and cylinders. Arcs are used to represent portions of these

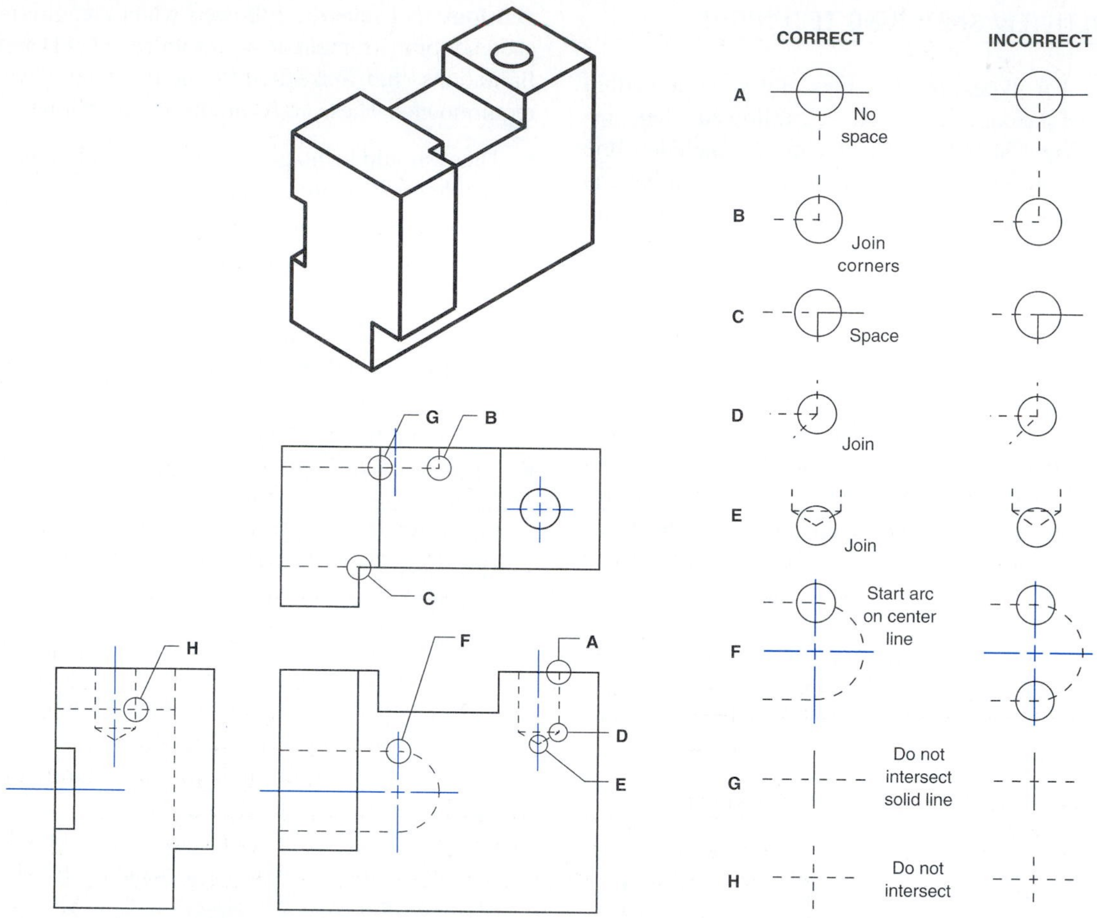

Figure 2.40 Drawing Conventions for Hidden Lines

VISIBLE LINE takes precedence over all other lines

HIDDEN LINE and CUTTING PLANE LINE take precedence over center lines

CENTER LINE does not have precedence

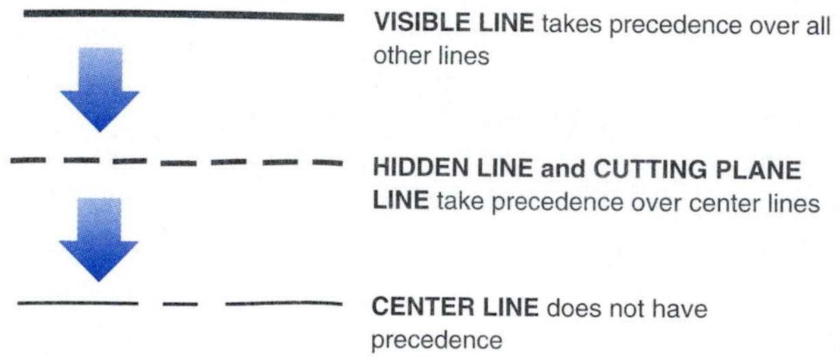

Figure 2.41 Precedence of Lines

The precedence of lines governs which lines are drawn when more than one line occupies the same position on a drawing. For example, a visible line has precedence over all other types of lines, and a hidden line and a cutting plane line have precedence over a center line.

elements, as well as rounded corners on objects. Whenever we are representing a hole, cylinder, or cone on a technical drawing, the conventional practice is to draw center lines, which are used to (1) locate the centers of circles and arcs; (2) represent the axes of cylinders, cones, and other curved surfaces; and (3) represent lines of symmetry.

Figure 2.43 is a multiview drawing of a cylinder. In the top view, horizontal and vertical center lines are drawn to locate the center of the circle. In the front view, a very thin center line (.035 mm) is drawn to show the location of the cylinder's axis. The small dashes that cross at the center of the circle extend approximately 8 mm or ⅜″ from the edges of the object. The short segment of the center line is approximately 3 mm or ⅛″ long. The long segment can be from 20 mm to 40 mm, or ¾″ to 1-½″ long. For very long cylinders, the center line is drawn as a series of long and short segments.

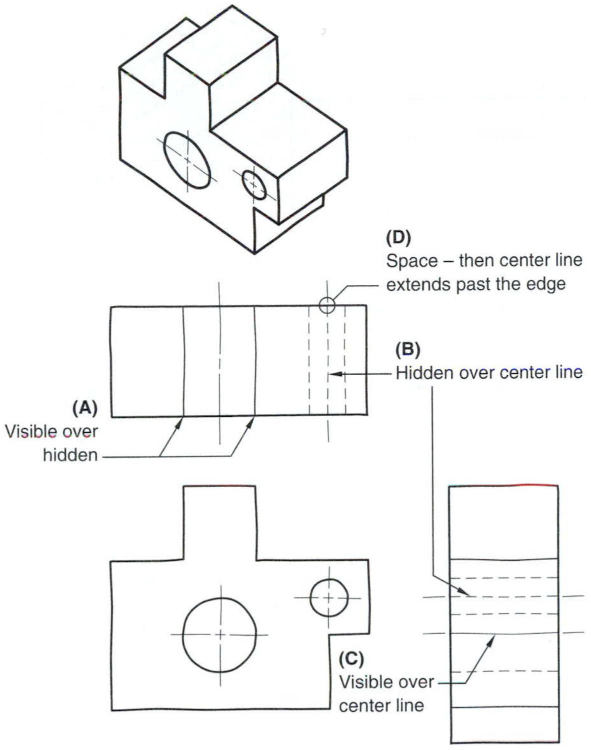

Figure 2.42 An Engineering Drawing Showing How the Precedence of Lines Is Applied

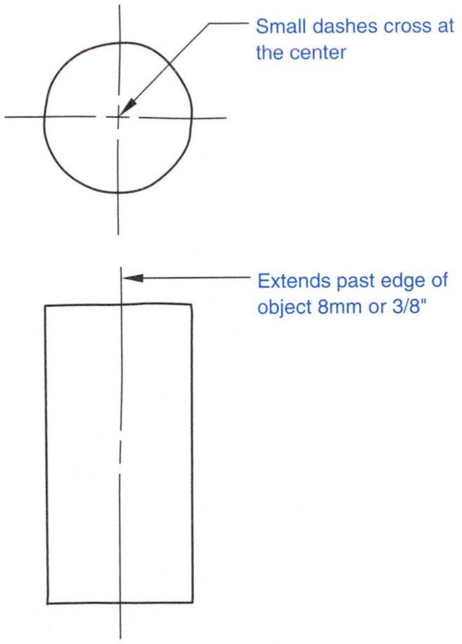

Figure 2.43 An Engineering Drawing of a Cylinder, Showing the Application of Center Lines

Figure 2.44 shows some applications and the associated conventions for placing center lines. Notice that center lines are used in both the circular and horizontal views of a hole. When adding center lines to the circular view of a very small hole, a solid center line may be used rather than a dashed line, as shown in part C. Part D shows how center lines are used to locate the centers of holes around a bolt circle. Part E shows how center lines, along with phantom lines, are used to show a path of motion.

2.6 MULTIVIEW SKETCHES

Multiview drawings can have from one to three or more views of an object. However, multiview sketches rarely have more than three views.

Multiview sketches are important in that they provide more accurate geometric information than a pictorial sketch, without requiring the time that a formal multiview drawing would take. If dimensions are provided, they are usually only for a particular feature(s) and are often only approximations, since these sketches are used early in the design process before final specifications have been made.

As is the goal with all sketches, multiview sketches should be done quickly and clearly. Straightedges, such as triangles and T-squares, should be avoided since they will only slow you down and will compel you toward a level of finish that is inappropriate in sketches. In addition, you should draw only as many views as are necessary to show the features accurately. An initial analysis of the features should tell you if one, two, or three views are needed to clearly show the elements of interest.

2.6.1 One-View Sketches

The simplest multiview sketch represents just the front view of an object. Though not truly a *multiview,* it is still meant to represent only two dimensions of the object, which is the basic concept of multiview drawings. This sketch can be produced using the techniques shown in Figure 2.20.

2.6.2 Two-View Sketches

Occasionally, an object can be completely described using only two views. As an example, Figure 2.45 shows a symmetrical part that can be described using two views. If the front view is as shown in the pictorial, the top and side views would be the same. Nothing would be gained by drawing both the top and side views, so only one of these views is sketched.

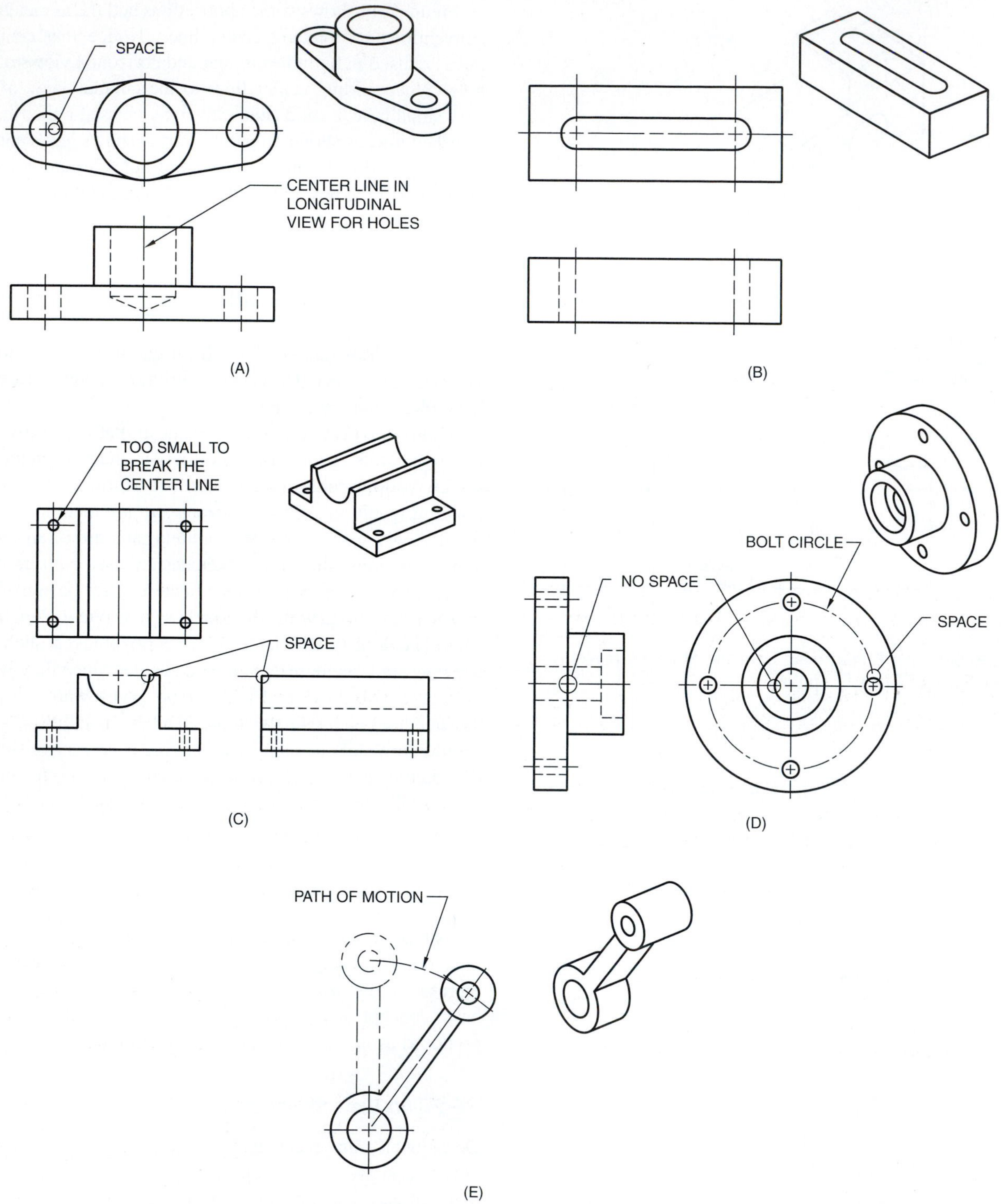

Figure 2.44 **Center Line Conventions**

These engineering drawings show various applications for center lines. Study each example to learn how center lines are used.

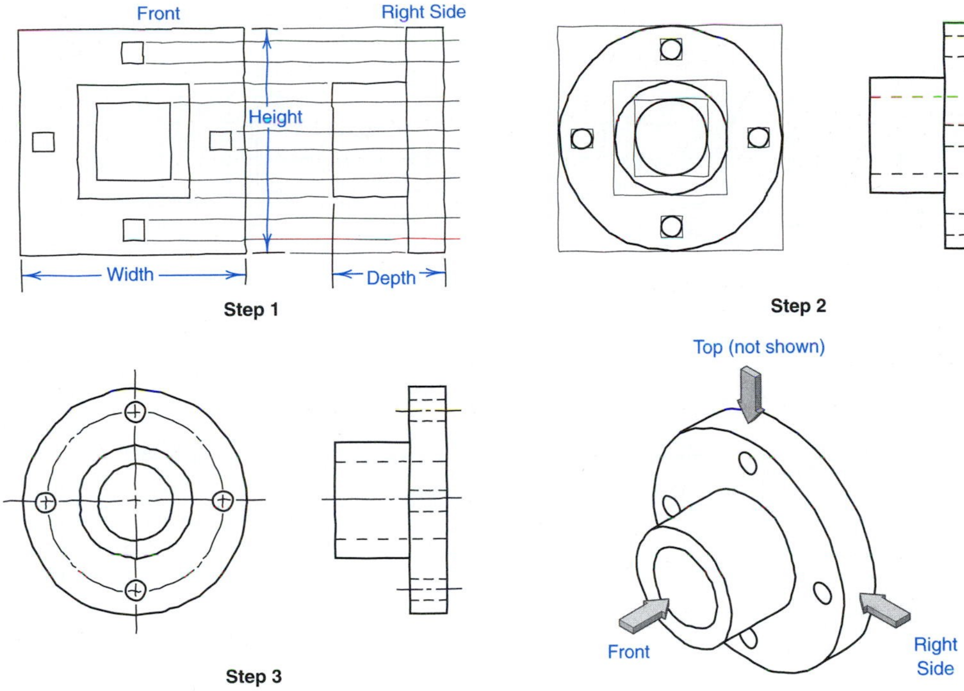

Figure 2.45 **Creating a Two-View Sketch**
A two-view sketch is created by blocking in the details, then adding center lines, circles, arcs, and hidden lines.

Creating a Two-View Sketch

Figure 2.45 and the following steps describe how to create a two-view sketch.

Step 1. In the front view, block in the squares with sides equal to the diameters of the circles. Since both the front and right side views show the height dimension, construction lines can be used to project the height of the squares onto the right side view. Block in the rectangles representing the details for the side view.

Step 2. Using the squares and center lines as guides, sketch the circles for each hole and the cylinder, in the front view. Using the construction lines as guides, sketch the hidden lines for the holes, in the side view.

Step 3. Darken all visible, center, and hidden lines.

2.6.3 Three-View Sketches

When an object is more complex, three views are needed. The object used in Figure 2.46 was chosen because it has many of the most common features you will be sketching, such as holes, arcs, lines, and planes.

Creating a Three-View Sketch

Figure 2.46 and the following steps show how to create a three-view sketch.

Step 1. Begin by blocking in the front, top, and right side views of the object, using the overall width, height, and depth. Sketch the front view first, then use construction lines to project the width dimension from the front view to the top view. Also, use construction lines to project the height dimension from the front view to the right side view. Leave room between the views so the sketch does not look crowded and there is room to add text for notes and dimensions. The spaces between views should be approximately the same. Make sure the depth dimension is equal in the top and side views by measuring the distance using a scale or dividers.

Step 2. Lightly block in the major features seen in each view. For example, the drilled holes are blocked in on the views where they look round. The angled edge and the rounded corner are most clearly seen in the top view. Begin darkening these major features.

Step 3. Quite often, features will appear in two and sometimes all three of the views, and construction lines can be used to project the location or size of a feature from one view

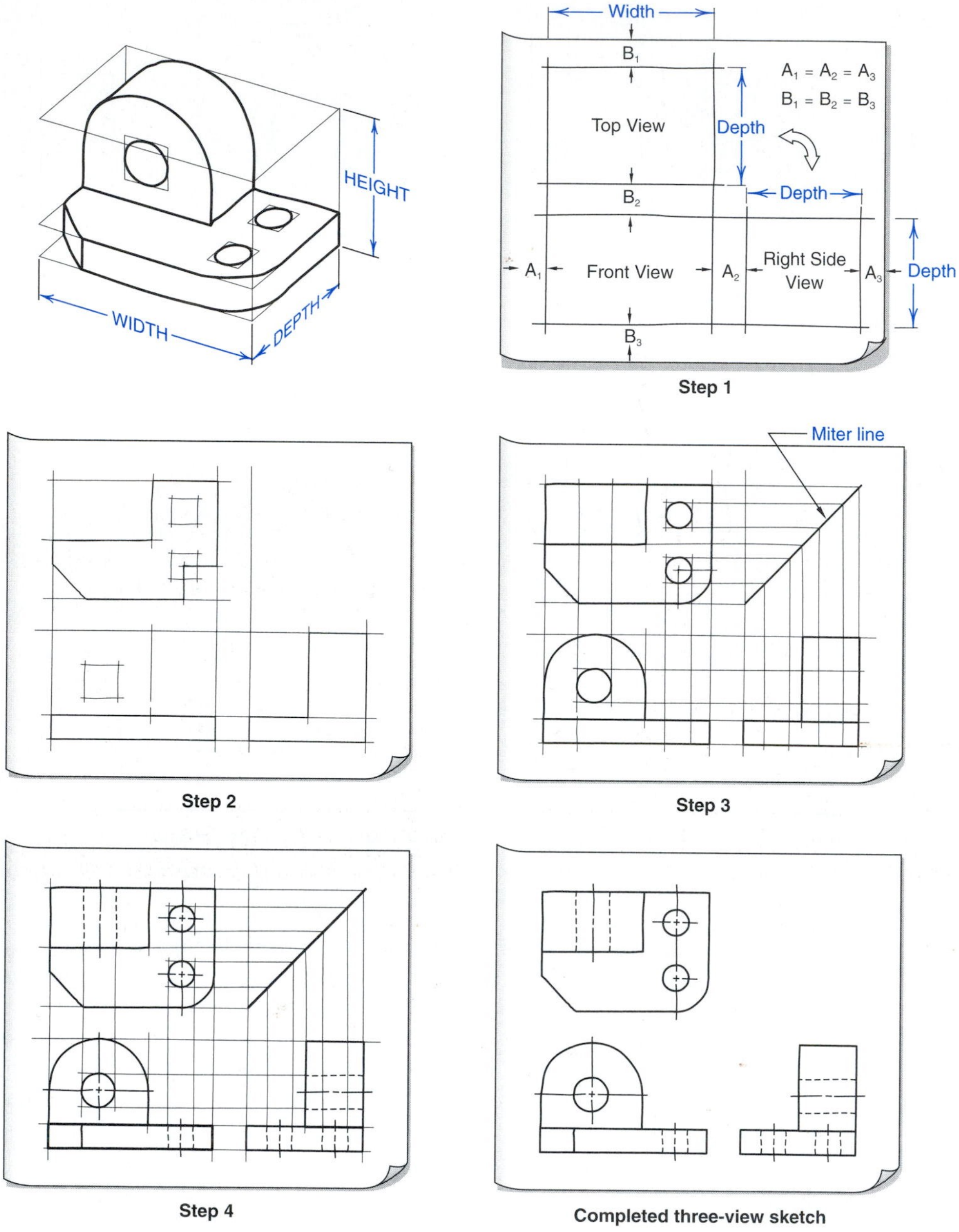

Figure 2.46 **Creating a Three-View Sketch**
A three-view sketch is created by blocking in the overall width, height, and depth dimensions for each view, then blocking in the details, using a miter line to transfer depth dimensions between the top and side views, and then adding the final details, such as circles, arcs, center lines, and hidden lines.

to another. Remember that each view always shares one dimension with an adjoining view. The depth dimension can be shared between the top and right side views with a special construction line called a **miter line.** The miter line is drawn at a 45-degree angle and is used as a point of intersection for lines coming to and from the right side and top views. For example, the width of a hole in the top view can be projected down to the front view. Then the location of the hole can be passed across the miter line to the right side view.

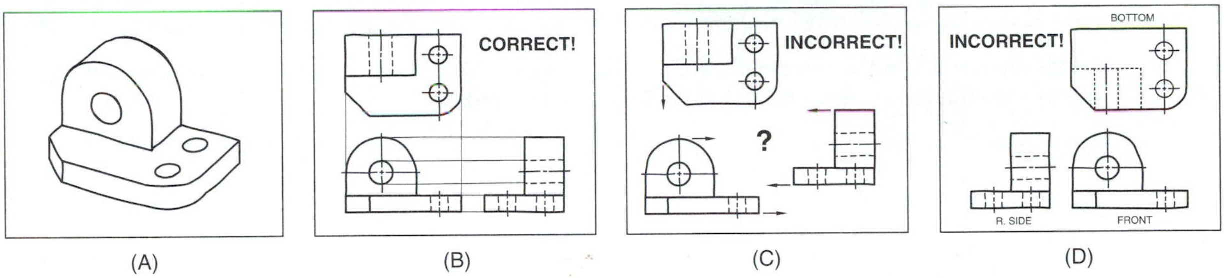

Figure 2.47 **View alignment**

A three-view sketch must be laid out properly, following standard practices. The width dimensions on the front and top views are aligned vertically, and the height dimensions on the front and side views are aligned horizontally.

Step 4. Finish adding the rest of the final lines. Be careful to do all hidden lines and center lines for the holes. Darken all final lines.

As with the two-view drawing, there are conventional practices that must be followed when arranging the three views on the paper (Figure 2.47). Make sure that all three views stay aligned with their neighboring views (Figure 2.47B). If they do not, they will not be able to share dimensions via projection lines. As with the two-view drawing, there is a strict organization for the views: the top view goes directly above and is aligned with the front view, and the right side view goes directly to the right of and is aligned with the front view. Do not rearrange the top, front, or right side views or substitute other regular views in their places.

Figure 2.48 **Convergence in Photographs**

This photograph shows an object in perspective, with major lines of convergence overdrawn.

2.7 | PERSPECTIVE PROJECTION

Perspective projection is the projection method that represents three-dimensional objects on two-dimensional media in a manner closest to how we perceive the objects with our eyes. If you were to take a photograph, lay tracing paper over it, and sketch the objects in it, the result would look like a perspective projection (Figure 2.48). Like the other pictorial sketches, all three dimensions of the object are presented in a single image.

If you were to stand in the middle of a straight, flat road and look toward the horizon (Figure 2.49), the road would appear to narrow to a single point, called the vanishing point. Even though the road appears to go to a point, in reality it does not. You know that the edges of the road are parallel to each other and, as you travel down the road, the portion of the road that looked so small before will be full size. Objects, whether they are portions of a road, cars, or billboards, look smaller as they get farther away. Also, parallel lines—such as the two edges of the road—will appear to *converge* (come together) as they recede from you. Through the use of construction lines in your perspective sketch, you can control the convergence of the parallel edges of an object, as well as the proportional sizes of objects as they recede in the distance (Figure 2.50).

Figure 2.50 shows the labels for the important elements of a perspective sketch. Most important is the **horizon line (HL),** which is an imaginary line in the distance, where objects are infinitely small and parallel lines converge. The point on the horizon line where parallel lines converge is called the **vanishing point (VP).** Where the portion of the object closest to the observer rests on the ground plane is called the **ground line (GL),** as shown in Figures 2.50 and 2.51.

▼ Practice Problem 2.2

Create an isometric sketch of the three-view drawing using the isometric grid shown. Double the size of the isometric sketch by making one square grid equal to two isometric grids.

Figure 2.49
Humans view their environment in perspective, where lines appear to converge to a single point. Even though it looks as though the edges of a road converge on the horizon, we know they don't. By permission of Johnny Hart and Creators Syndicate, Inc.

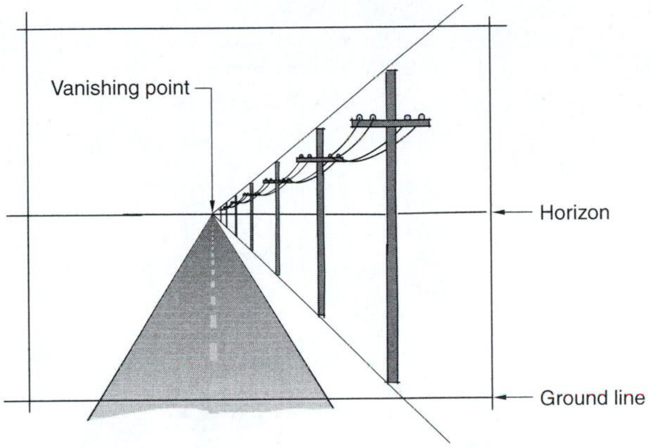

Figure 2.50 Principal Elements in Perspective Sketches
Principal elements of a perspective sketch include the horizon, vanishing point, and ground line. All elements in a perspective sketch are drawn to a single vanishing point.

The horizon line shown in Figures 2.50 and 2.51 also represents the observer's eye level. The relationship of the ground line to the horizon line reflects the height of the observer relative to the object. Figure 2.52 shows the different types of views created by raising or lowering the ground line, and each view type has a specific use, as shown in Table 2.1.

The human's eye view is the most commonly used for sketching everyday objects. The ground's eye is useful for one- to three-story buildings, and the worm's eye view is used for taller structures (Figure 2.53).

Figures 2.49 through 2.52 show perspective projections with only one vanishing point. In Figures 2.51 and 2.52, the parallel edges of only one of the object's three dimensions converge to the vanishing point. If you want more of the dimensions to converge, more vanishing points are needed. Perspective views are classified according to the number of vanishing points used to create the drawing.

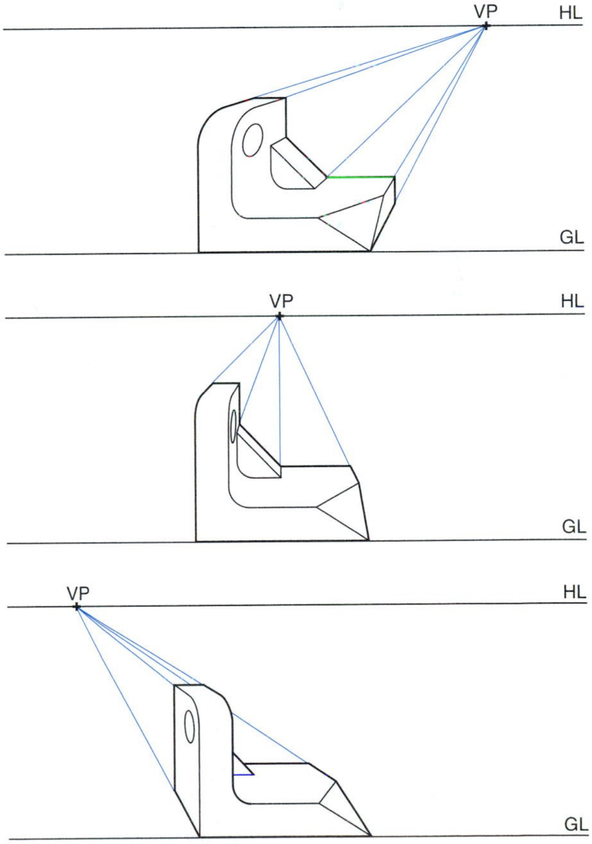

Figure 2.51 Moving the Vanishing Point Right to Left
The view of the object can be dramatically changed by moving the vanishing point along the horizon line. Points to the right of center will reveal details about the right side of the object; points to the left of center will reveal details about the left side of the object.

Table 2.1 Relationship between the Ground line and Horizon Line

View Name	Relationship	Effect
Bird's eye view	Ground line well below the horizon line.	Observer is much taller than the object.
Human's eye view	Ground line six feet below the horizon line.	Observer is somewhat taller than the object.
Ground's eye view	Ground line at the same level as the horizon line.	Observer is the same height as the object.
Worm's eye view	Ground line well above the horizon line.	Observer is much shorter than the object.

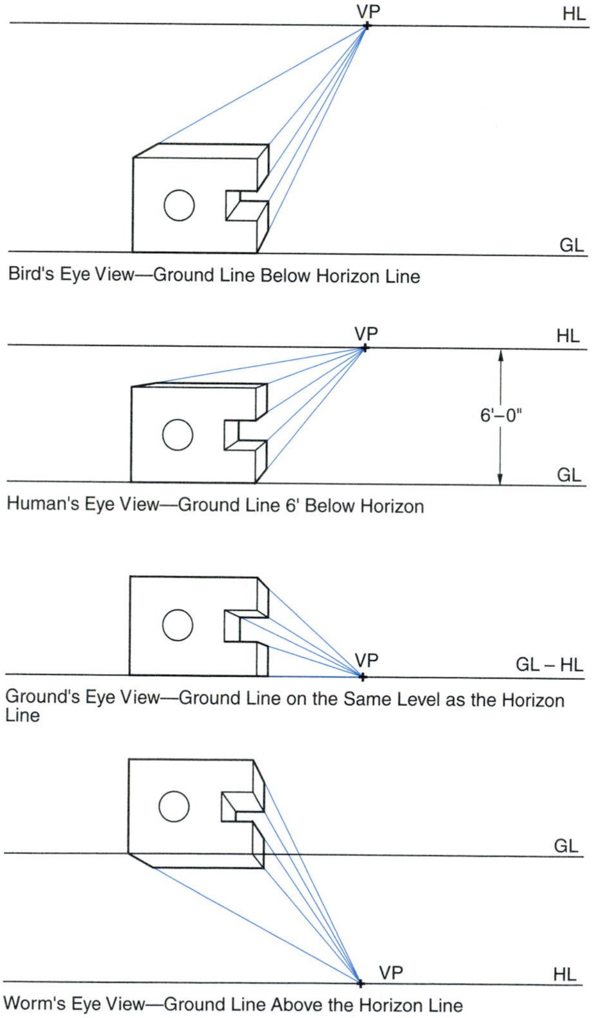

Figure 2.52 Moving the Vanishing Point Vertically

The view of the object can be dramatically changed by moving the ground line relative to the horizon line.

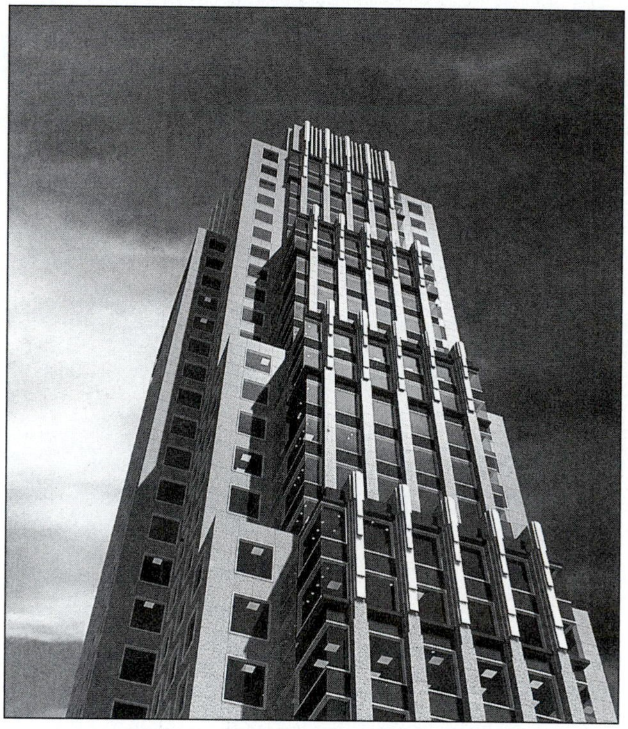

Figure 2.53 A Worm's Eye Perspective View of a Building, Created with CAD

This view simulates what the building would look like when viewed looking up from the ground. Courtesy of Michael Sechman and Associates, Oakland, California.

Figure 2.54 shows one-, two-, and three-point perspective sketches. Although three-point perspective is the most *realistic,* one- and two-point perspectives are usually adequate and are simpler to sketch. By varying the number and positions of the vanishing points, and the position of the ground line relative to the horizon line, it is possible to create virtually any perspective view of an object.

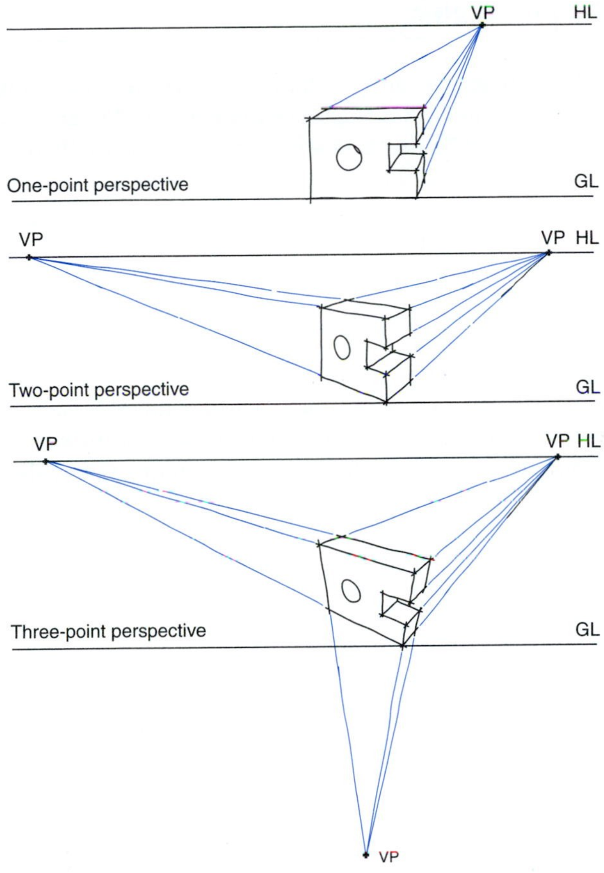

Figure 2.54 One-, Two-, and Three-Point Perspective Sketches of the Same Object

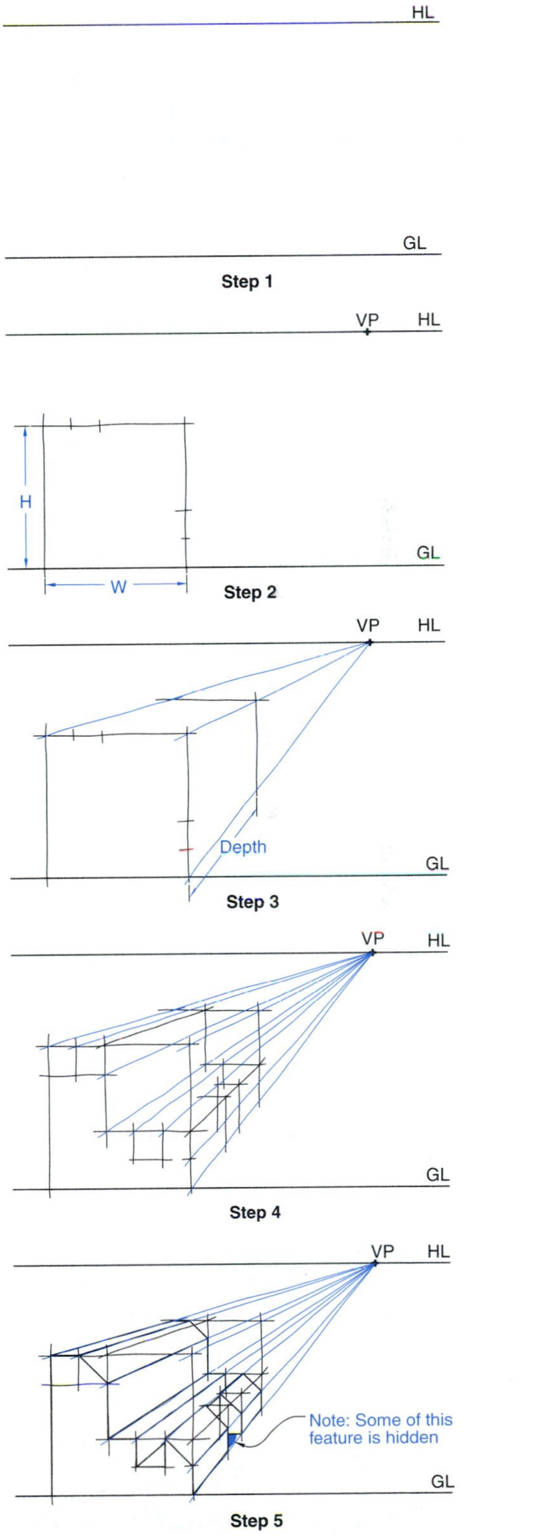

Figure 2.55 Constructing a One-Point Perspective Sketch, Using Five Steps

2.7.1 One-Point Perspective Sketch

One-point perspective sketches are used to quickly create a pictorial representation of an object or structure.

Creating a One-Point Perspective Sketch

The following steps describe how to create a one-point perspective sketch of the guide block shown in Figure 2.55.

Step 1. For a one-point perspective sketch, begin by determining the type of view you want (human's eye, ground's eye, etc.); then place the corresponding horizon and ground lines on the paper.

Step 2. Establish the relationship of the vanishing point to the object, such as to the right of the object. With the vanishing point marked, box in a front view of the object with construction lines, marking the height and width dimensions of the features of the object.

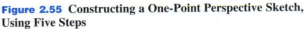

Step 3. From three corners of the bounding box (points 1, 2, and 3), draw converging lines to the vanishing point. The fourth corner would be hidden, so it is not drawn. Based on the depth of the object, determine where the back of the object would be on the converging lines, and draw a horizontal and a vertical line to mark the back of the object.

Step 4. A bounding box in all three dimensions is now established for the object. Continue refining the object by sketching more bounding boxes representing other features. Remember that bounding boxes drawn on the front and back faces of the object will be square, while those placed on the top and sides will not. On the right side, the vertical edges will all be parallel, while on the top the horizontal edges will be parallel. Those edges going back in depth on either face will converge at the vanishing point.

Step 5. Finally, sketch the dark lines representing the object, paying close attention to those edges that converge.

2.8 LETTERING

All engineering and technical drawings, including sketches, contain some notes or dimensions. All text on a drawing must be neat and legible so it can be easily read on both the original drawing and a reproduction, such as a blueprint or photocopy. Although precise lettering is not required, legibility is very important.

Until the invention of printing by Johann Gutenburg in the 15th century, all text was hand lettered, using a very personalized style. With the invention of printing, text styles became more standardized. Although some early technical drawings were embellished with personalized text, current ANSI standards suggest the use of single-stroke Gothic characters for maximum readability (Figure 2.56).

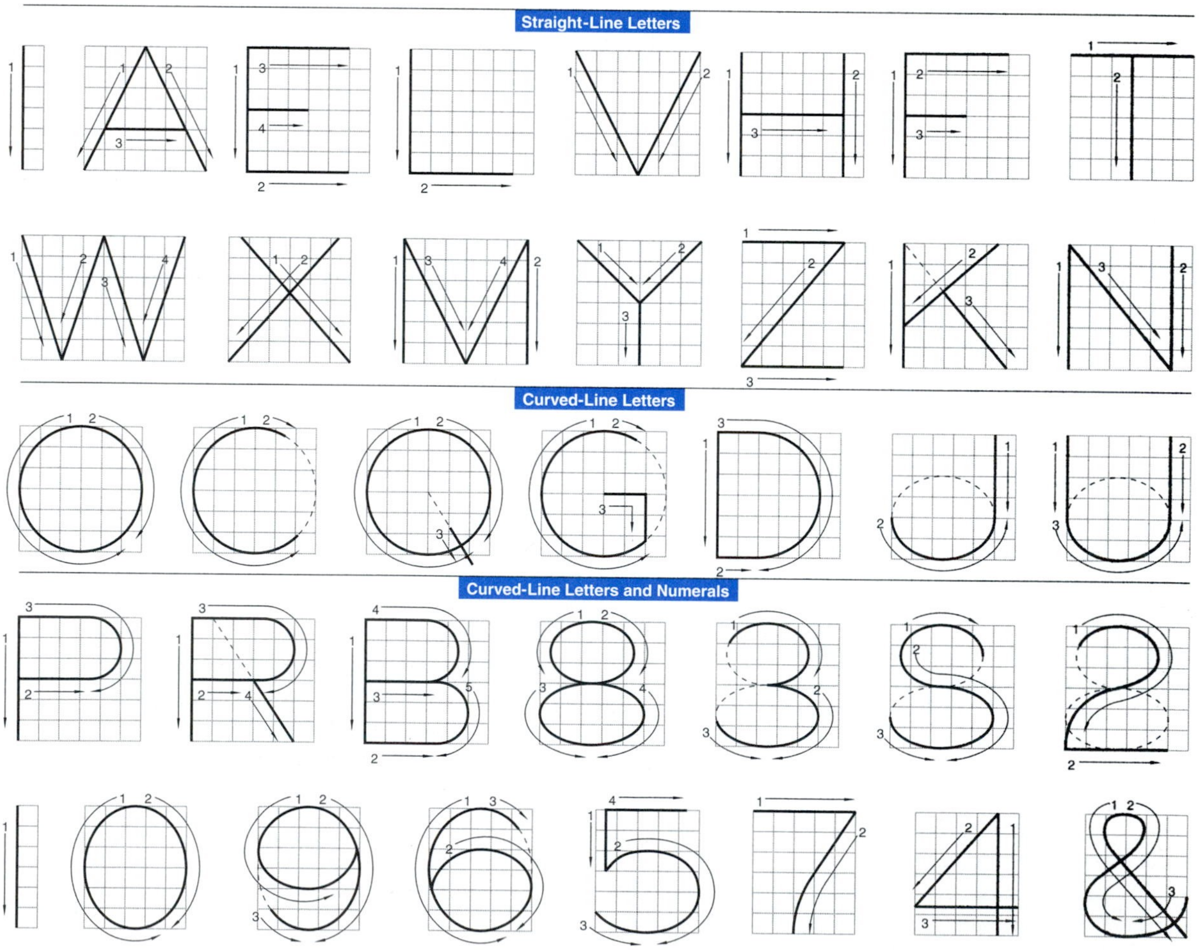

Figure 2.56 Vertical Gothic Letter and Numeral Design, with Suggested Sequence of Strokes That Can Be Used as a Guide for Hand Lettering a Technical Drawing

Questions for Review

1. Define and describe the uses for technical sketching.
2. Define an ideation sketch, and explain how it differs from a document sketch.
3. List the four types of sketches, grouped by projection methods. Sketch an example of each type.
4. Describe the major differences between parallel and perspective projection.
5. Define multiview drawing, and make a sketch of one.
6. Define principal view.
7. Describe the precedence of lines.
8. List the two important uses for text on a drawing.

Further Reading

Edwards, B. *Drawing on the Right Side of the Brain.* New York: St. Martin's Press, 1979.

Hanks, K., and L. Belliston. *Draw!* Los Altos, CA: William Kaufmann, 1977.

Hanks, K., L. Belliston, and D. Edwards. *Design Yourself!* Los Altos, CA: William Kaufmann, 1978.

Knowlton, K. W. *Technical Freehand Drawing and Sketching.* New York: McGraw-Hill, 1977.

Martin, L. C. *Design Graphics.* New York: Macmillan, 1968.

Problems

Use the gridded problem sheets provided at the end of this section to complete the problems that follow.

Hints for Isometric Sketching

- Identify the major features and overall dimensions of the object.
- Use clean, crisp strokes.
- Do not use straightedges or scales when sketching.
- Start by drawing a bounding box, using construction lines.
- Only measure dimensions along the primary axes.
- Do not directly transfer angles from a multiview to a pictorial.
- Use light construction lines to locate vertices and edges.
- Sketch faces roughly in this order:
  1. Normal faces on the perimeter of the bounding box.
  2. Normal faces in the interior of the bounding box.
  3. Inclined faces.
  4. Oblique faces.
- Darken all object lines.

Hints for Multiview Sketching

- Identify the major features and overall dimensions of the object.
- Use clean, crisp strokes.
- Do not use straightedges or scales when sketching.
- Start by drawing bounding boxes and a miter line, using construction lines.
- Align the views.
- Use light construction lines to locate vertices and edges.
- Only measure dimensions along the primary axes.
- Map inclined and oblique faces between all three views.
- Follow the precedence of lines.
- Double-check to make sure there are no missing hidden or center lines.
- Darken all visible, hidden, and center lines.

2.1 Use upside-down sketching to create the word sketch as shown in Figure 2.57.

2.2 Use upside-down sketching to create the table shown in Figure 2.58.

2.3 Use upside-down sketching to create the chair shown in Figure 2.59.

2.4 Use contour sketching to create the series of cubes and cylinders shown in Figure 2.60.

2.5 Use contour sketching to create the overlapping shapes shown in Figure 2.61.

2.6 Use contour sketching to create the optical illusions shown in Figure 2.62.

2.7 Create a negative space sketch of the paper clips shown in Figure 2.63.

2.8 Using visible, center, and hidden line styles, practice sketching the following types of lines/features using size A or A4 plain or grid paper:

- Straight lines.
- 90- and 180-degree arcs.
- Circles.
- Ellipses.

2.9 Refer to Figures 2.64A–D, Multiview Sketching Problems. Sketch freehand multiview representations of the given objects using either size A or A4 plain or grid paper.

Figure 2.57 Problem 2.1 Upside-Down Sketch of the Word SKETCH

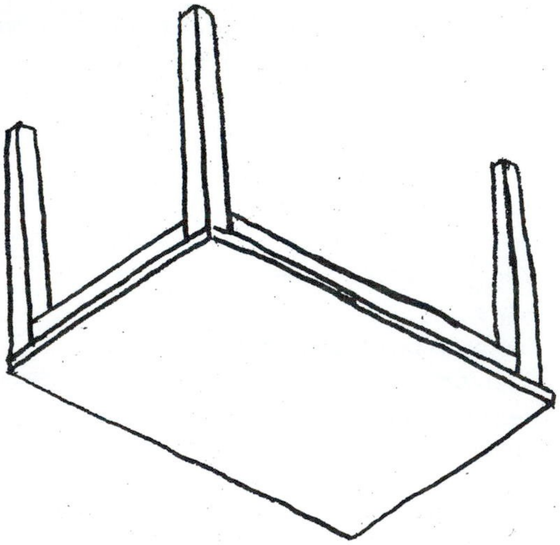

Figure 2.58 Problem 2.2 Upside-Down Sketch of a Table

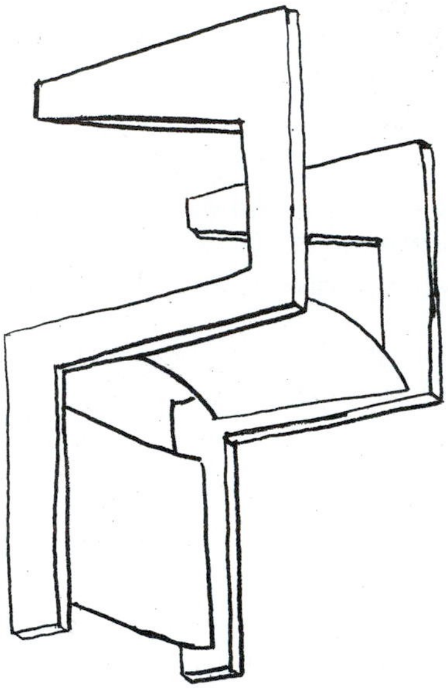

Figure 2.59 Problem 2.3 Upside-Down Sketch of a Chair

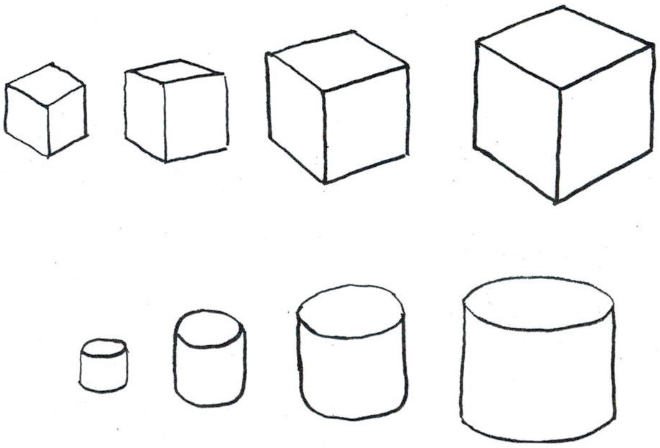

Figure 2.60 Problem 2.4 Contour Sketching of Cubes and Cylinders

2.10 Refer to Figures 2.65A–C, Pictorial Sketching Problems. Sketch freehand pictorials of the given objects using either size A or A4 plain or grid paper.

2.11 Draw a pictorial sketch of one of the multiviews in Problem 2.10. Pass the pictorial sketch to another student in the class, but do not specify which multiview you used. Have that student draw a multiview from your sketch. Compare that with the original multiview. Do they look the same? Reverse the process by starting with one of the objects in Problem 2.9.

2.12 Using one object from either Problem 2.9 or 2.10, sketch isometric pictorials from three different viewpoints on size A or A4 plain or grid paper.

2.13 Refer to Figures 2.64 and 2.65. Create one-point perspective sketches.

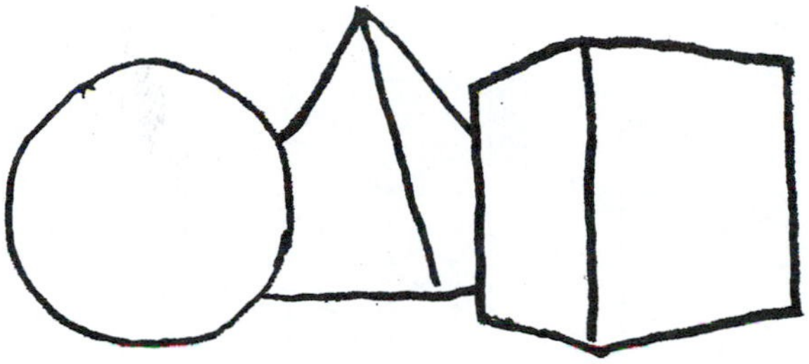

Figure 2.61 Problem 2.5 Contour Sketch of Overlapping Shapes

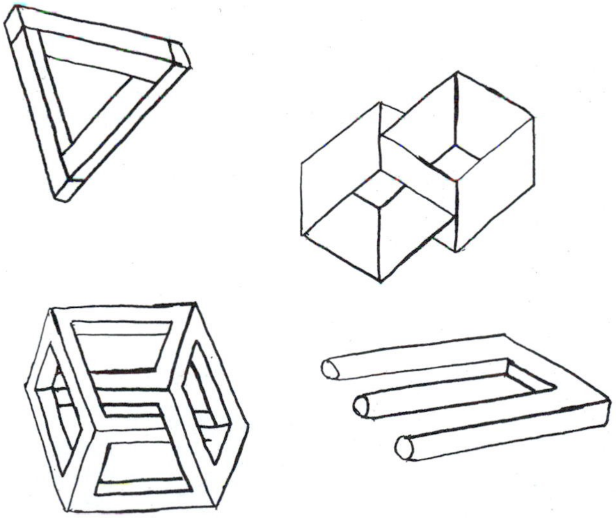

Figure 2.62 Problem 2.6 Contour Sketching of Optical Illusions

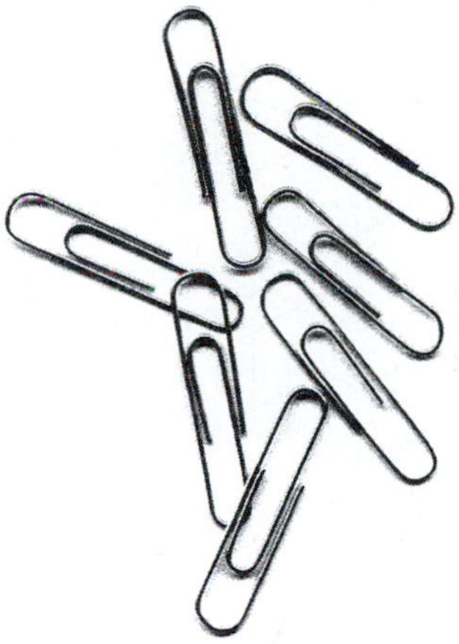

Figure 2.63 Problem 2.7 Negative Space Sketching

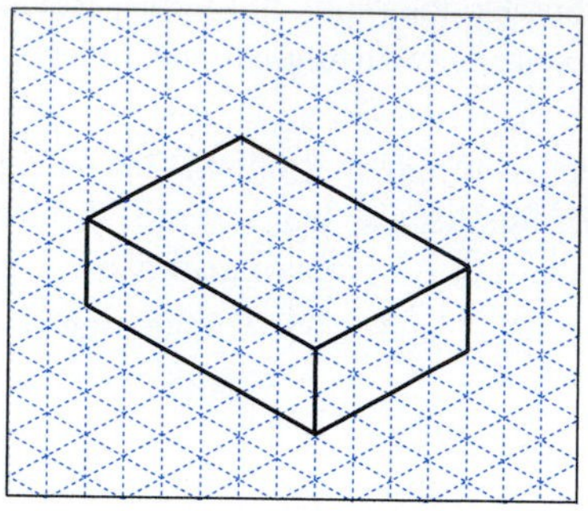

(1)

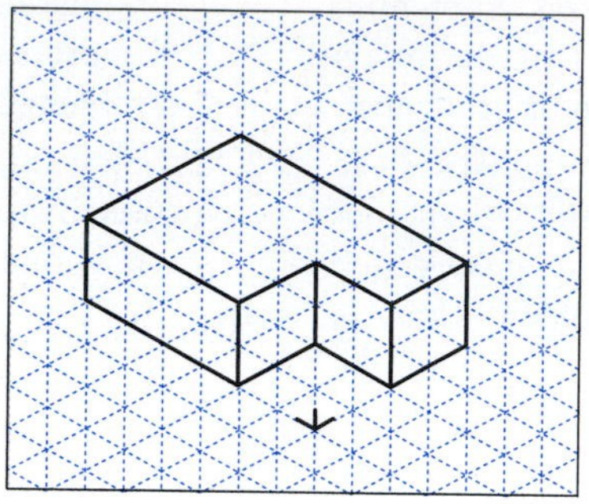

(2)

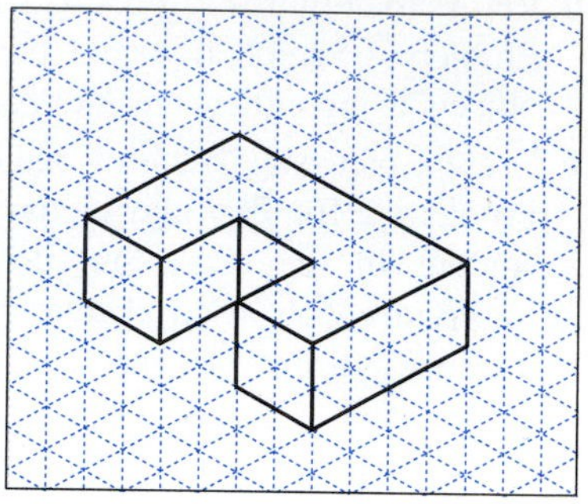

(3)

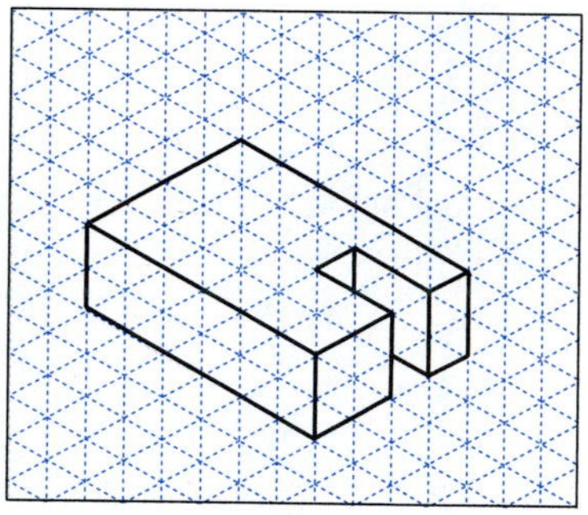

(4)

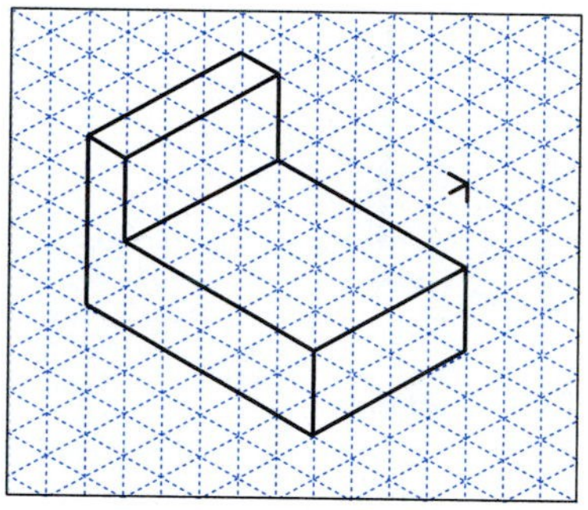

(5)

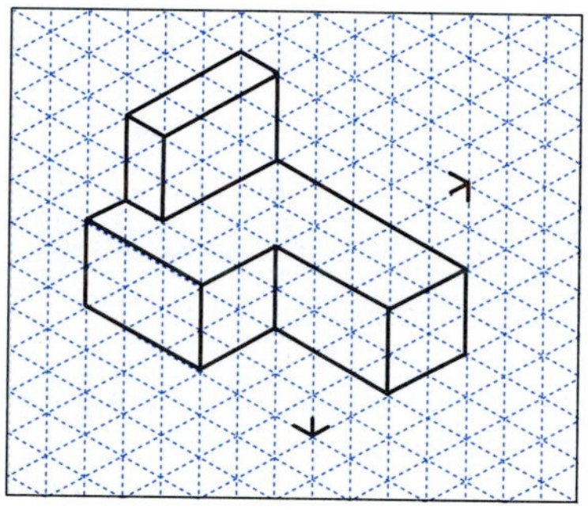

(6)

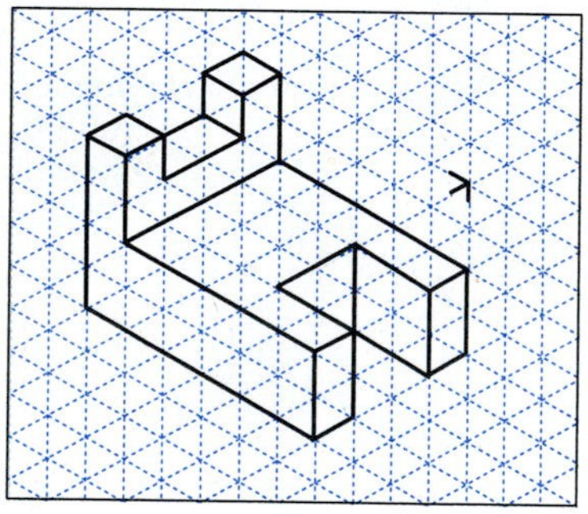

(7)

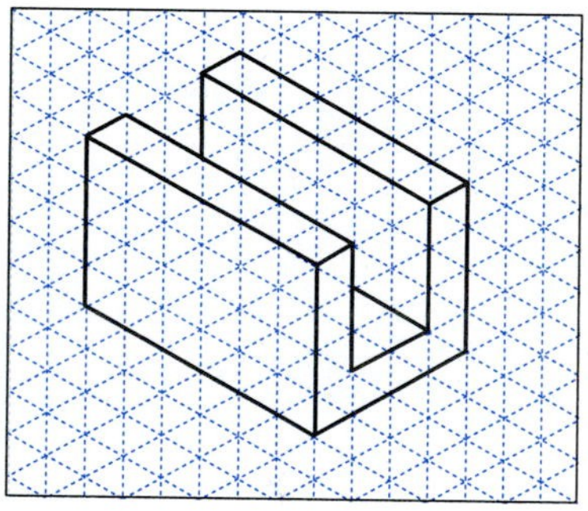

(8)

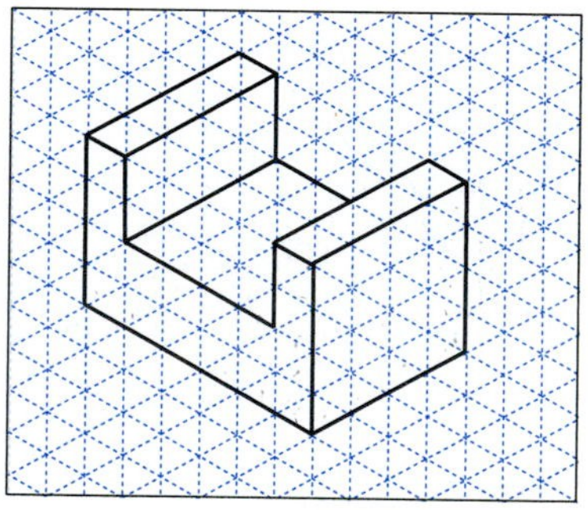

(9)

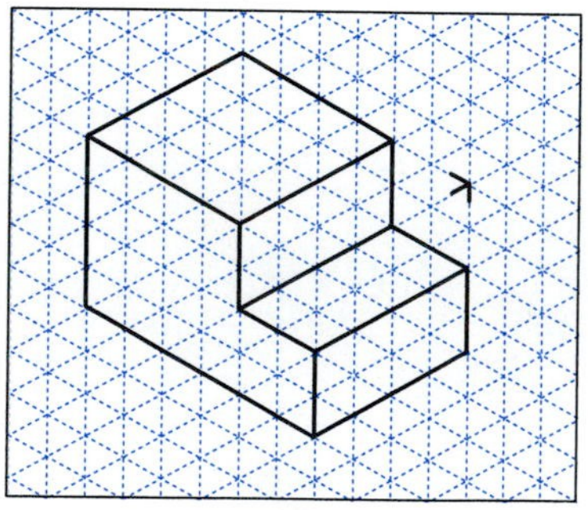

(10)

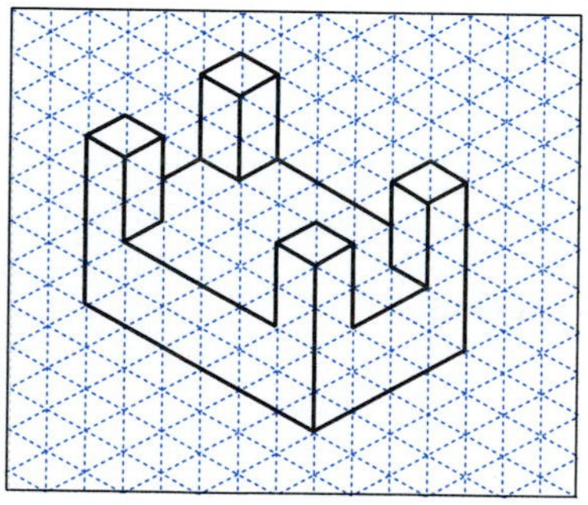

(11)

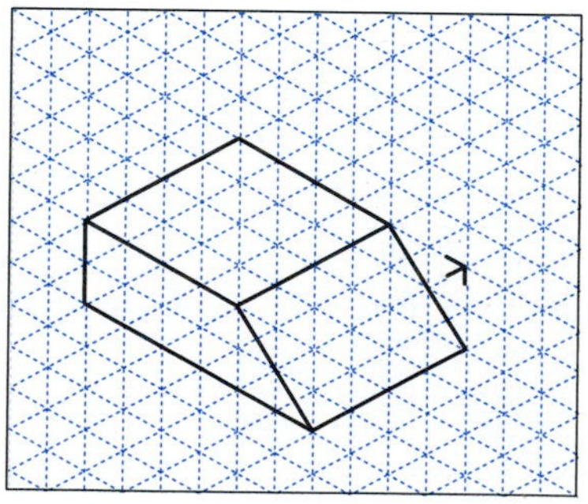

(12)

Figure 2.64A Problem 2.9 Multiview Sketching Problems

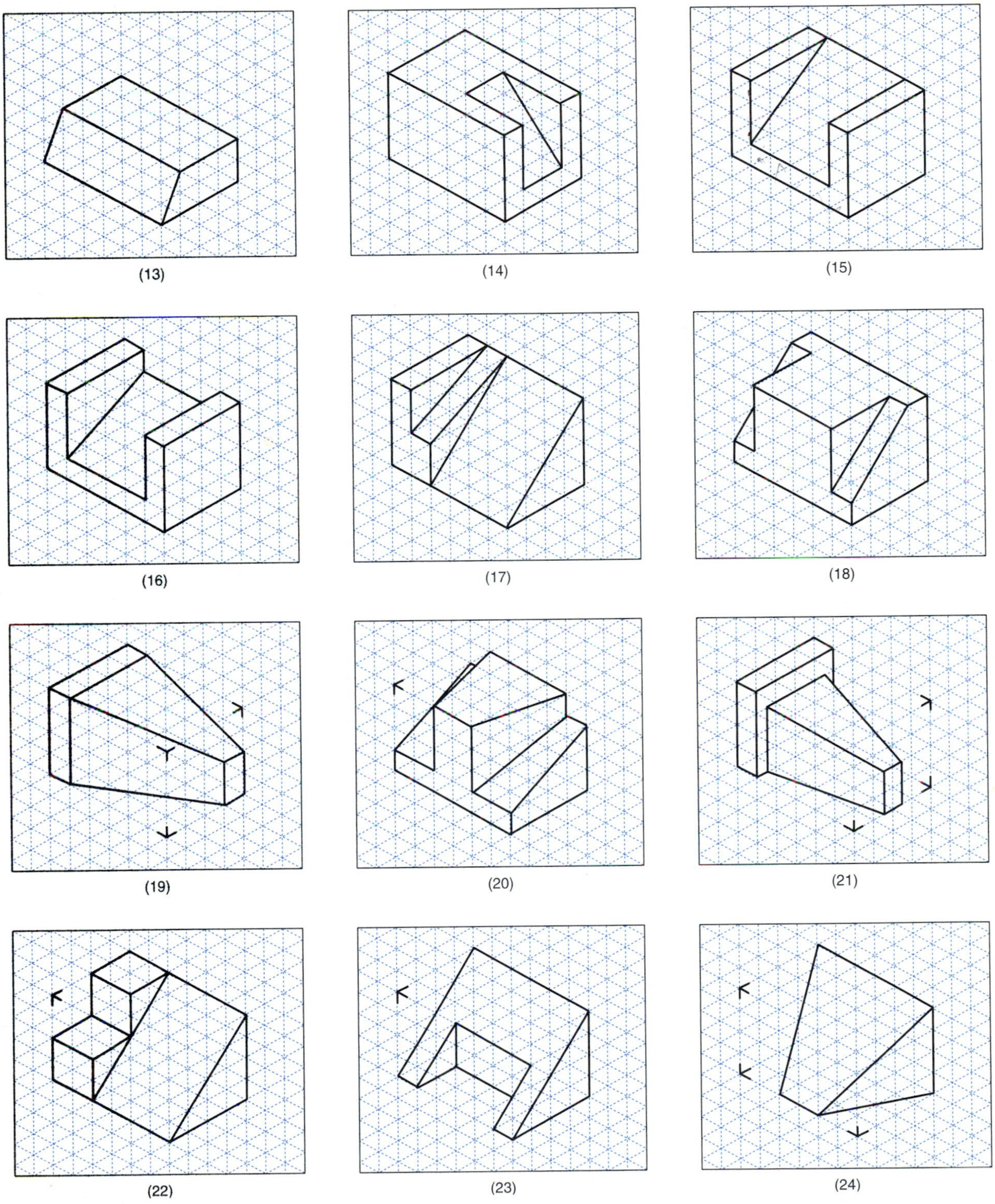

Figure 2.64B **Problem 2.9 Multiview Sketching Problems**

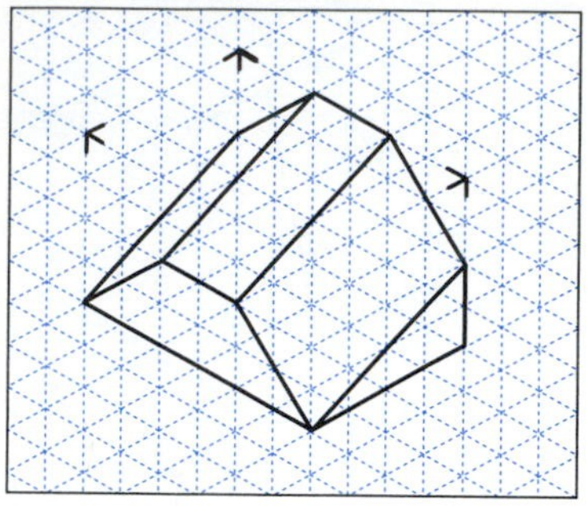

(25)

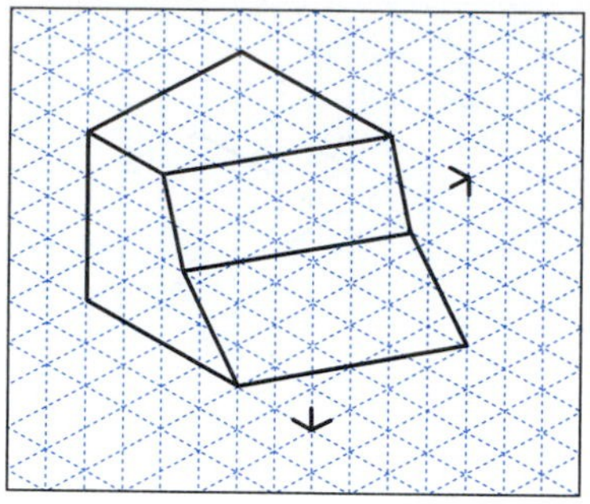

(26)

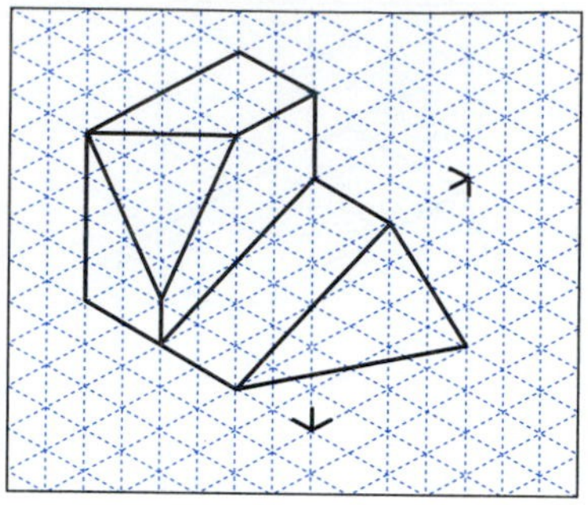

(27)

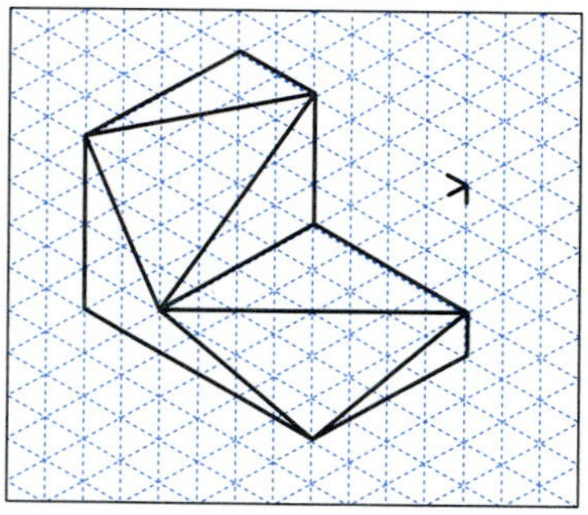

(28)

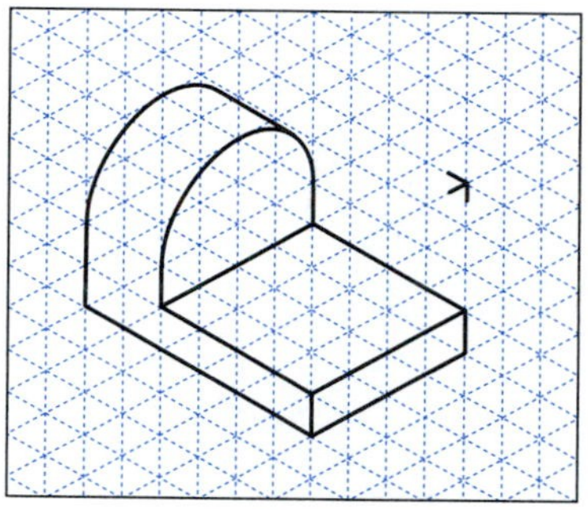

(29)

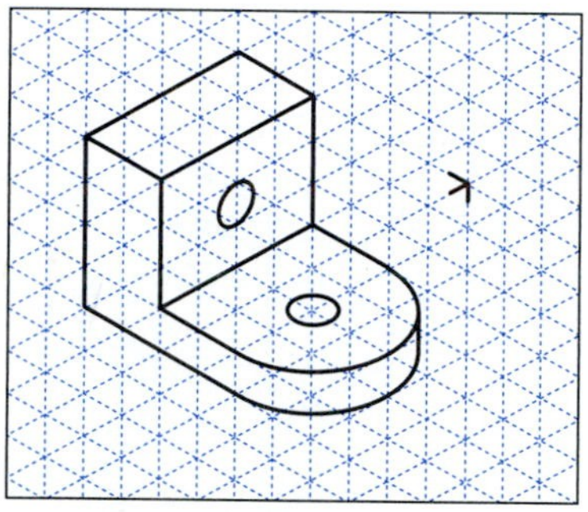

(30)

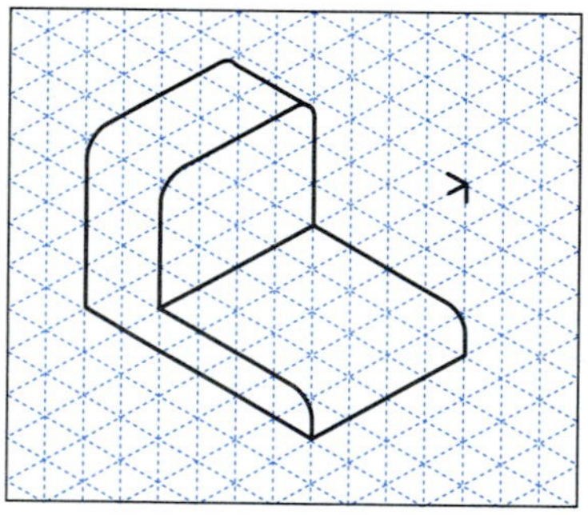

(31)

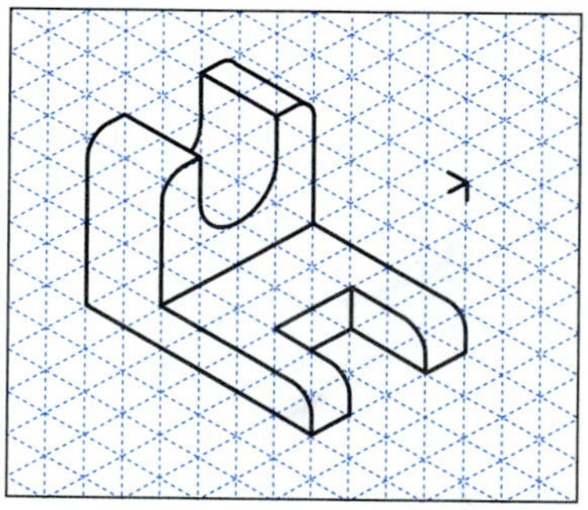

(32)

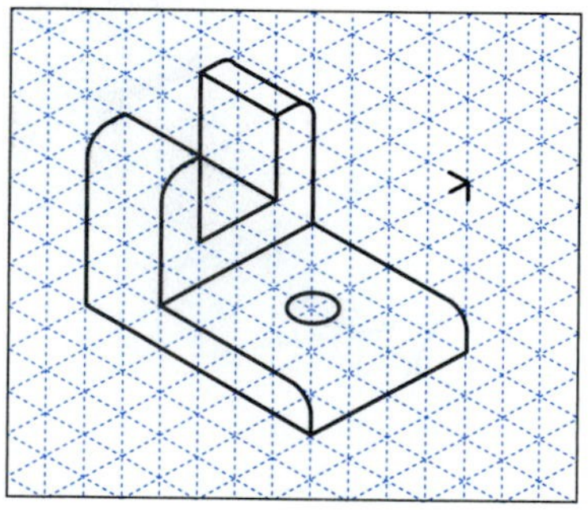

(33)

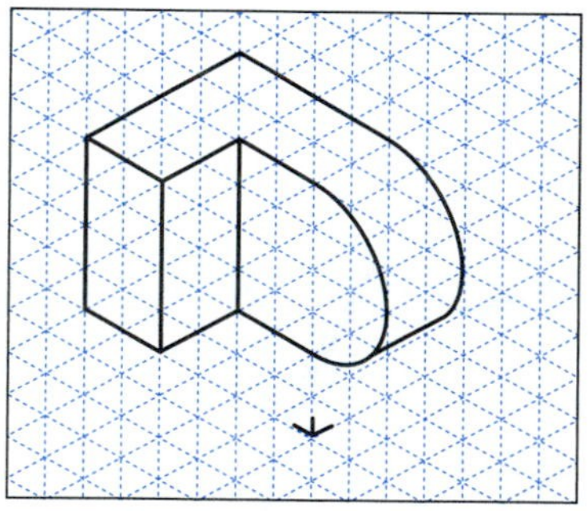

(34)

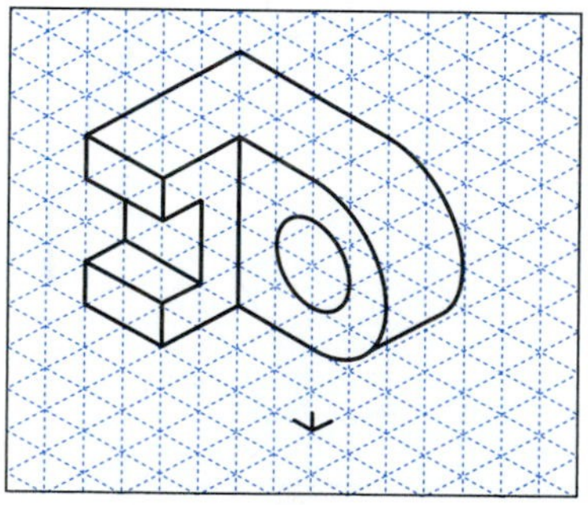

(35)

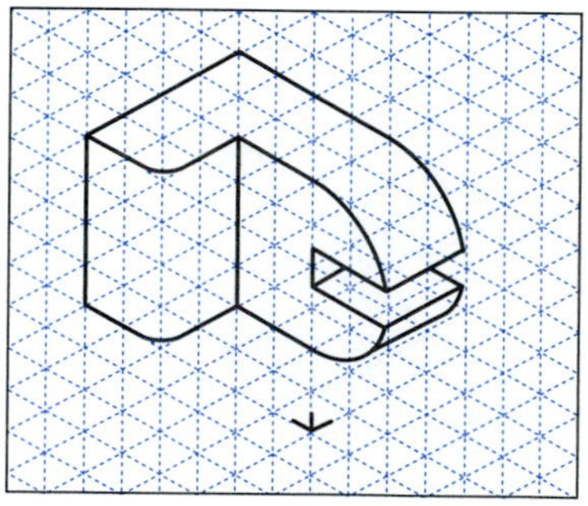

(36)

Figure 2.64C **Problem 2.9 Multiview Sketching Problems**

Assume all holes to be through.

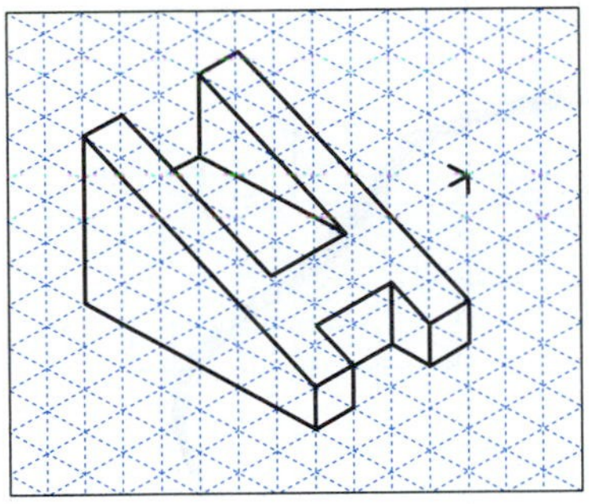

(37)

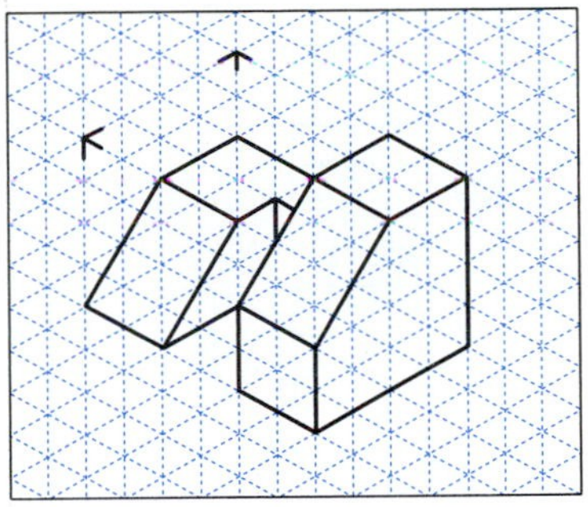

(38)

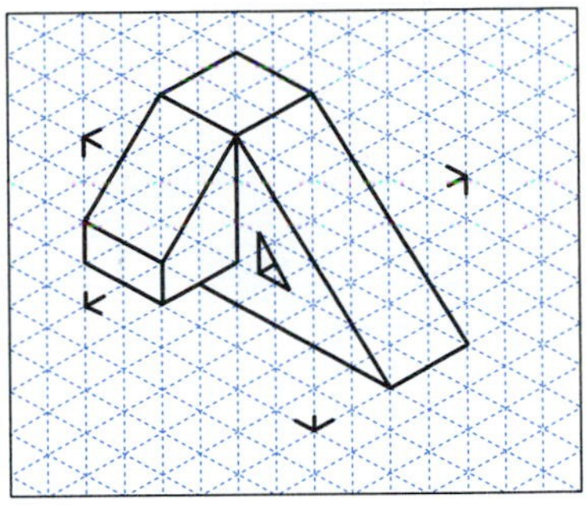

(39)

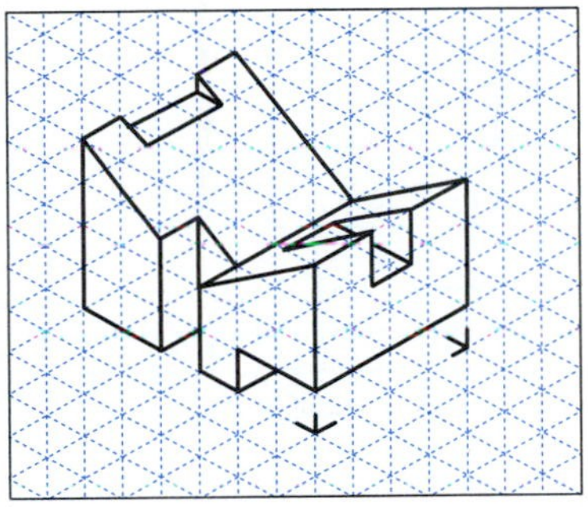

(40)

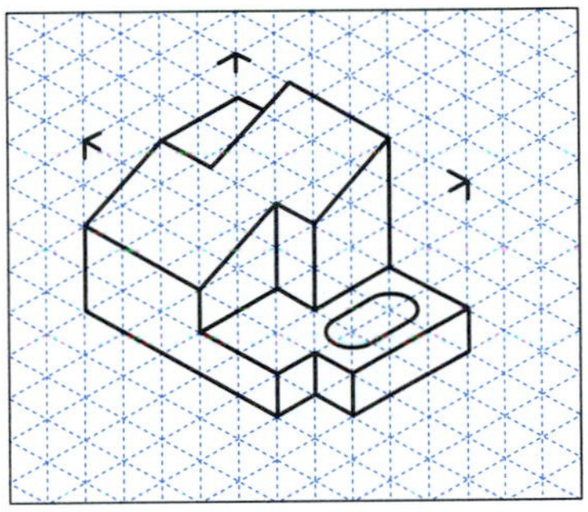

(41)

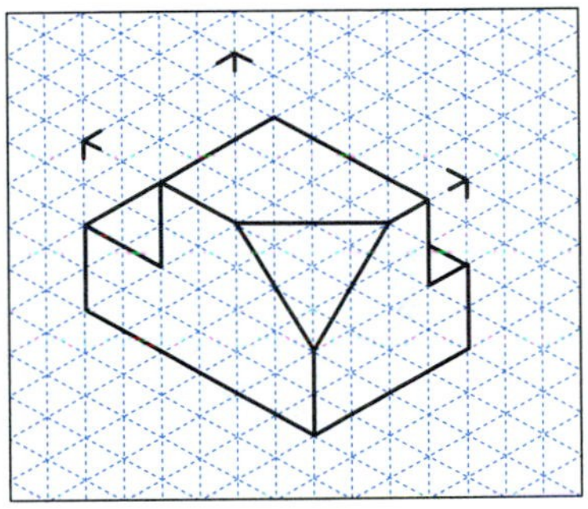

(42)

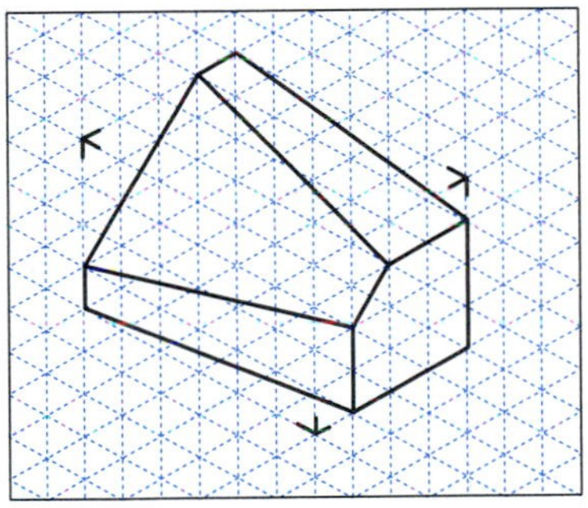

(43)

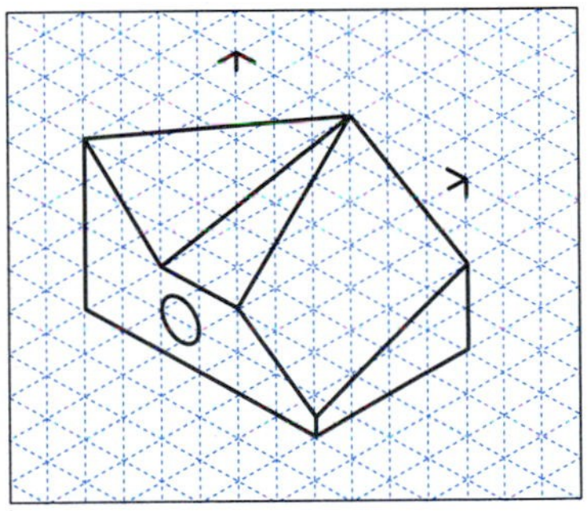

(44)

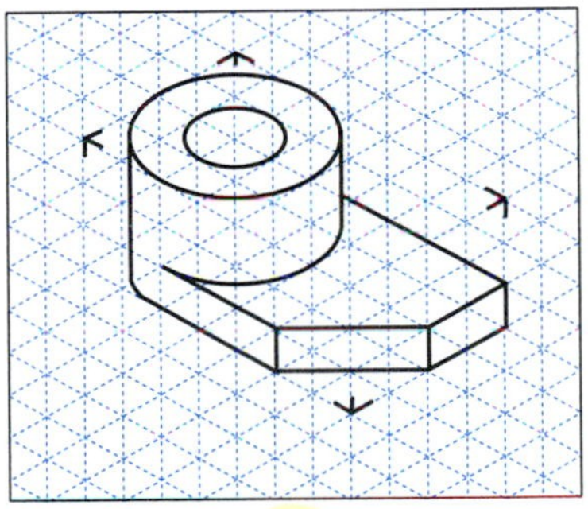

(45)

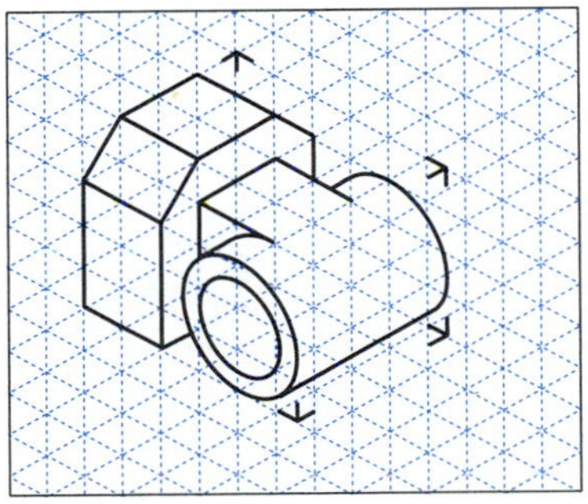

(46)

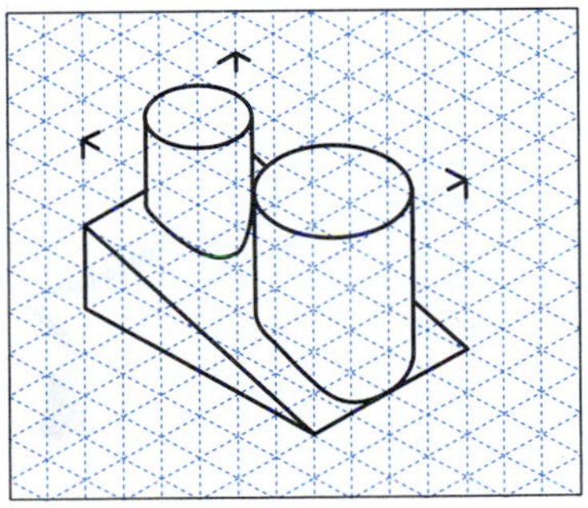

(47)

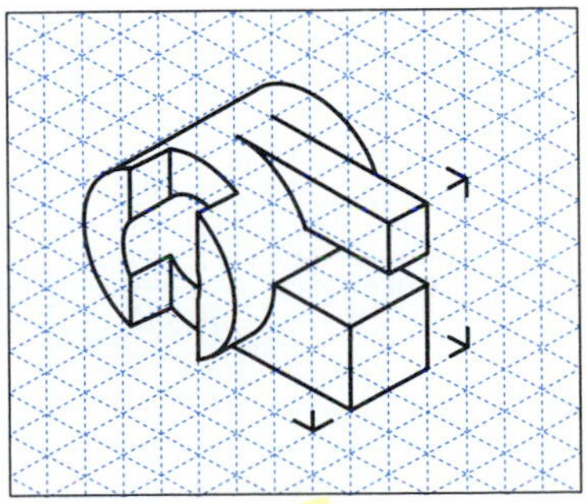

(48)

Figure 2.64D **Problem 2.9 Multiview Sketching Problems**
Assume all holes to be through.

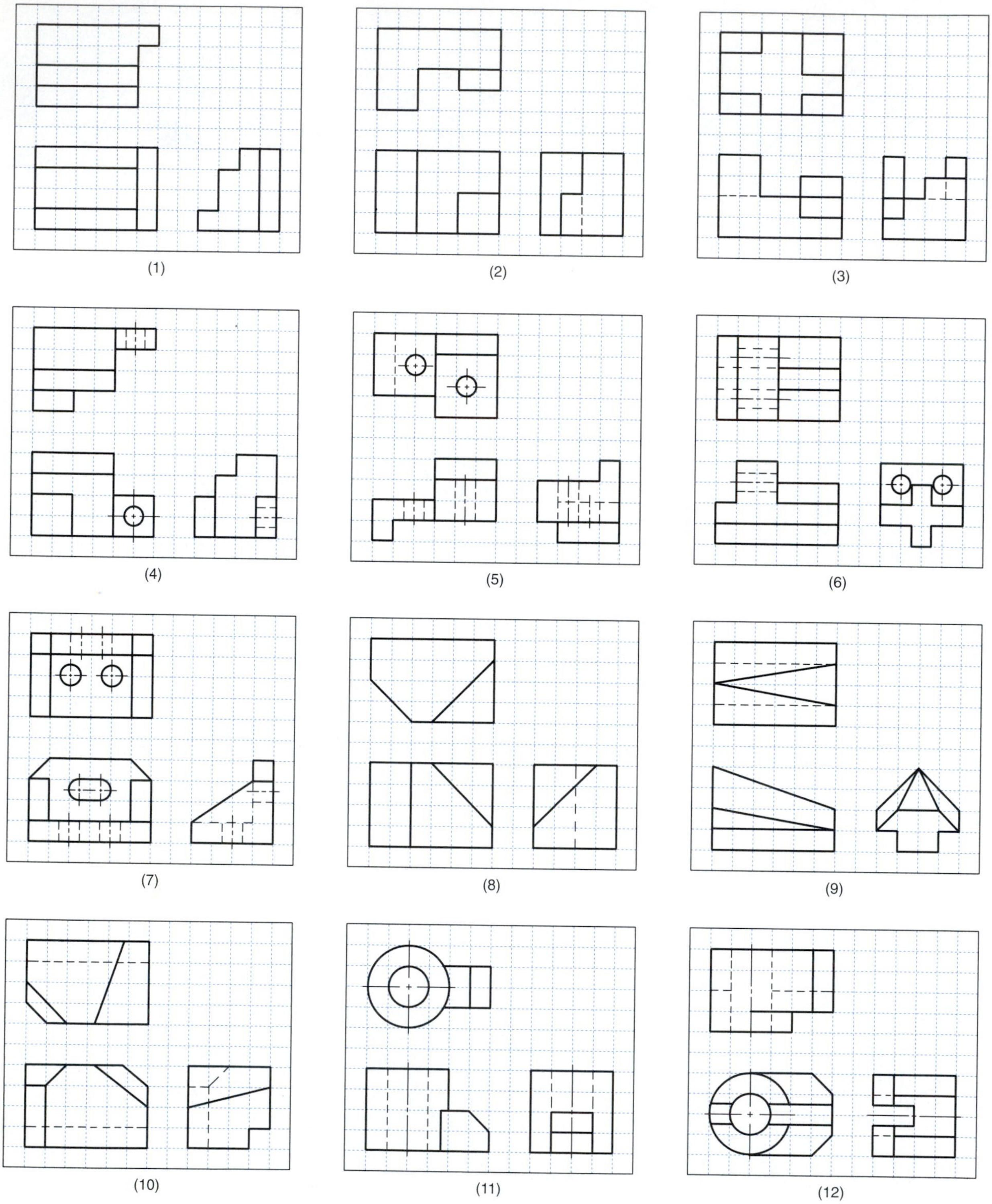

Figure 2.65A Problem 2.10 Pictorial Sketching Problems

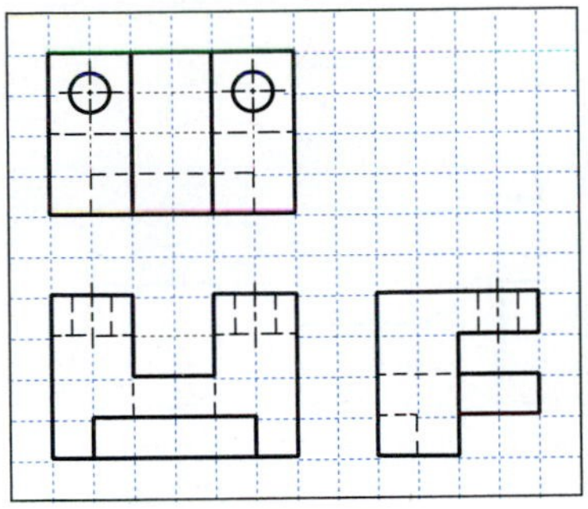

(13)

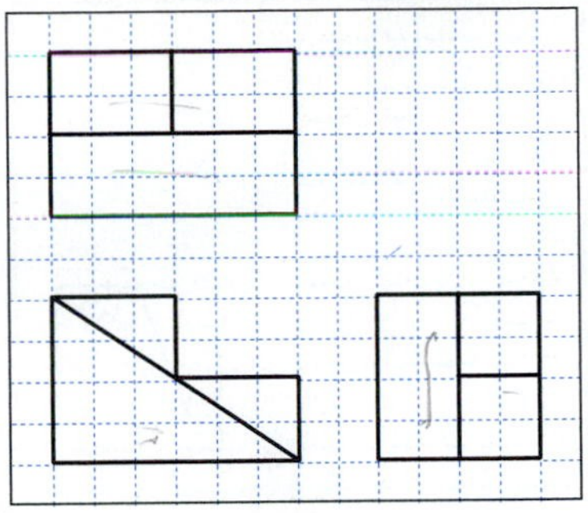

(14)

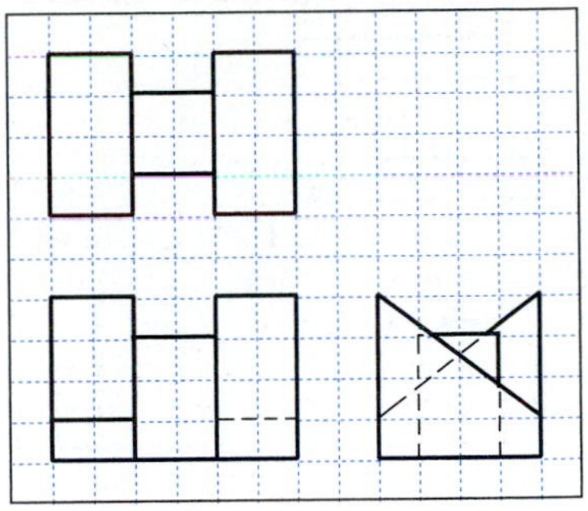

(15)

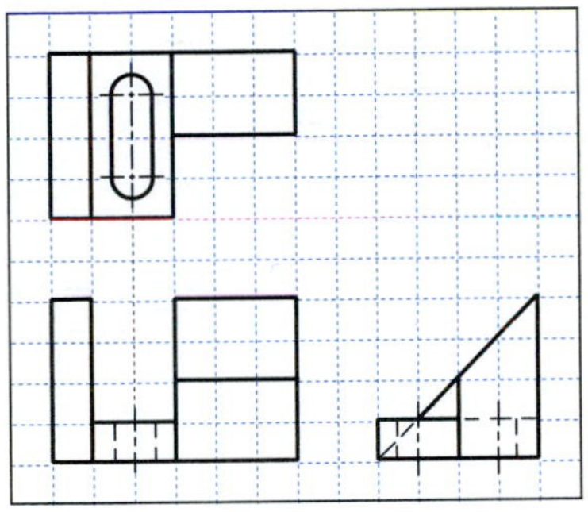

(16)

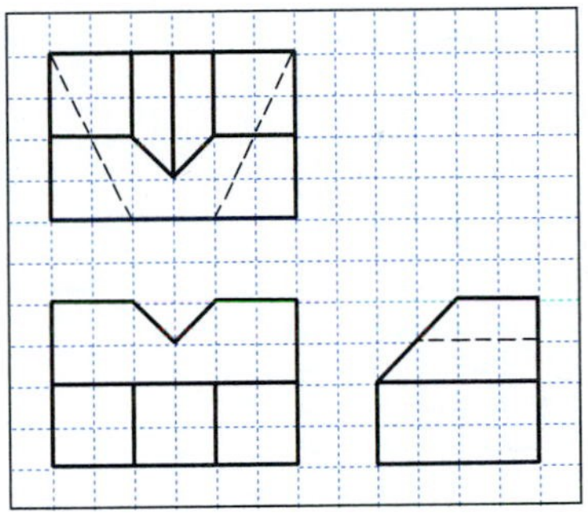

(17)

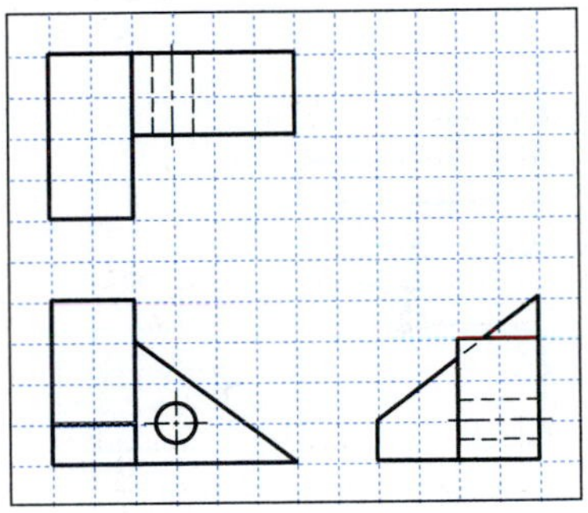

(18)

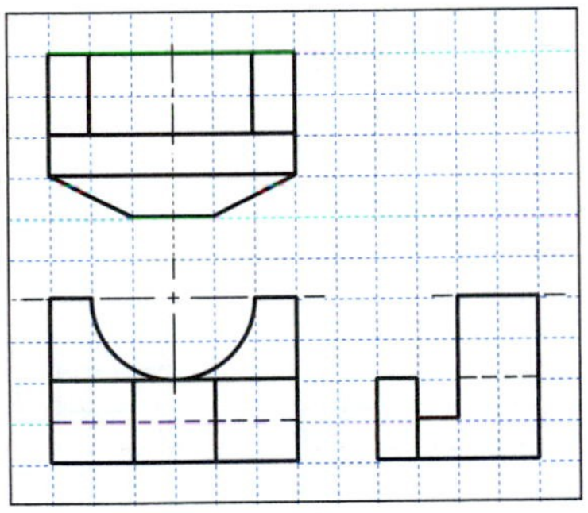

(19)

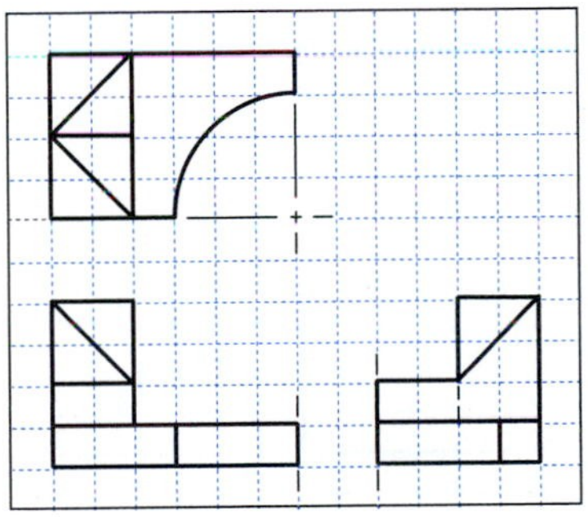

(20)

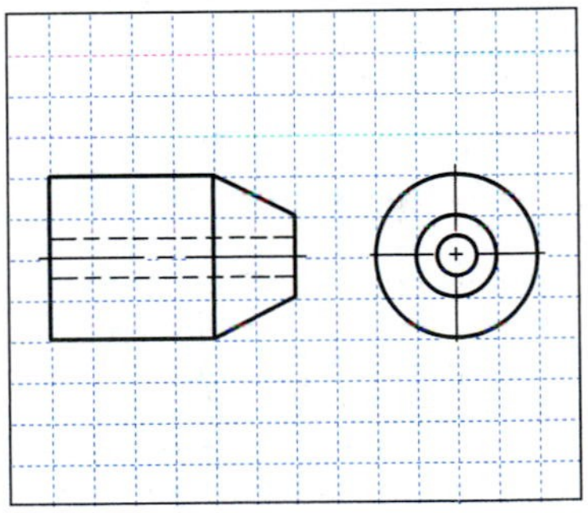

(21)

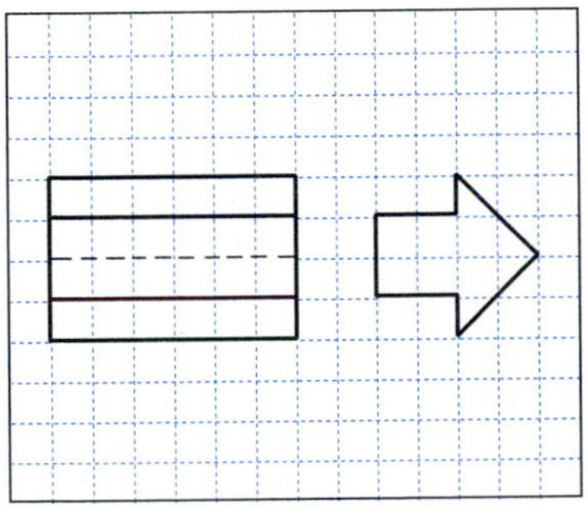

(22)

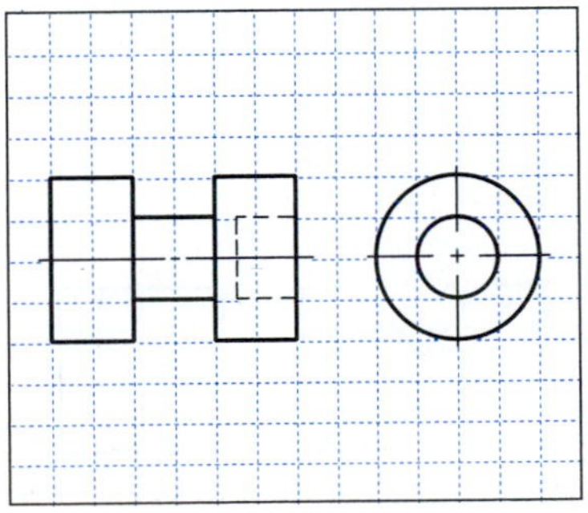

(23)

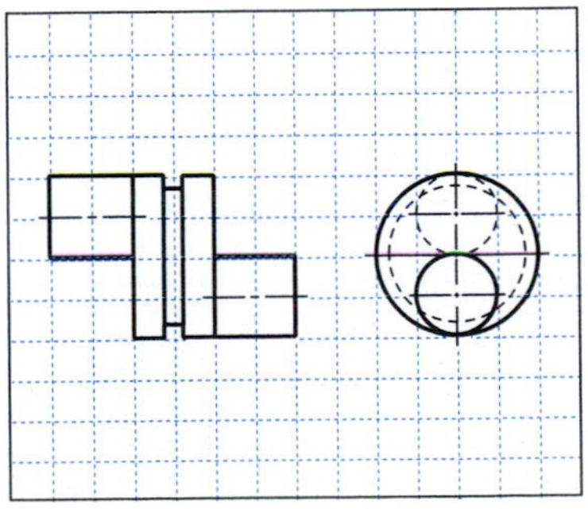

(24)

Figure 2.65B **Problem 2.10 Pictorial sketching problems**

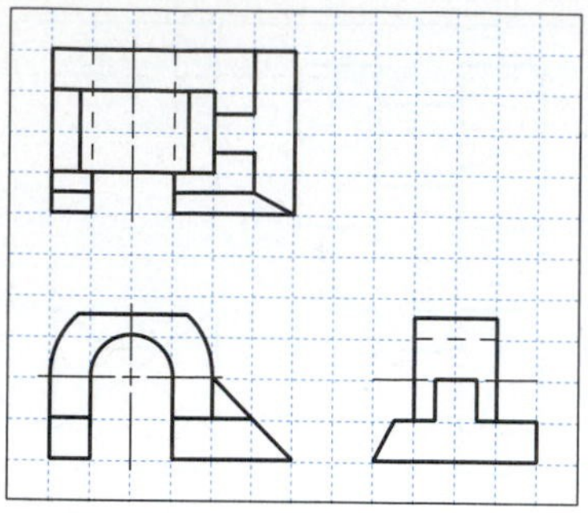

(25)

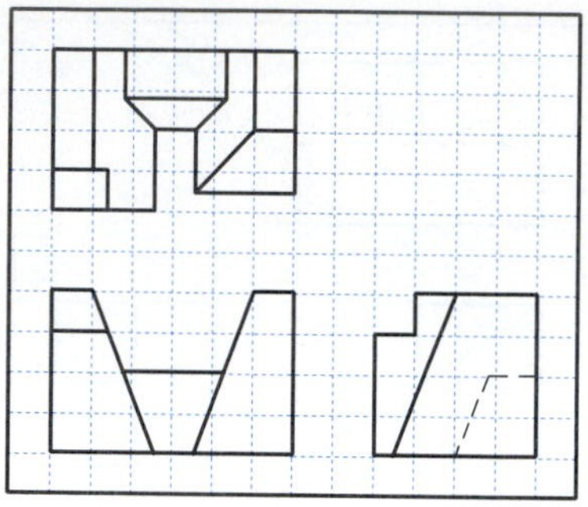

(26)

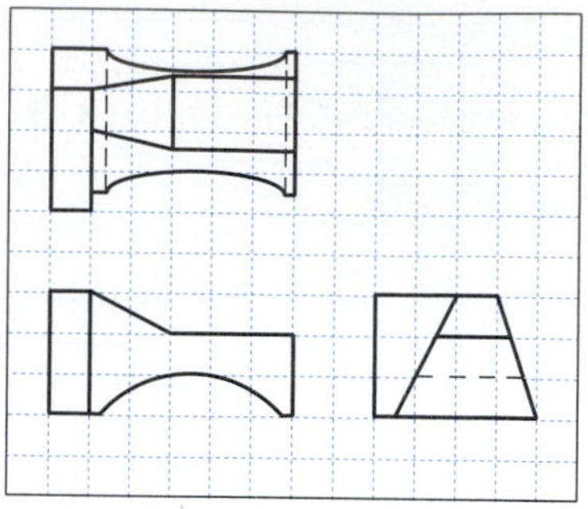

(27)

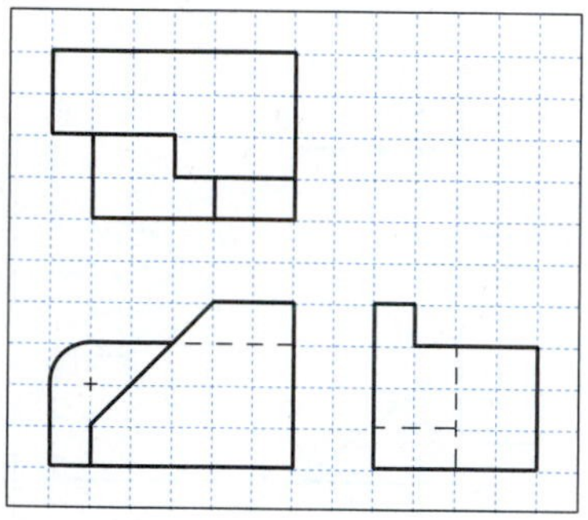

(28)

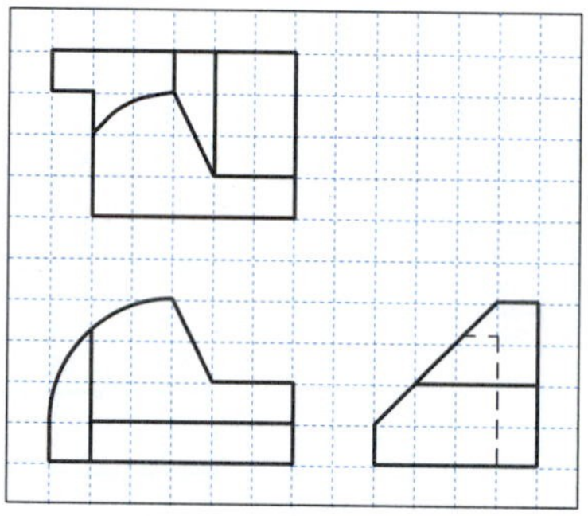

(29)

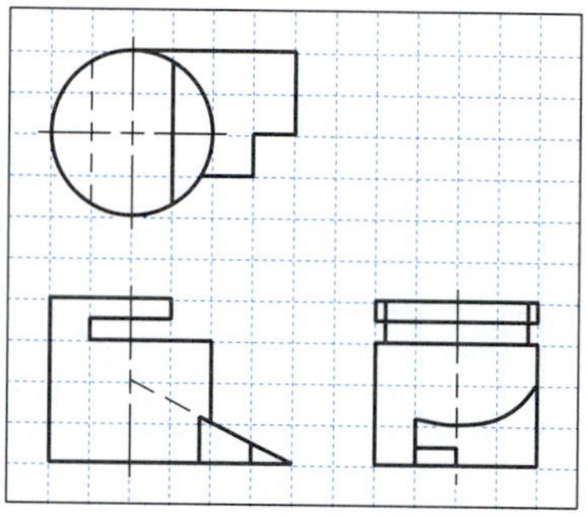

(30)

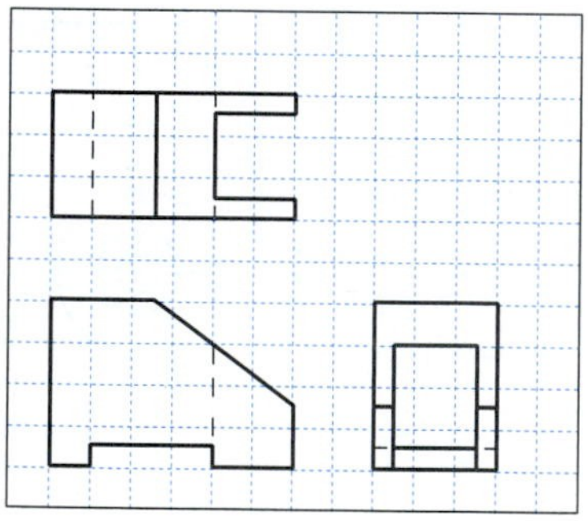

(31)

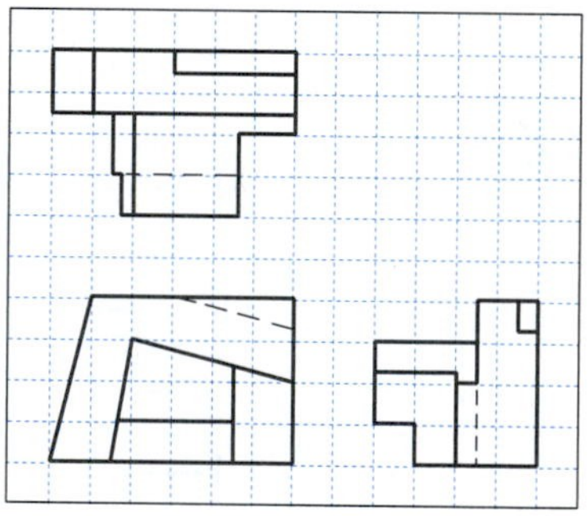

(32)

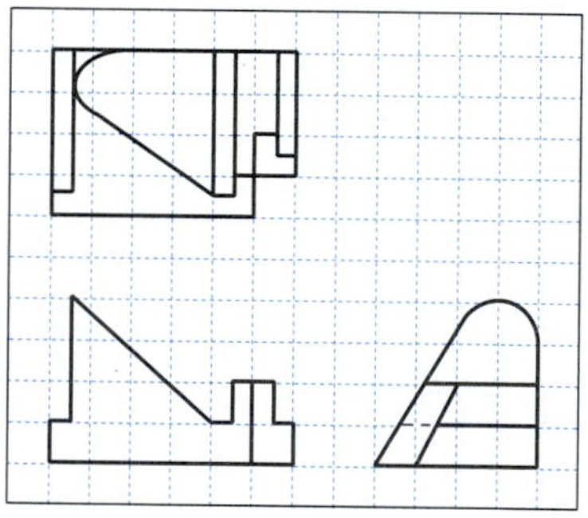

(33)

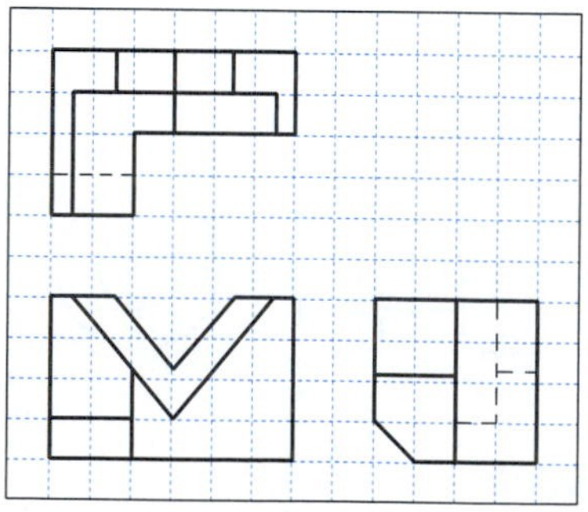

(34)

Figure 2.65C **Problem 2.10 Pictorial Sketching Problems**

Problems 2.1 through 2.5

Refer to Figures 2.57 through 2.63. Create those sketches assigned by your instructor.

Sketch Number: ____________________

Name: ____________________

Div/Sec: ____________________

Date: ____________________

Problems 2.1 through 2.5

Refer to Figures 2.57 through 2.63. Create those sketches assigned by your instructor.

Sketch Number: ____________________

Name: ____________________

Div/Sec: ____________________

Date: ____________________

Problem 2.8

Using visible, center, and hidden line styles, practice sketching straight lines, 90- and 180-degree arcs, circles, and ellipses.

Sketch Number: ______________

Name: ______________

Div/Sec: ______________

Date: ______________

Problem 2.9

Refer to Figure 2.64. Sketch multiviews of the pictorial views assigned. Divide the long dimension of the paper with a thick, dark line.

Sketch Number: ____________

Name: ____________

Div/Sec: ____________

Date: ____________

Problem 2.10

Refer to Figure 2.65. Sketch freehand pictorials of the multiviews assigned.

Sketch Number: ____________

Name: ____________

Div/Sec: ____________

Date: ____________

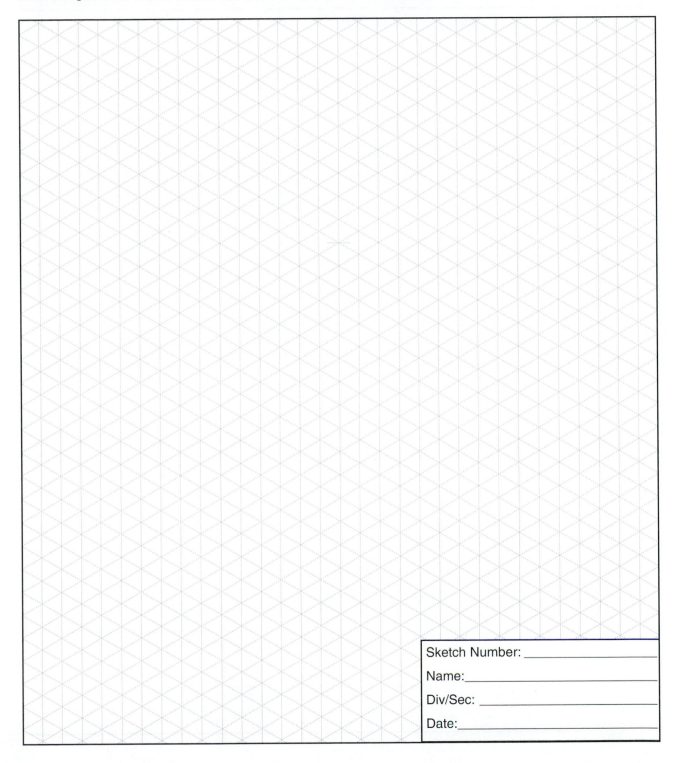

Problem 2.12

Using one object from either Problem 2.9 or 2.10, sketch isometric pictorials from three different viewpoints.

Sketch Number: ____________________

Name: ____________________

Div/Sec: ____________________

Date: ____________________

Problem 2.12

Using one object from either Problem 2.9 or 2.10, sketch isometric pictorials from three different viewpoints.

Sketch Number: ____________________

Name: ____________________

Div/Sec: ____________________

Date: ____________________

Problem 2.12

Using one object from either Problem 2.9 or 2.10, sketch isometric pictorials from three different viewpoints.

Sketch Number: ____________________

Name: ____________________

Div/Sec: ____________________

Date: ____________________

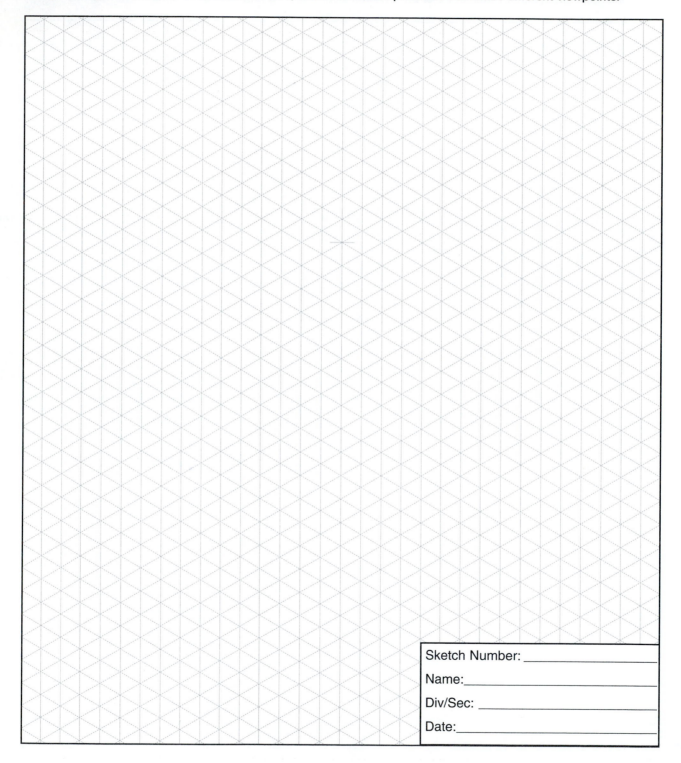

Problem 2.13

Refer to Figures 2.64 and 2.65. Create those sketches assigned by your instructor.

Sketch Number: ________________

Name: ________________

Div/Sec: ________________

Date: ________________

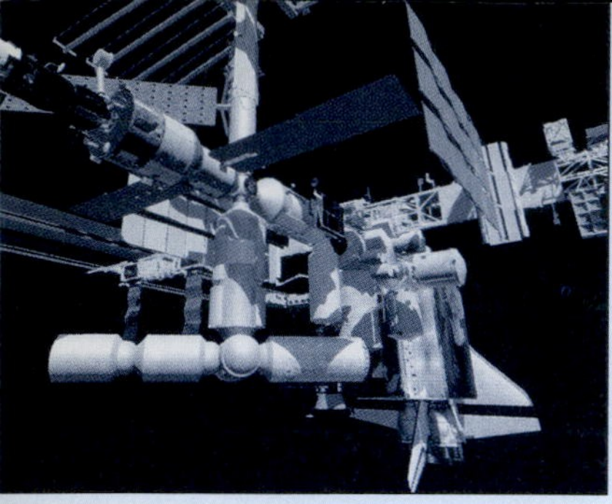

Chapter

Section and Auxiliary Views

3

OBJECTIVES

After completing this chapter, you will be able to:

1. Apply the concept of cutting planes to create section views.
2. Represent cutting plane lines and section lines, using conventional practices.
3. Create full, half, offset, removed, revolved, broken-out, auxiliary, and assembly section views, using conventional practices.
4. Create conventional breaks for different materials and cross sections.
5. Represent ribs, webs, and thin features in section, using conventional practices.
6. Represent aligned sectioned features, using conventional practices.
7. Create auxiliary views.

3.1 SECTIONING BASICS

Section views are an important aspect of design and documentation, and are used to improve clarity and reveal interior features of parts and structures (Figure 3.1). Section views are also used in the ideation and refinement stages of engineering design to improve the communications and problem-solving processes. **Sectional drawings** are multiview technical drawings that contain special views of a part or parts, views that reveal interior features. A primary reason for creating a section view is the elimination of hidden lines so that a drawing can be more easily understood or visualized. Figure 3.2 shows a regular multiview drawing and a sectioned multiview drawing of the same part in the front view; the hidden features can be seen after sectioning.

Traditional section views are based on the use of an imaginary cutting plane that cuts through the object to reveal interior features (Figure 3.3). This imaginary cutting

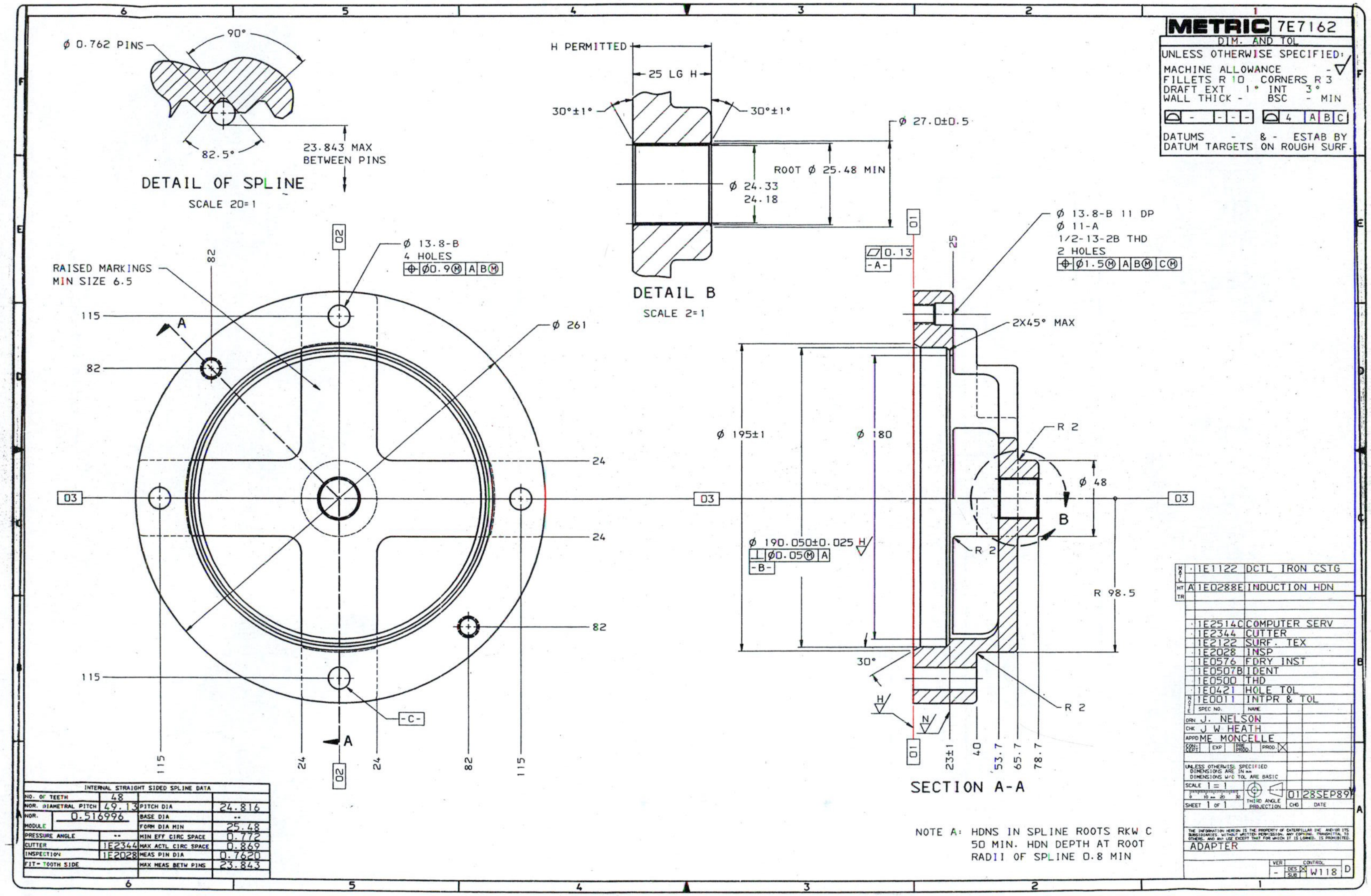

Figure 3.1

A typical multiview technical drawing shows the right side view in full section and removed section details.

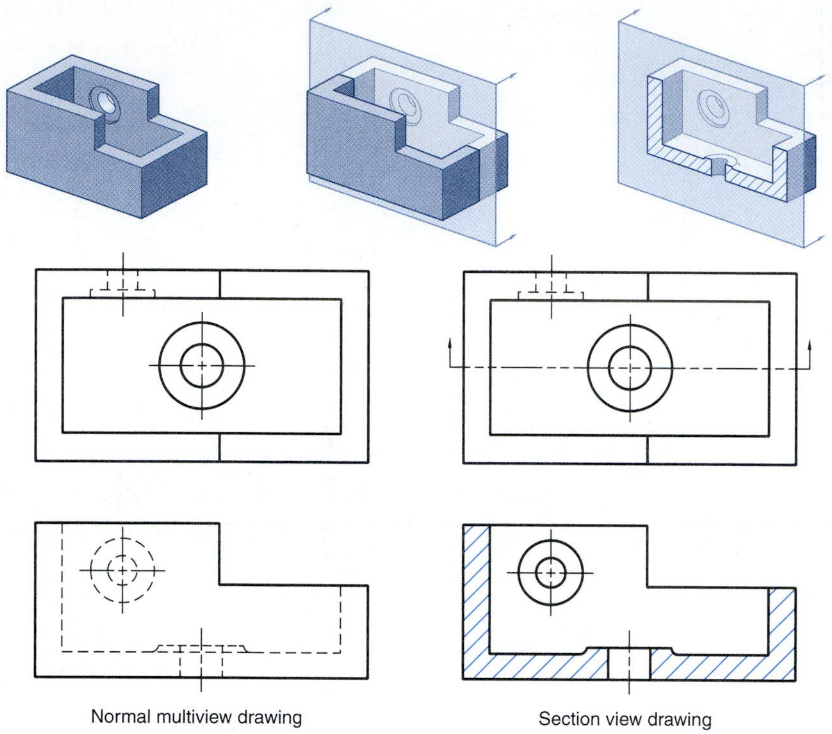

Figure 3.2 Section View Reveals Hidden Features
A section view will typically reveal hidden features so that the object is more easily visualized.

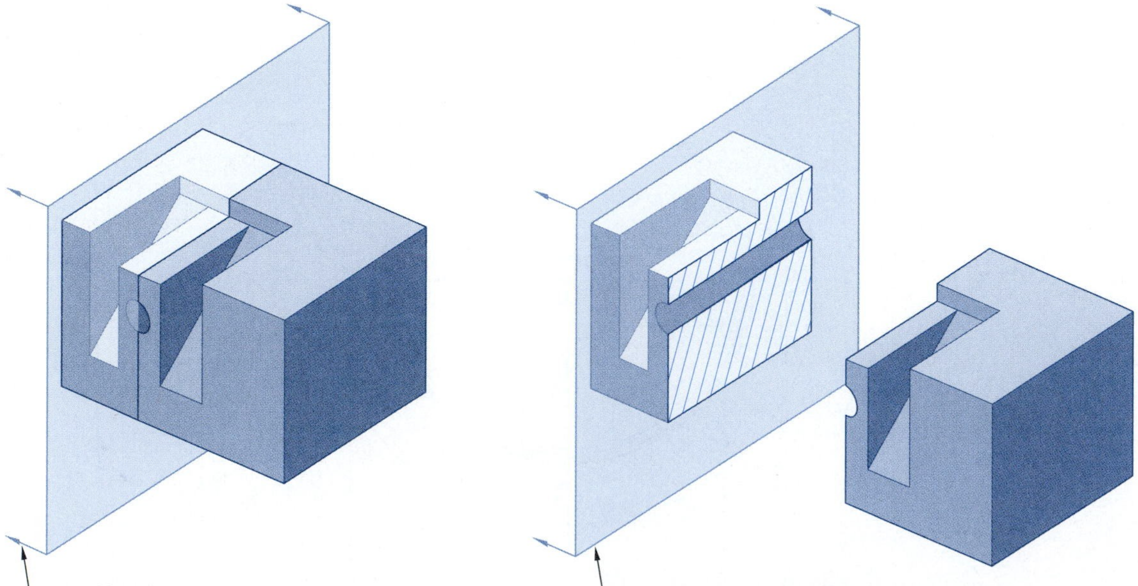

Figure 3.3 Cutting Planes
Imaginary cutting planes used to create section views are passed through the object to reveal interior features.

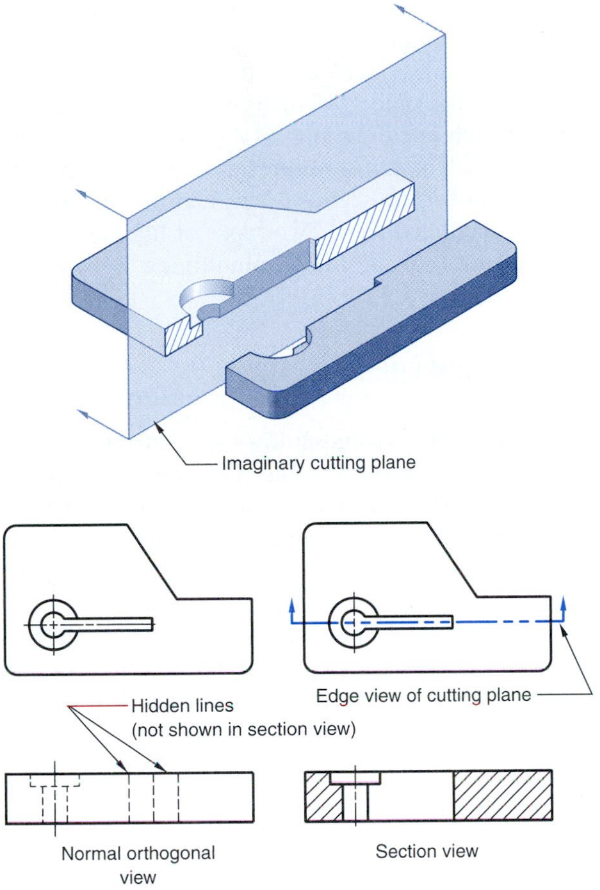

Figure 3.4 Treatment of Hidden Lines
Normally, hidden lines are omitted from section views.

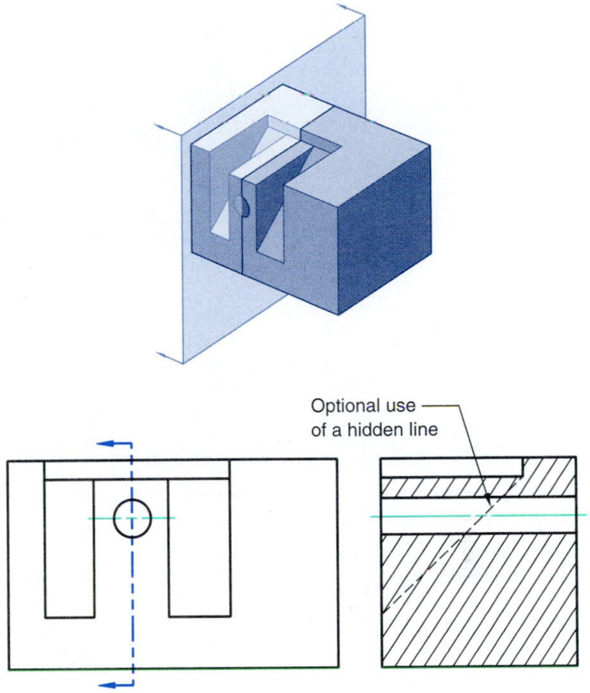

Figure 3.5 Optional Use of Hidden Lines
Hidden lines can be shown in section views. They are usually used to eliminate the need for another view.

plane is controlled by the designer and can (1) go completely through the object (full section), (2) go halfway through the object (half section), (3) be bent to go through features that are not aligned (offset section), or (4) go through part of the object (broken-out section).

An important reason for using section views is to reduce the number of hidden lines in a drawing. A section view reveals hidden features without the use of hidden lines (Figure 3.4). Adding hidden lines to a section view complicates the drawing, defeating the purpose of using a section. There are times, however, when a minimum number of hidden lines are needed to represent features other than the primary one shown by the section (Figure 3.5).

3.1.1 Visualization of Section Views

Figure 3.6 is a multiview drawing of a part that is difficult to visualize in its three-dimensional form because of the many hidden lines. A section view is created by passing an imaginary cutting plane vertically through the center of the part. Figure 3.6 shows an isometric representation of the part after it is sectioned. This isometric section view shows the interior features of the part more clearly and is an aid for visualizing the 3-D form.

In Figure 3.6, the cutting plane arrows in the top view point up to represent the direction of sight for producing a front view in full section. The direction of the arrow can also be thought of as pointing toward the half of the object being kept. The front half of the object is "removed" to reveal the interior features of the part.

All the surfaces touched by the cutting plane are marked with section lines. Because all the surfaces are the same part, the section lines are identical and are drawn in the same direction. The center line is added to the counterbored hole to complete the section view.

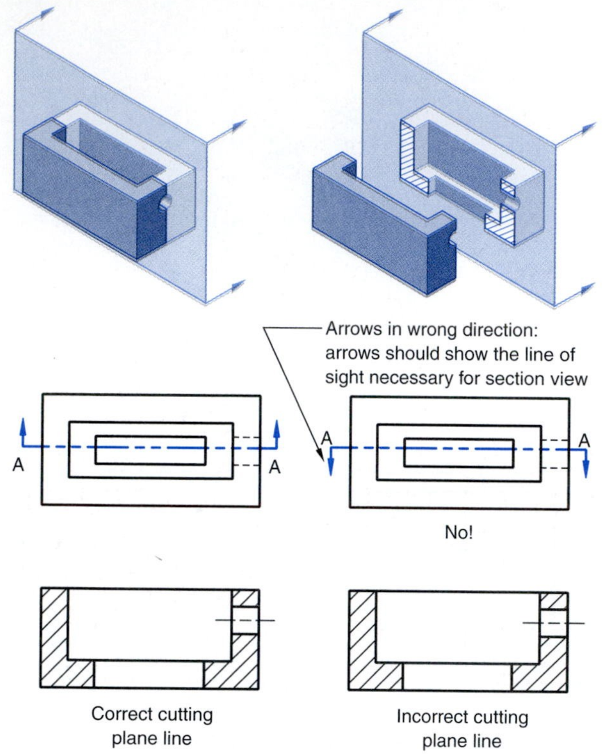

Figure 3.6 Placement of Cutting Plane Lines

The cutting plane line is placed in the view where the cutting plane appears on edge.

Figure 3.7 Standard Cutting Plane Linestyles

Standard cutting plane linestyles are thick lines terminated with arrows.

3.2 CUTTING PLANE LINES

Cutting plane lines, which show where the cutting plane passes through the object, represent the *edge view* of the cutting plane and are drawn in the view(s) adjacent to the section view. In Figure 3.6, the cutting plane line is drawn in the top view, which is adjacent to the sectioned front view. Cutting plane lines are *thick* (0.6 mm or 0.032 inch) dashed lines that extend past the edge of the object ¼″ or 6 mm and have line segments at each end drawn at 90 degrees and terminated with arrows. The arrows represent the direction of the line of sight for the section view, and they point away from the sectioned view. Two types of lines are acceptable for cutting plane lines in multiview drawings, as shown in Figure 3.7.

3.3 SECTION LINE PRACTICES

Section lines or **cross-hatch lines** are added to a section view to indicate the surfaces that are cut by the imaginary cutting plane. Different section line symbols can be used to represent various types of materials. However, there are so many different materials used in design that the general symbol (i.e., the one used for cast iron) may be used for most purposes on technical drawings. The actual type of material required is then noted in the title block or parts list or entered as a note on the drawing. The angle at which section lines are drawn is usually 45 degrees to the horizontal, but this can be changed for adjacent parts shown in the same section. Also, the spacing between section lines is uniform on a section view.

3.3.1 Material Symbols

The type of section line used to represent a surface varies according to the type of material. However, the **general-purpose section line** symbol used in most section view drawings is that of *cast iron.* Figure 3.8 shows some of the ANSI standard section line symbols used for only a few materials; there are literally hundreds of different materials used in design. Occasionally, a general section line symbol is used to represent a group of materials, such as steel. The specific type of steel to be used will be indicated in the title block or parts list. Occasionally, with assembly section views, material symbols are used to identify different parts of the assembly.

3.3.2 Drawing Techniques

The general-purpose cast iron section line is drawn at a 45-degree angle and spaced 1/16 inch (1.5 mm) to ⅛ inch (3 mm) or more, depending on the size of the drawing.

Section lines should not run parallel or perpendicular to the visible outline (Figures 3.9A and B). If the visible outline to be sectioned is drawn at a 45-degree angle, the

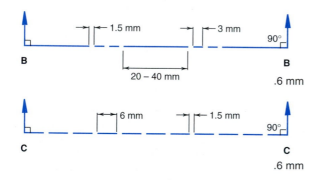

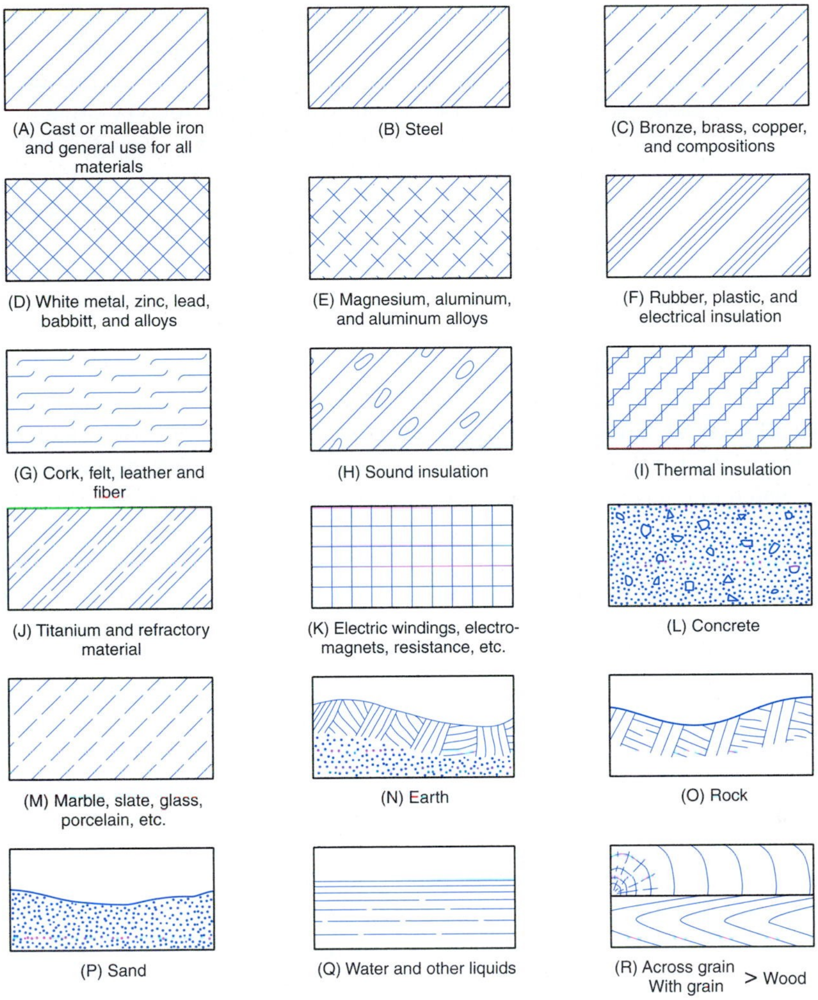

Figure 3.8 **ANSI Standard Section Lines for Various Materials**

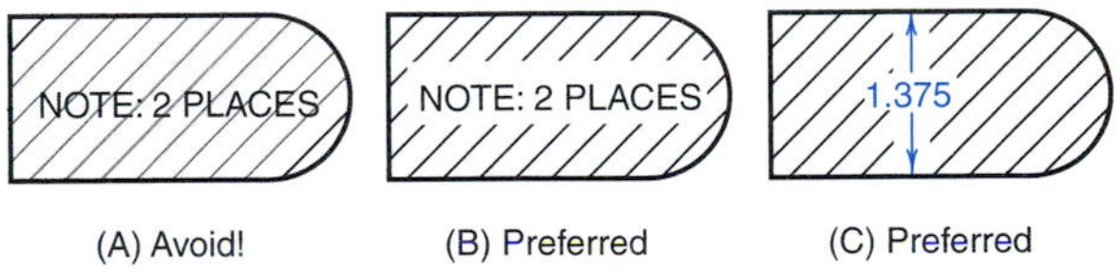

Figure 3.10 **Notes in Section Lined Areas**

Section lines are omitted around notes and dimensions.

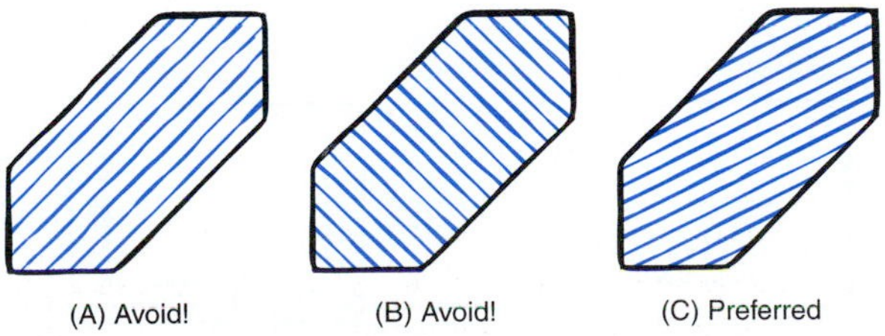

Figure 3.9 **Section Line Placement**

Avoid placing section lines parallel or perpendicular to visible lines.

section lines are drawn at a different angle, such as 30 degrees (Figure 3.9C).

Avoid placing dimensions or notes within the section-lined areas (Figure 3.10). If the dimension or note must be placed within the sectioned area, omit the section lines in the area of the note, as shown in Figures 3.10B and C.

3.4 SECTION VIEW TYPES

There are many different types of section views used on technical drawings:

Full section
Half section
Broken-out section
Revolved section
Removed section
Offset section
Assembly section
Auxiliary section

Up to this point, only full sections have been shown in the figures. The selection of the type of section view to be used is based upon which one most clearly and concisely represents the features of interest. For example, there are times when it may be more appropriate to use a half-section view, for symmetrical objects, or a broken-out section, for small areas. The various types of section views are described in the following paragraphs.

3.4.1 Full Sections

A **full-section** view is made by passing the imaginary cutting plane completely through the object, as shown in Figure 3.11A. Figure 3.11B shows the orthographic views of the object, and Figure 3.11C shows the full-section view. All the hidden features intersected by the cutting plane are represented by visible lines in the section view. Surfaces touched by the cutting plane have section lines drawn at a 45-degree angle to the horizontal. Hidden lines are omitted in all section views unless they must be used to provide a clear understanding of the object.

The top view of the section drawing shows the cutting plane line, with arrows pointing in the direction of the line of sight necessary to view the sectioned half of the object. In a multiview drawing, a full-section view is placed in the same position that an unsectioned view would normally occupy; that is, a front section view would replace the traditional front view.

3.4.2 Half Sections

Half sections are created by passing an imaginary cutting plane only *halfway* through an object. Figure 3.12A shows the cutting plane passing halfway through an object and *one quarter* of the object being removed. Figure 3.12B shows the normal orthographic views before sectioning, and Figure 3.12C shows the orthographic views after sectioning. Hidden lines are omitted on both halves of the section view. Hidden lines may be added to the unsectioned half for dimensioning or for clarity.

External features of the part are drawn on the unsectioned half of the view. A center line, not an object line, is used to separate the sectioned half from the unsectioned half of the view. The cutting plane line shown in the top view of Figure 3.12C is bent at 90 degrees, and one arrow is drawn to represent the line of sight needed to create the front view in section.

Half-section views are used most often on parts that are symmetrical, such as cylinders. Also, half sections are sometimes used in assembly drawings when external features must be shown.

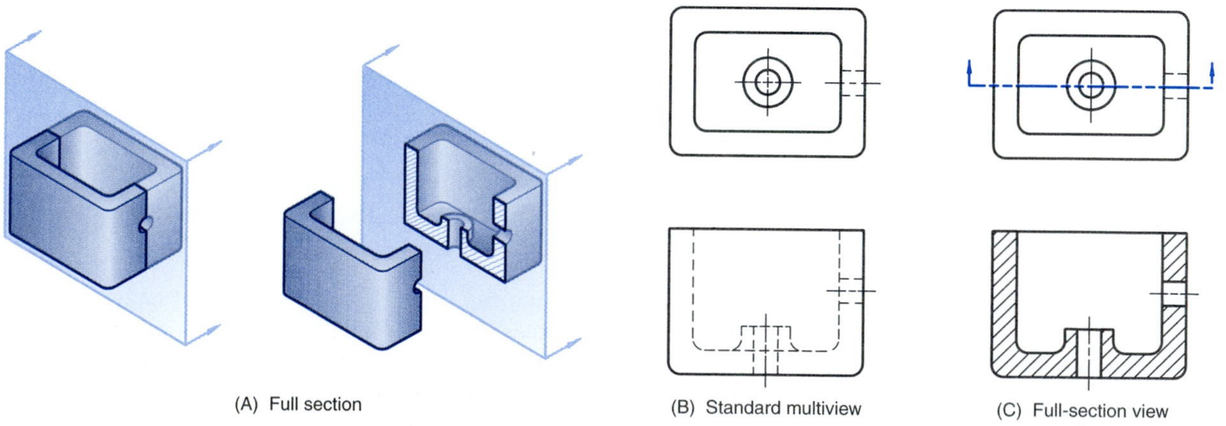

Figure 3.11 Full-Section View
A full-section view is created by passing a cutting plane fully through the object.

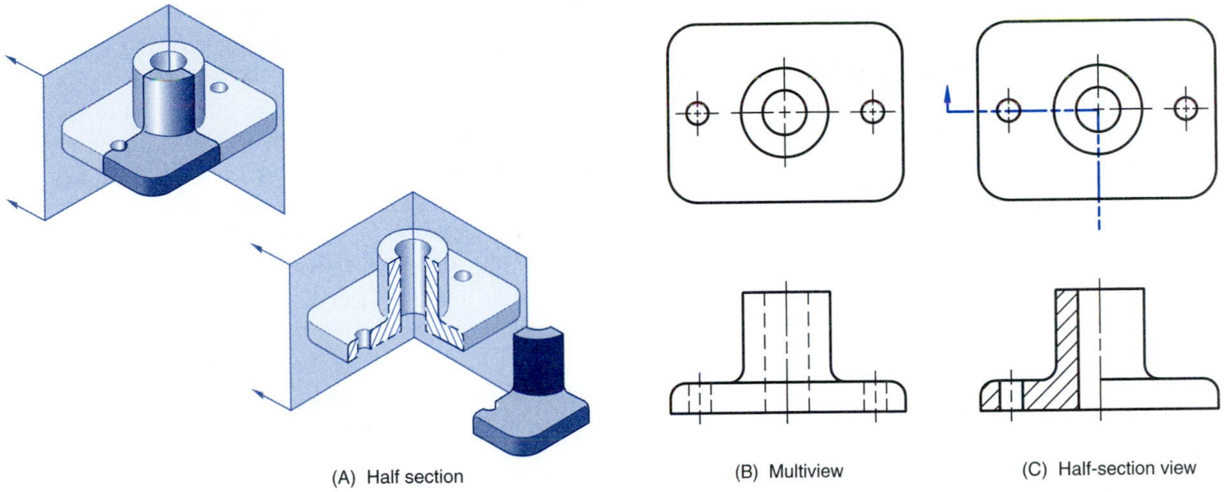

Figure 3.12 Half Section
A half-section view is created by passing a cutting plane halfway through the object.

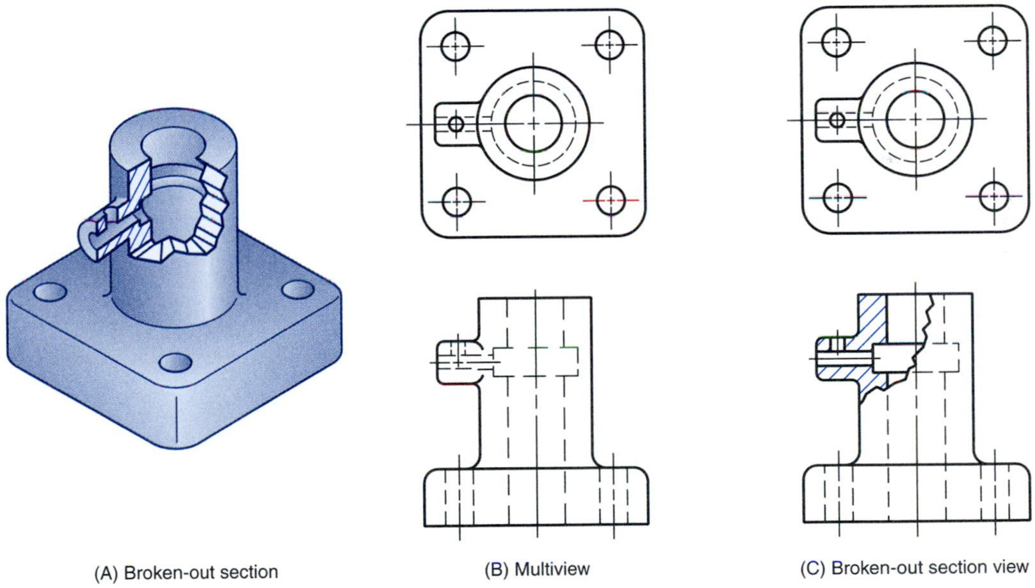

Figure 3.13 Broken-Out Section
A broken-out section view is created by breaking off part of the object to reveal interior features.

3.4.3 Broken-Out Sections

A **broken-out section** is used when only a portion of the object needs to be sectioned. Figure 3.13A shows a part with a portion removed or broken away. A broken-out section is used instead of a half or full section view to save time.

A break line separates the sectioned portion from the unsectioned portion of the view. A break line is drawn freehand to represent the jagged edge of the break. No cutting plane line is drawn. Hidden lines may be omitted from the unsectioned part of the view unless they are needed for clarity, as shown in Figure 3.13C.

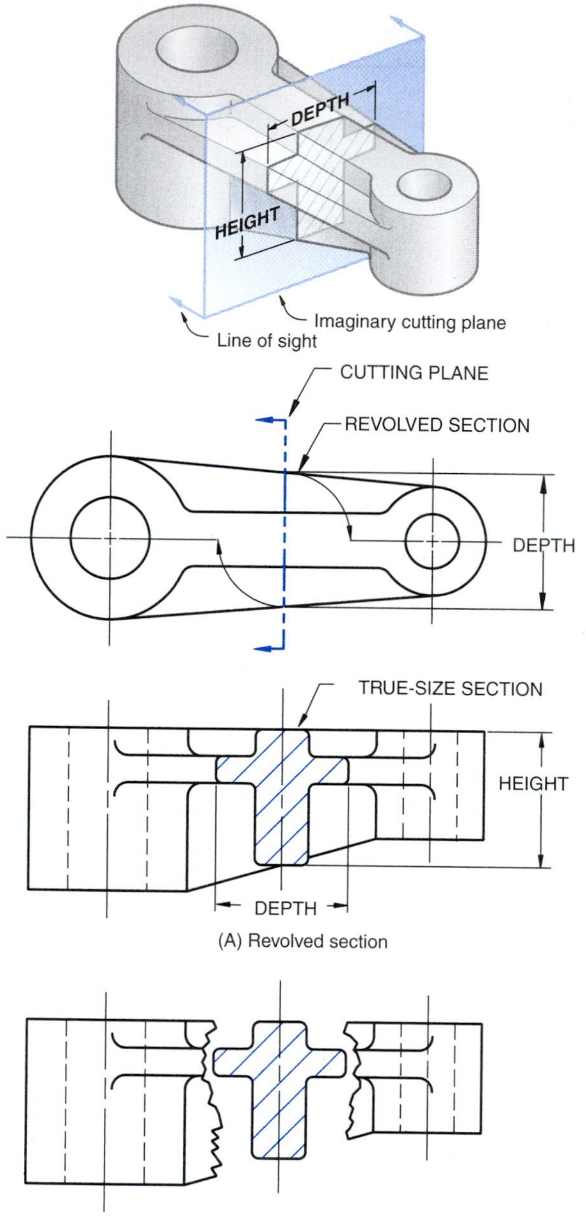

Figure 3.14 **Revolved Section**
A revolved section view is created by passing a cutting plane through the object, then revolving the cross section 90 degrees.

3.4.4 Revolved Sections

A **revolved section** is made by revolving the cross-section view 90 degrees about an axis of revolution and superimposing the section view on the orthographic view, as shown in Figure 3.14A. Visible lines adjacent to the revolved view can either be drawn or broken out using conventional breaks, as shown in Figure 3.14B. When the revolved view is superimposed on the part, the original lines of the part behind the section are deleted. The cross section is drawn true shape and size, not distorted to fit the view. The axis of revolution is shown on the revolved view as a center line.

A revolved section is used to represent the cross section of a bar, handle, spoke, web, aircraft wing, or other elongated feature. Revolved sections are useful for describing a cross section without having to draw another view. In addition, these sections are especially helpful when a cross section varies or the shape of the part is not apparent from the given orthographic views.

3.4.5 Removed Sections

Removed section views do not follow standard view alignments as practiced in multiview drawings. Removed sections are made in a manner similar to revolved sections by passing an imaginary cutting plane perpendicular to a part then revolving the cross section 90 degrees. However, the cross section is then drawn adjacent to the orthographic view, not on it (Figure 3.15). Removed sections are used when there is not enough room on the orthographic view for a revolved section.

Removed sections are used to show the contours of complicated shapes, such as the wings and fuselage of an airplane, blades for jet engines or power plant turbines, and other parts that have continuously varying shapes.

Removed sections can also be drawn to a larger scale for better representation of the details of the cross section and for dimensioning. The scale used for the removed section view is labeled beneath the view. Sometimes removed sections are placed on center lines adjacent to the axis of revolution.

3.4.6 Offset Sections

An **offset section** has a cutting plane that is bent at one or more 90-degree angles to pass through important features (Figure 3.16A). Offset sections are used for complex parts that have a number of important features that cannot be sectioned using a straight cutting plane. In Figure 3.16, the cutting plane is first bent at 90 degrees to pass through the hole and then bent at 90 degrees to pass through the slot. The front portion of the object is "removed" to create a full-section view of the part. The cutting plane line is drawn with 90-degree offsets, as shown in Figure 3.16A.

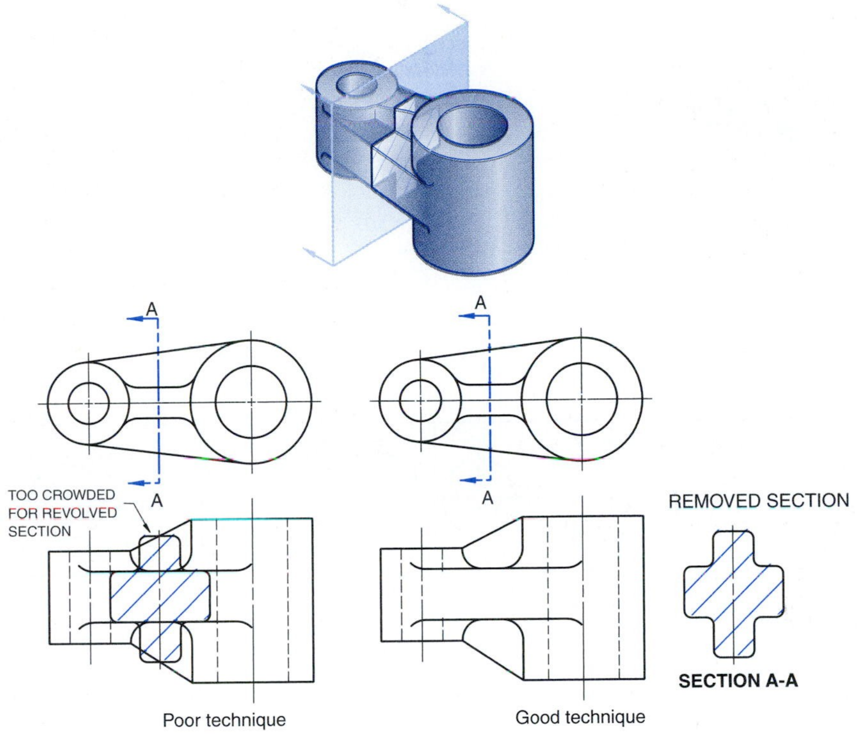

Figure 3.15 Removed Section
A removed section view is created by making a cross section, then moving it to an area adjacent to the view.

As shown in Figure 3.16B, the change of plane that occurs when the cutting plane is bent at 90 degrees is not represented with lines in the section view.

3.4.7 Assembly Sections

Assembly sections are typically orthographic, pictorial, full, or half-section views of parts as assembled. Leonardo da Vinci was one of the first to create assembly sections, using them to illustrate pictorially the complex machine designs he developed.

Section assembly drawings follow special conventions. Standard parts, such as fasteners, dowels, pins, washers, springs, bearings, and gears, and nonstandard parts, such as shafts, are *not* sectioned; instead, they are drawn showing all of their exterior features. For example, in Figure 3.17, fasteners would be cut in half by the cutting plane, yet they are not cross-hatched with section lines.

Typically, the following features are not section lined in a mechanical assembly section drawing:

Shafts	Ribs
Bearings, roller or ball	Spokes
Gear teeth	Lugs
Threaded fasteners	Washers
Nuts and bolts	Keys
Rivets	Pins

Adjacent parts in assembly sections are cross-hatched at different angles so that they are more easily identified (Figure 3.18). Different material symbols can also be used for this purpose. Also, if a part in an assembly section is separated by some distance, the section lines are still drawn in the same direction.

CAD 3-D modeling software can create models of each part, the individual models can be sectioned, and the section views can be placed together, resulting in a 3-D assembly section. An alternative involves using a feature that adds translucency to parts to reveal interior assemblies (Figure 3.19). The models can be used to check for interference of parts or can be analyzed by dynamically showing the parts in action. The models can also be rotated to produce an orthographic assembly section view.

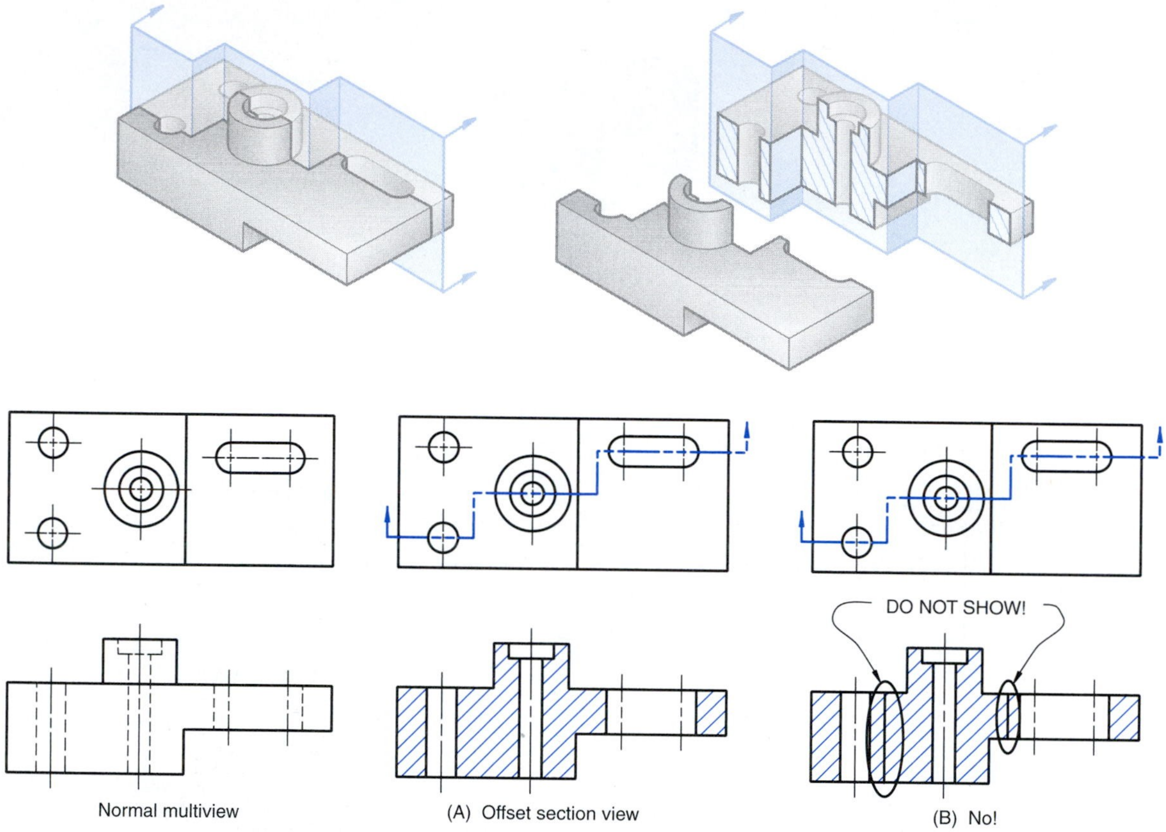

Figure 3.16 Offset Section
An offset section view is created by bending the cutting plane at 90-degree angles to pass through important features.

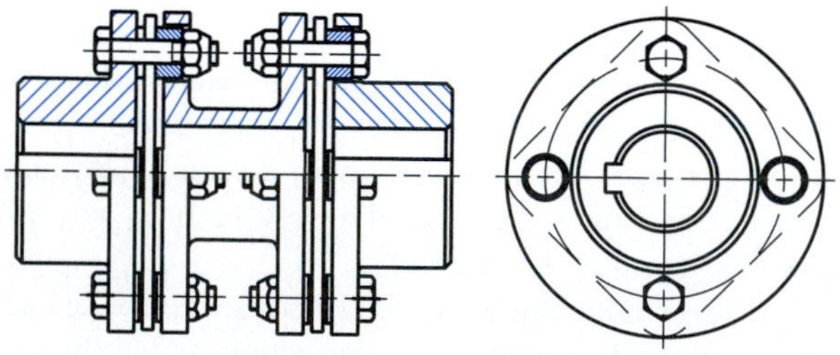

Figure 3.17 Standard Parts Not Section Lined
Standard parts, such as fasteners and shafts, are not section lined in assembly sections, even if they are cut by the cutting plane.

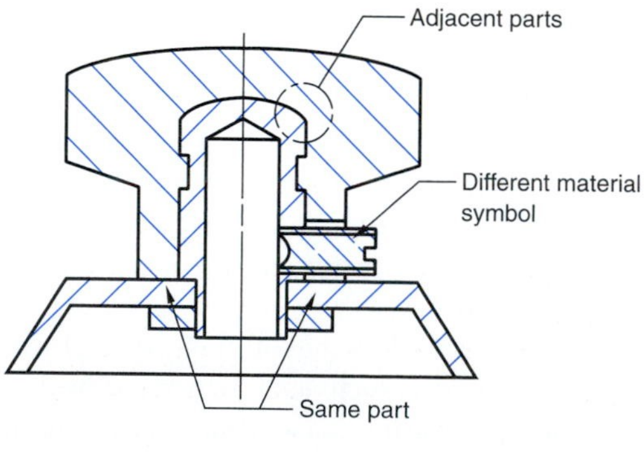

Figure 3.18 Section Lining Adjacent Parts
Adjacent parts in an assembly section are section lined at different angles so that individual parts can be more easily identified.

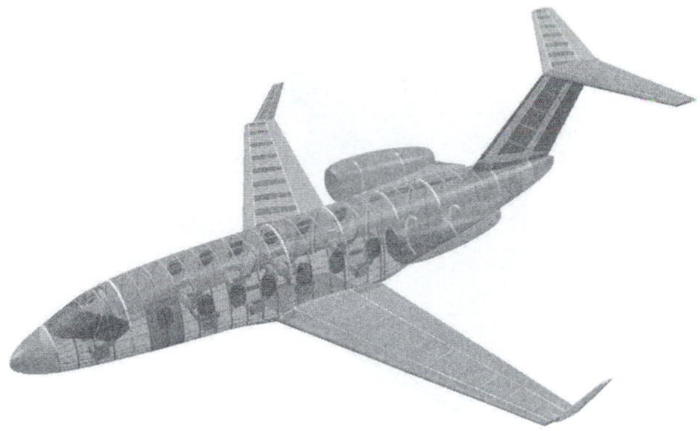

Figure 3.19 Translucency of a CAD Model
With a 3-D CAD model, translucency can be used instead of cutting planes to reveal interior features. Courtesy of Hewlett-Packard.

3.5 SPECIAL SECTIONING CONVENTIONS

Conventional practices have been established to handle section views of special situations, such as the alignment of holes, ribs, and spokes. These practices are described in the following paragraphs.

3.5.1 Ribs, Webs, and Other Thin Features

Ribs, webs, spokes, lugs, gear teeth, and other thin features are not section lined when the cutting plane passes parallel to the feature. A **rib** or **web** is a thin, flat part that acts as a support (Figure 3.20). Adding section lines to these features would give the false impression that the part is thicker than it really is. Figure 3.21 shows a cutting plane that passes parallel to and through a web (SECTION B–B). Figure 3.21B shows the view drawn using conventional practice, which leaves the web unsectioned. Figure 3.21A shows an incorrect representation of the section view, with the web having section lines. This view gives the false impression that the web has substantial thickness.

Leaving thin features unsectioned only applies if the cutting plane passes parallel to the feature. If the cutting plane passes perpendicular or crosswise to the feature (cutting plane A–A), section lines are added as shown in Figure 3.21C.

Occasionally, section lines are added to a thin feature so that it is not mistaken or read as an open area. Figure 3.22A shows a part with webs, which are flush with the rim and the hub. In Figure 3.22B, the part could be interpreted as being round and without webs. To section-line a thin feature, use alternate lines, as shown in Figure 3.22C. However, if the feature is not lost, as shown in Figure 3.23, then section lines should not be added.

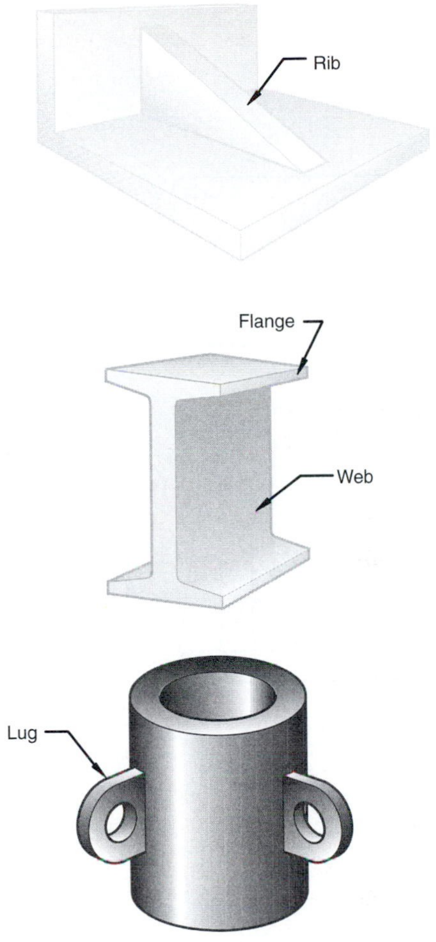

Figure 3.20 Ribs, Webs, and Lugs
Ribs, webs, and lugs are special types of features commonly found in mechanical components. These types of features require special treatment in section views.

3.5.2 Aligned Sections

Aligned sections are special types of orthographic drawings used to revolve or align special features of parts in order to clarify them or make them easier to represent in section. Aligned sections are used when it is important to include details of a part by "bending" the cutting plane. The cutting plane and the feature are imagined to be aligned or revolved before the section view is created. In other words, the principles of orthographic projection are violated in order to more clearly represent the features of the object.

Normally, the alignment is done along a horizontal or vertical center line, and the realignment is always less than 90 degrees (Figure 3.24). The aligned section view gives a clearer, more complete description of the geometry of the part. The cutting plane line may be bent to pass

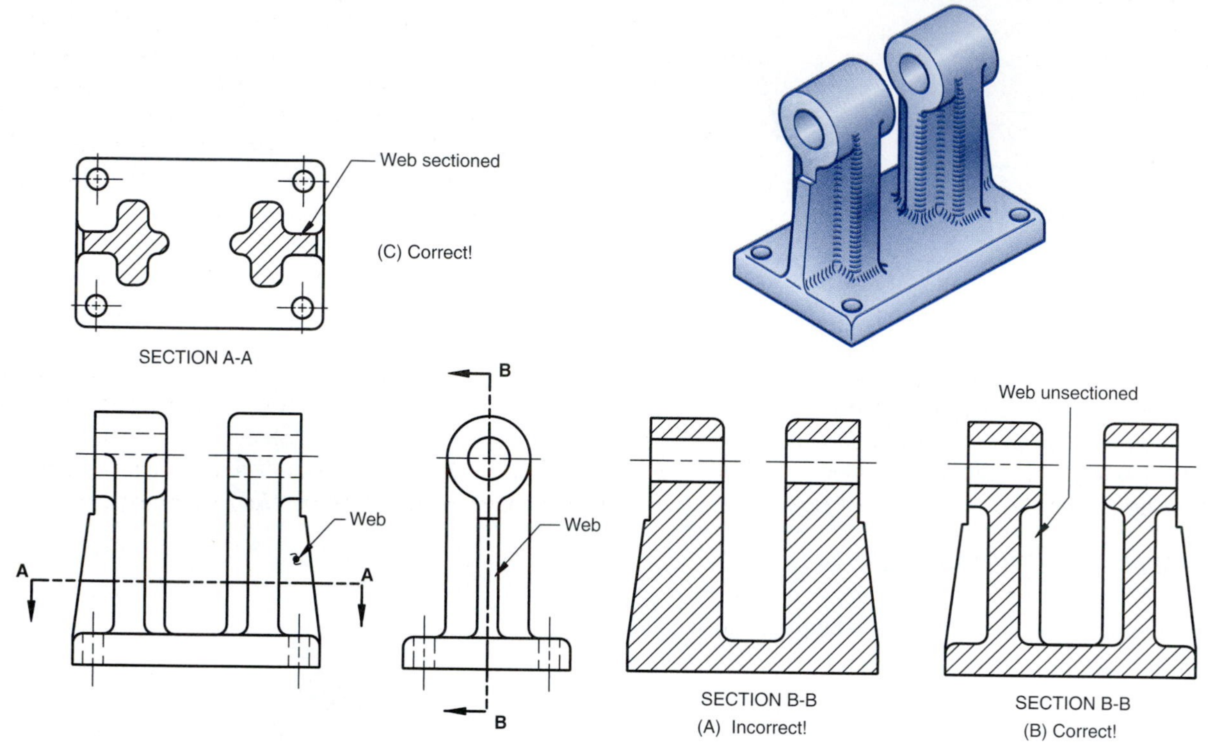

Figure 3.21 **Conventional Practices for Webs in Section**

Thin features, such as webs, are left unsectioned when cut parallel to the feature by the cutting plane.

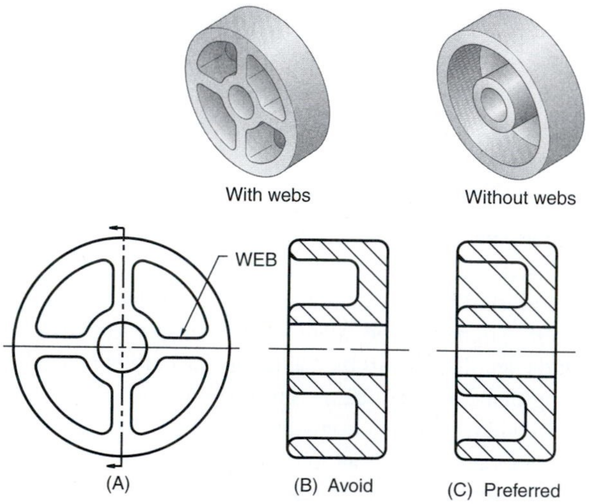

Figure 3.22 **Alternate Method of Representing a Web in Section**

Thin features are section lined with alternate lines if it clarifies the geometry of the object.

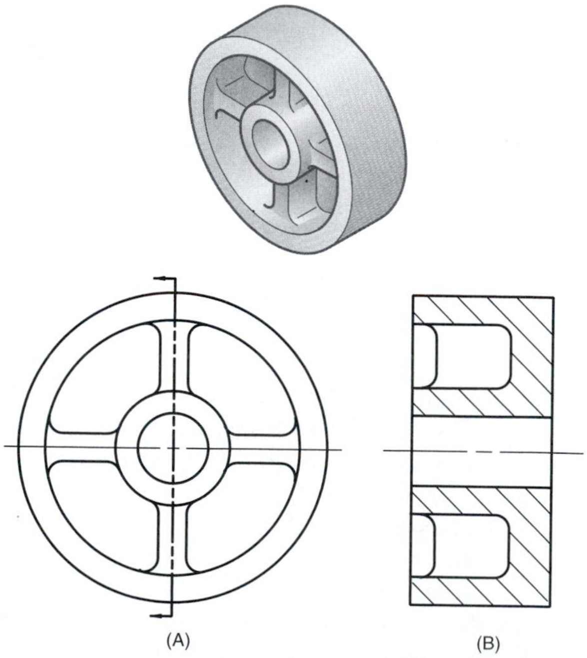

Figure 3.23 **Omitting Section Lines on Webs**

When the feature is not lost, section lines are omitted.

through all of the nonaligned features in the unsectioned view (Figures 3.25 through 3.27).

Conventional practice also dictates the method for representing certain features that are perpendicular to the line of sight. Figure 3.25 shows a part with a feature called a *spoke,* which is bent at a 30-degree angle from the vertical center line. True projection would show the spoke foreshortened in the side view. However, aligning the spoke with the vertical center line in the front view allows the section view to be created more easily, and this is the preferred method for representing the part. Also, even though the spoke has been aligned, standard practice requires that section lines not be placed on the spoke. Also, spoke A is not drawn in the section view, to save time and increase clarity.

Alignment of features is used for many applications. For example, the lug shown in Figure 3.26A is aligned in the section view but is not sectioned. A lug is considered a thin feature, so section lines are normally not used when the cutting plane is parallel to the lug's thickness. However, if the lug were positioned as shown in Figure 3.26B, then the lug would be drawn with section lines. Figure 3.27 shows how ribs are aligned before the section view is drawn. The standard practice is not to put section lines on ribs (Figure 3.27C).

3.5.3 Conventional Breaks

Conventional breaks are used for revolved section views or for shortening the view of an elongated part, such as a shovel handle or vehicle axle. Shortening the length of a part leaves room for the part to be drawn to a larger scale. The printed dimension specifies the true length (Figure 3.28).

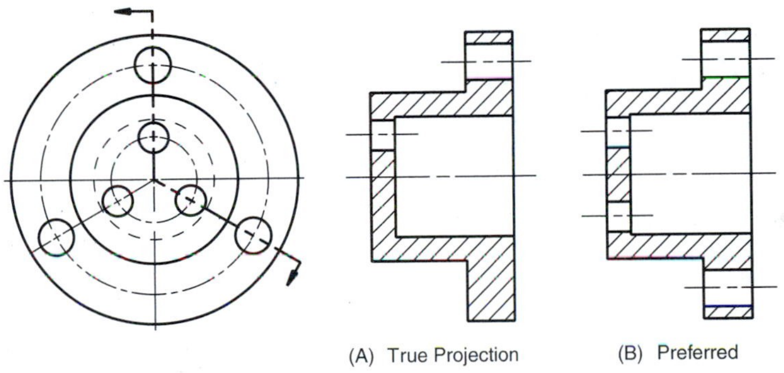

Figure 3.24 Aligned Section
Aligned section conversions are used to rotate the holes into position along the vertical center line.

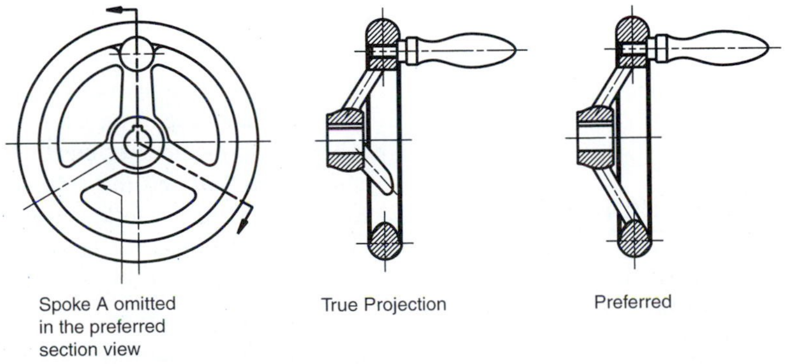

Figure 3.25 Aligning Spokes
Aligning spokes in section views is the conventional method of representation.

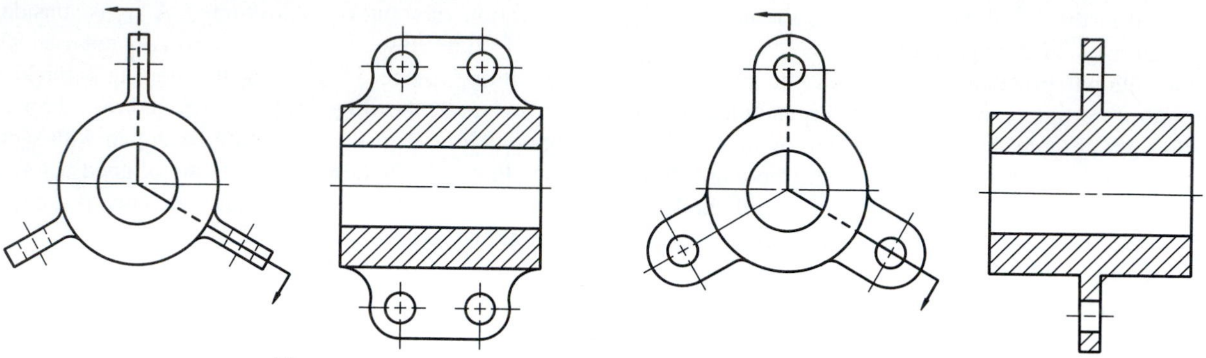

Figure 3.26 Aligning Lugs
Aligning lugs in section views is the conventional method of representation.

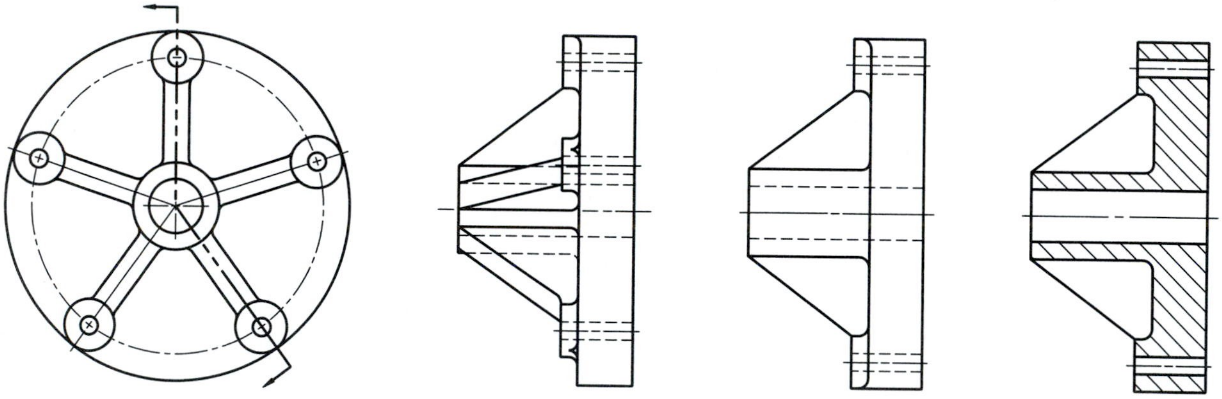

Figure 3.27 Aligning Ribs
Aligning ribs in section views is the conventional method of representation.

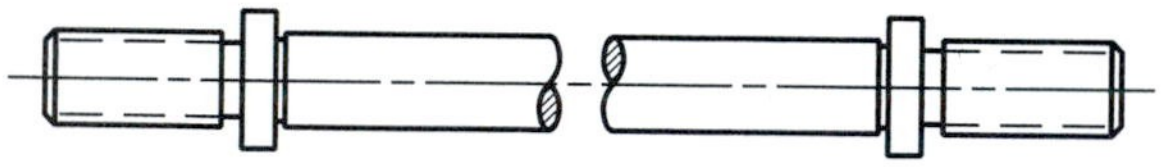

Figure 3.28 Conventional Break Symbols Can Shorten the Drawn Length of a Long Object

3.6 AUXILIARY VIEW PROJECTION THEORY

An **auxiliary view** is an orthographic view that is projected onto any plane other than one of the six principal views. Figure 3.29A shows three principal views of an object. Surface ABCD is in an inclined plane and is therefore never seen in true size or shape in any of these views. In a multiview drawing, a true size and shape plane is shown only when the *line of sight* (LOS) used to create the view is perpendicular to the projection plane. To show the true size and shape of surface ABCD, an auxiliary view can be created by positioning a line of sight perpendicular to the inclined plane, then constructing the new view (Figure 3.29B). Two methods of creating auxiliary views are the *fold-line method* and the *reference plane method.* These are discussed in the following sections. The use of construction techniques for auxiliary views has diminished with the use of 3-D CAD.

3.6.1 Fold-Line Method

In Figure 3.30, the object is suspended in a glass box to show the six principal views, created by projecting the object onto

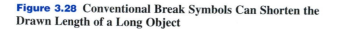

▼ Practice Problem 3.1

Sketch the front view as a full-section view.

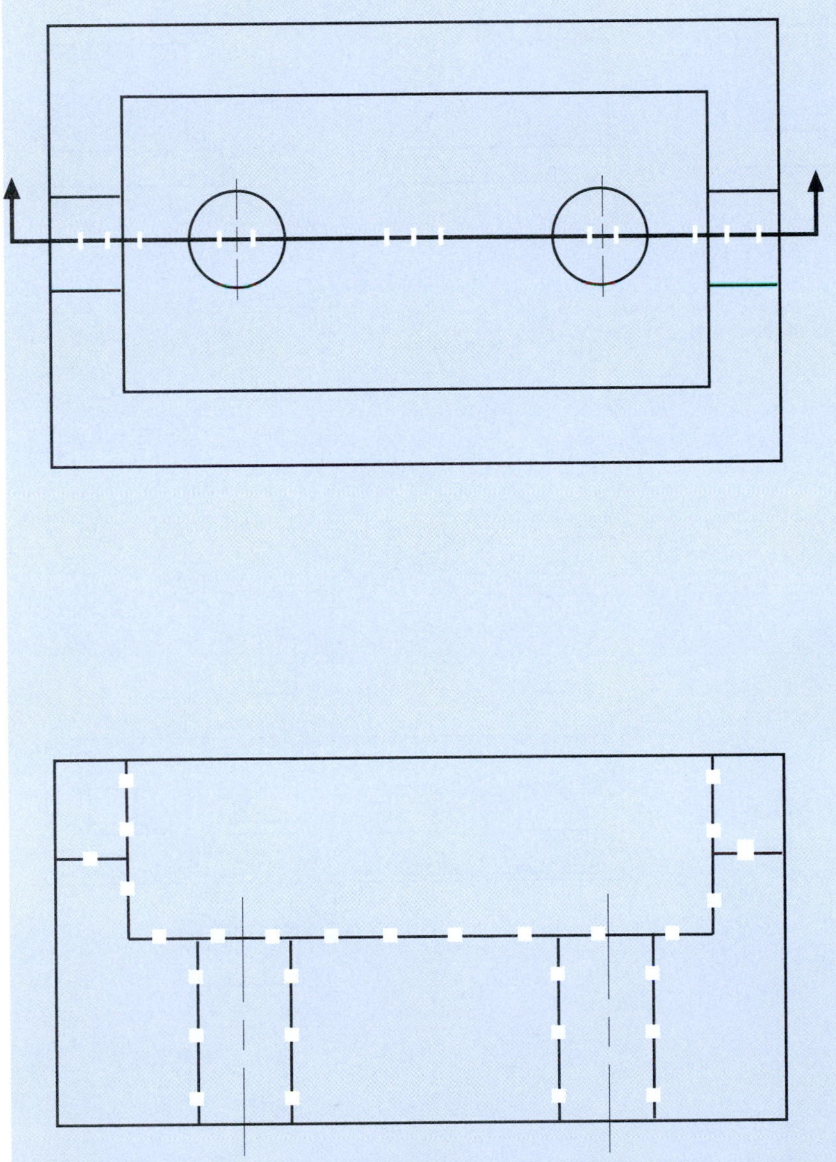

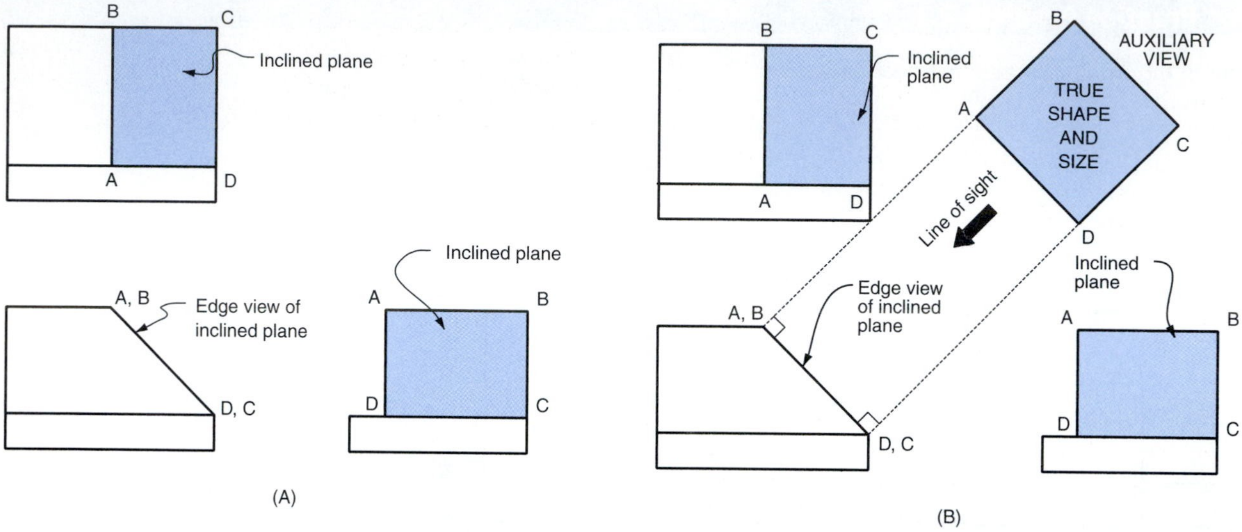

Figure 3.29 Auxiliary View
An auxiliary view of an inclined plane is not one of the principal views.

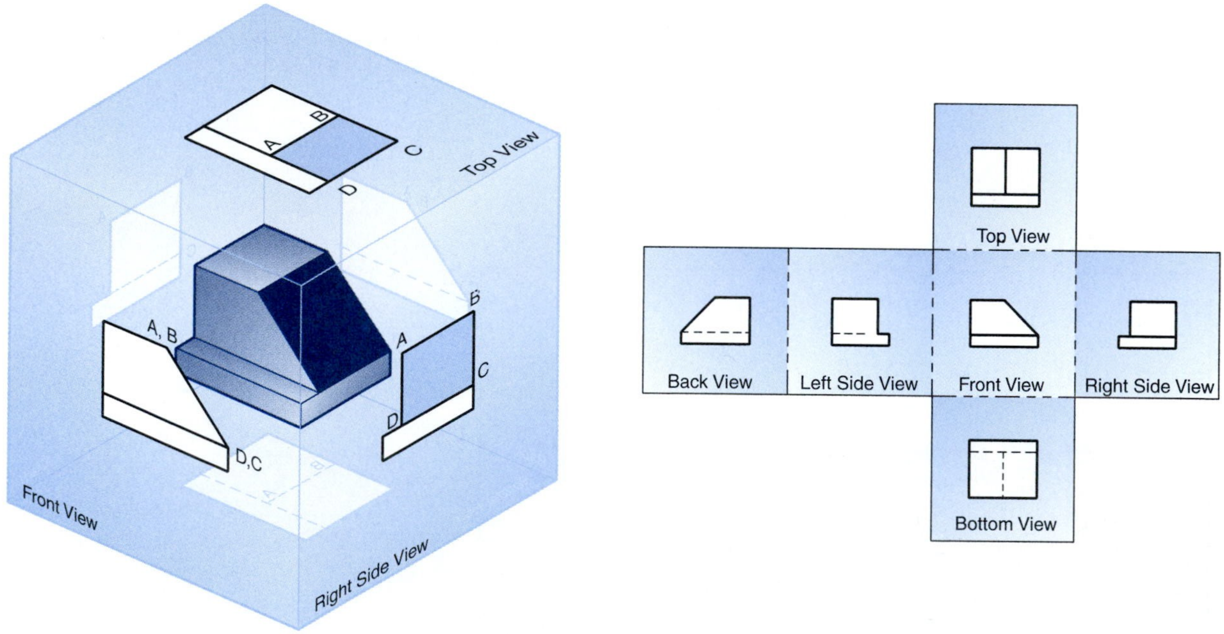

Figure 3.30 Object in Glass Box, and Resulting Six Views When the Box Is Unfolded

the planes of the box. The box is then unfolded, resulting in the six principal views. However, when the six views are created, surface ABCD never appears true size and shape; it always appears either foreshortened or on edge.

Figure 3.31 shows the object suspended inside a glass box, which has a special or *auxiliary plane* that is parallel to inclined surface ABCD. The line of sight required to create the auxiliary view is perpendicular to the new projection plane and to surface ABCD. The auxiliary plane is perpendicular to and hinged to the frontal plane, creating a *fold line* between the front view and the new auxiliary view.

In Figure 3.32, the auxiliary glass box is unfolded, with the fold lines between the views shown as phantom lines. In the auxiliary view, surface ABCD is shown true size and shape and is located at distance M from the fold line. The line AB in the top view is also located at distance M from its fold line. Changing the position of the object, such as moving it closer to the frontal plane, changes distance M (Figure 3.33).

3.6.2 Constructing an Auxiliary View

Figure 3.34 shows an auxiliary view that is projected from the front view of an object, using the fold-line method. Since plane ABCD is an inclined plane in the principal views, an auxiliary view is needed to create a true-size view of that plane. An auxiliary view of plane ABCD is created as described in the following steps.

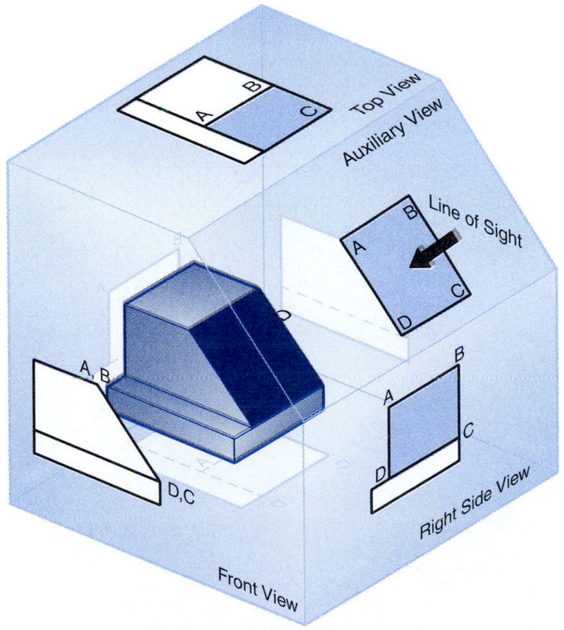

Figure 3.31 Object in Glass Box with Special Auxiliary Plane

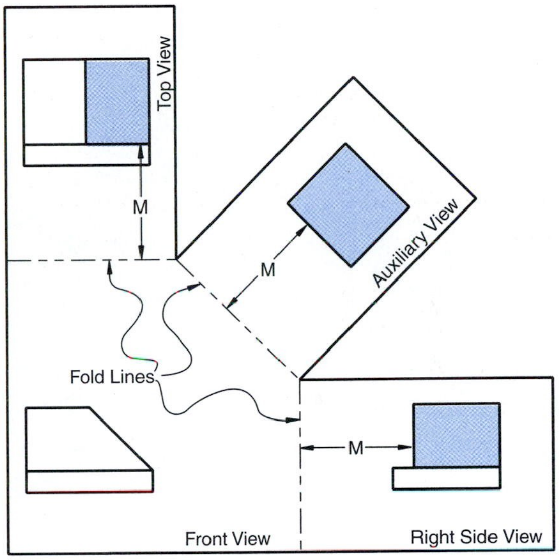

Figure 3.32 Unfolding the Glass Box to Create an Auxiliary View of the Inclined Plane

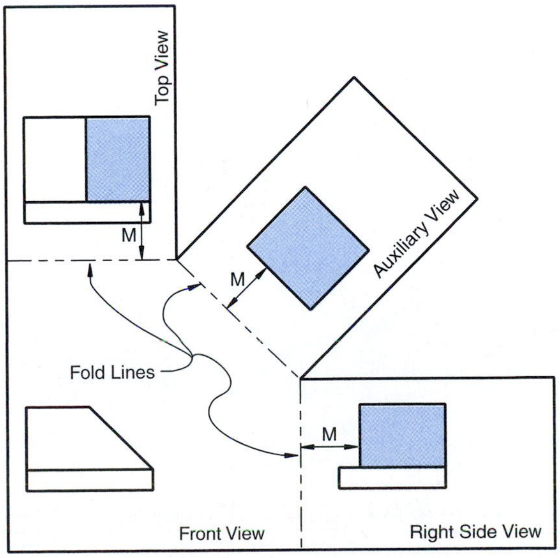

Figure 3.33 Object Distance from Fold Line

Object distance from the frontal plane determines the distance from the fold lines in the right side, auxiliary, and top views.

Only the inclined plane has been drawn in the auxiliary view; the rest of the object is not represented. When only the feature of interest is drawn in an auxiliary view and not the whole object, the view is called a partial auxiliary view. Most auxiliary views will be partial. Also, hidden features are not shown unless absolutely necessary.

Constructing an Auxiliary View

Step 1. Given the front, top, and right side views, draw fold line F–1 using a phantom line parallel to the edge view of the inclined surface. Place line F–1 at any convenient distance from the front view.

Step 2. Draw fold line F–H between the front and top views. Line F–H should be perpendicular to the projectors between

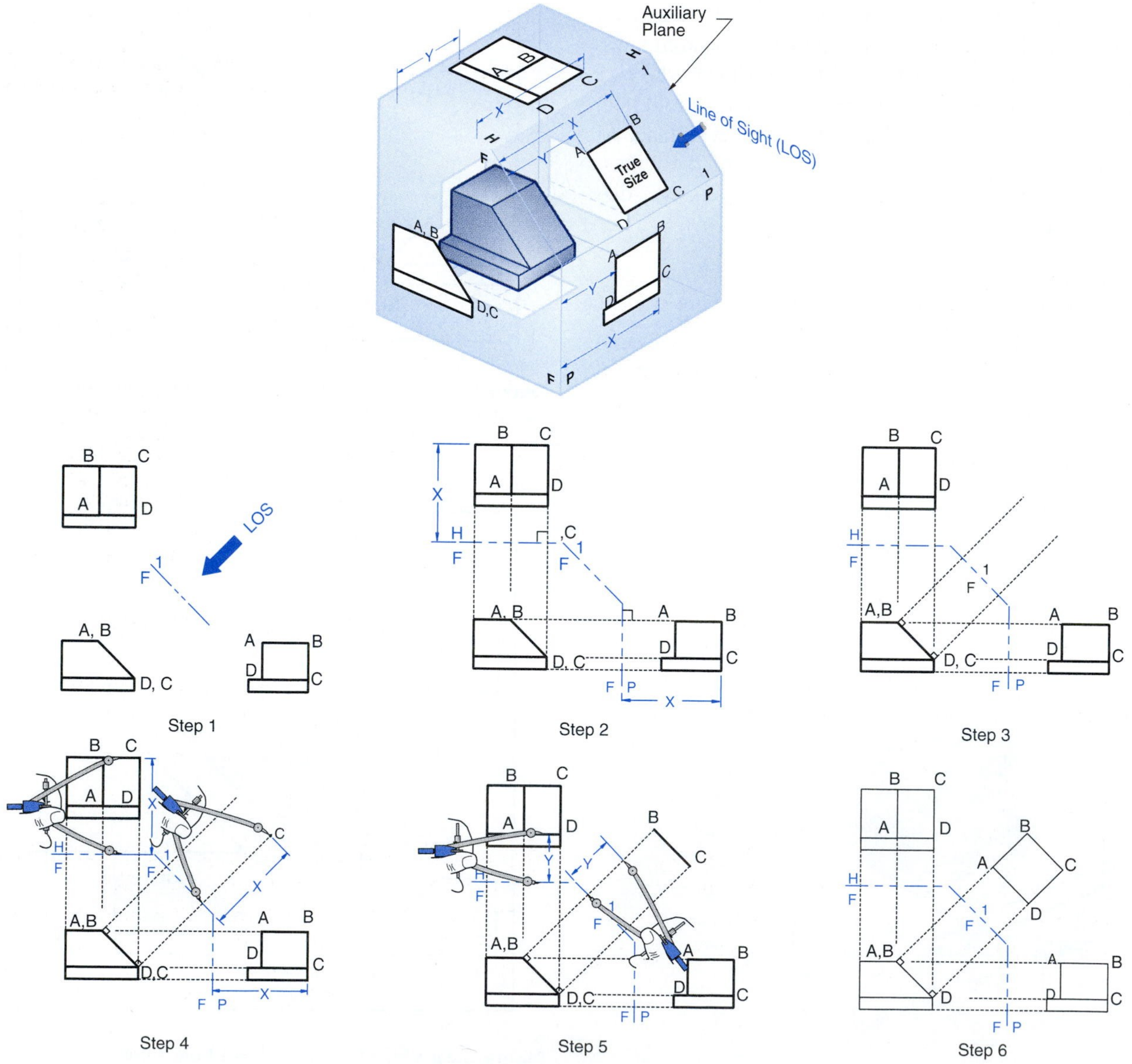

Figure 3.34 Constructing a Depth Auxiliary View to Determine the True Size and Shape of the Inclined Surface

the views and at a distance X from the rear edge of the top view. Draw fold line F–P between the front and right side views, perpendicular to the projectors between the two views and at the distance X from the rear edge of the right side view. The distance from fold line F–H to the top view must be equal to the distance from fold line F–P to the right side view. Draw parallel projectors between the principal views, using construction lines.

Step 3. Project the length of the inclined surface from the front view to the auxiliary view, using construction lines. The projectors are perpendicular to the edge view and projected well into the auxiliary view from corners A,B and D,C.

Step 4. Transfer the depth of the inclined surface from the top view to the auxiliary view by first measuring the *perpendicular* distance from fold line H–F to point C at the rear of the top view. This is distance X. Measure this same distance on the projectors in the auxiliary view, measuring from fold line F–1. The measurement used to locate point C could have been taken from the profile view.

Step 5. From point C in the auxiliary view, draw a line perpendicular to the projectors. Depth dimension Y is transferred from the top view by measuring the *perpendicular* distance from fold line H–F to point A (or D) in the top view and transferring that distance to the auxiliary view along the projectors *perpendicular* to fold line F–1. Draw a line at the transferred point A (or D) in the auxiliary view, perpendicular to the projectors.

Step 6. To complete the auxiliary view of the inclined surface, darken lines AB and DC.

3.6.3 Partial Auxiliary Views

In auxiliary views, it is normal practice *not* to project hidden features or other features that are not part of the inclined surface. When only the details for the inclined surface are projected and drawn in the auxiliary view, the view is called a **partial auxiliary view.** A partial auxiliary view saves time and produces a drawing that is much more readable. Figure 3.35 shows a partial and a full auxiliary view of the same object. The full auxiliary view is harder to draw, read, and visualize. In this example, some of the holes have to be drawn as ellipses in the full auxiliary view. Sometimes a break line is used in a partial auxiliary view. When drawing break lines, do not locate them coincident with a visible or hidden line.

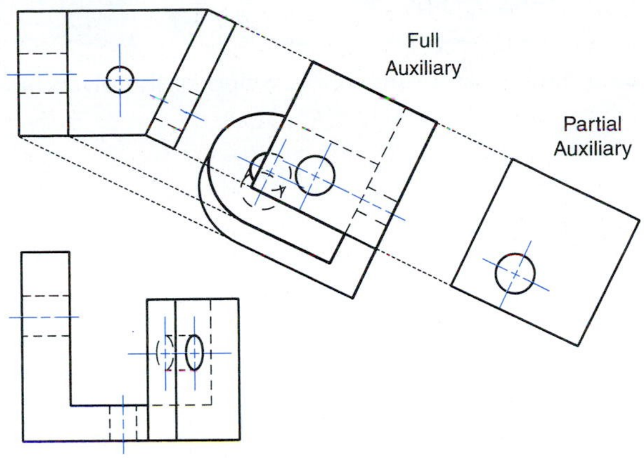

Figure 3.35 **A Full Auxiliary View, Including Hidden Lines, and a Partial Auxiliary View with No Hidden Lines**

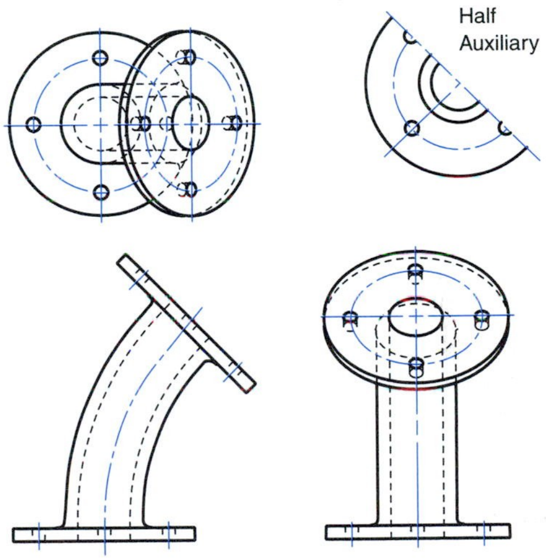

Figure 3.36 **A Half Auxiliary View of a Symmetrical Feature**

3.6.4 Half Auxiliary Views

Symmetrical objects can be represented in **half auxiliary views;** that is, only half of the object is drawn. The construction of a half auxiliary view is the same as described earlier for full auxiliary views. Figure 3.36 shows an object represented as a half auxiliary view.

▼ Practice Problem 3.2

Create a three-view and auxiliary view of the given object on the square grid.

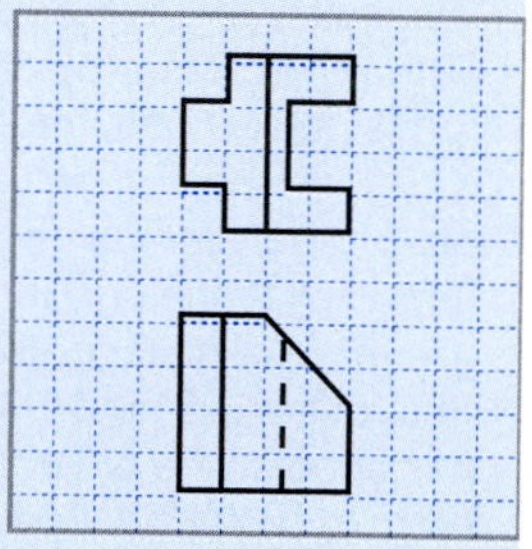

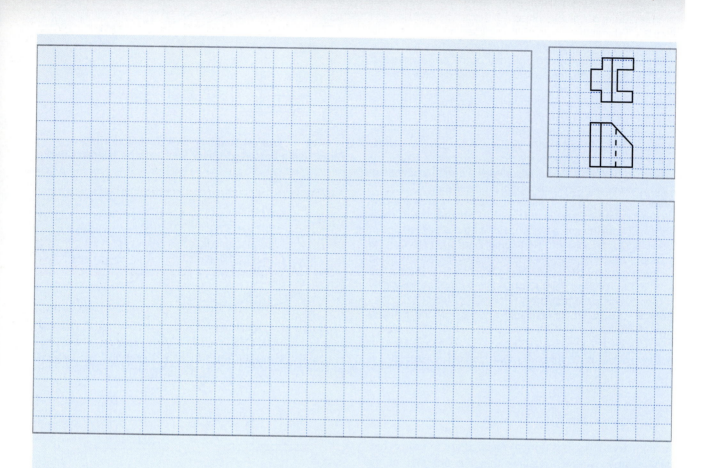

Questions for Review

1. Define section drawings.
2. Describe how 3-D CAD is used to create section views.
3. Describe how cutting plane lines are used.
4. Sketch the two standard types of cutting plane lines.
5. List three applications of section views.
6. What convention is used for hidden lines in a section view?
7. What convention is used for hidden lines on the unsectioned half of a half-section view?
8. Define section lines.
9. Sketch the material symbol used to represent steel.
10. Describe the difference between a revolved and a removed section view.
11. List some of the standard parts that are not sectioned in an assembly section. Explain why.
12. What type of line is used to separate the sectioned half from the unsectioned half of a half section view?
13. Define auxiliary views.
14. What is a partial auxiliary view?

Problems

Use the gridded sheets provided at the end of this section to complete the problems that follow.

3.1 For the Figure 3.37 section view problems, sketch (or draw with instruments or use a CAD software package) the orthographic view containing the cutting plane line and appropriate section view. Assume each grid represents 6 mm.

3.2 Sketch, or draw with instruments or CAD, the necessary views, including a section view, of the objects shown in Figures 3.38 through 3.45.

3.3 Using instruments or CAD, sketch or draw the two given views and a partial auxiliary view with hidden lines of the largest inclined surface in Figure 3.46.

3.4 Using instruments or CAD, sketch or draw the two given views and a complete auxiliary view without hidden lines of the largest inclined surface in the Figure 3.47 objects.

3.5 For Figures 3.48–3.51, sketch (or draw with instruments or use a CAD software package) two adjacent orthographic views of the object and a projected auxiliary view showing the object such that the largest inclined or largest oblique plane is in true shape.

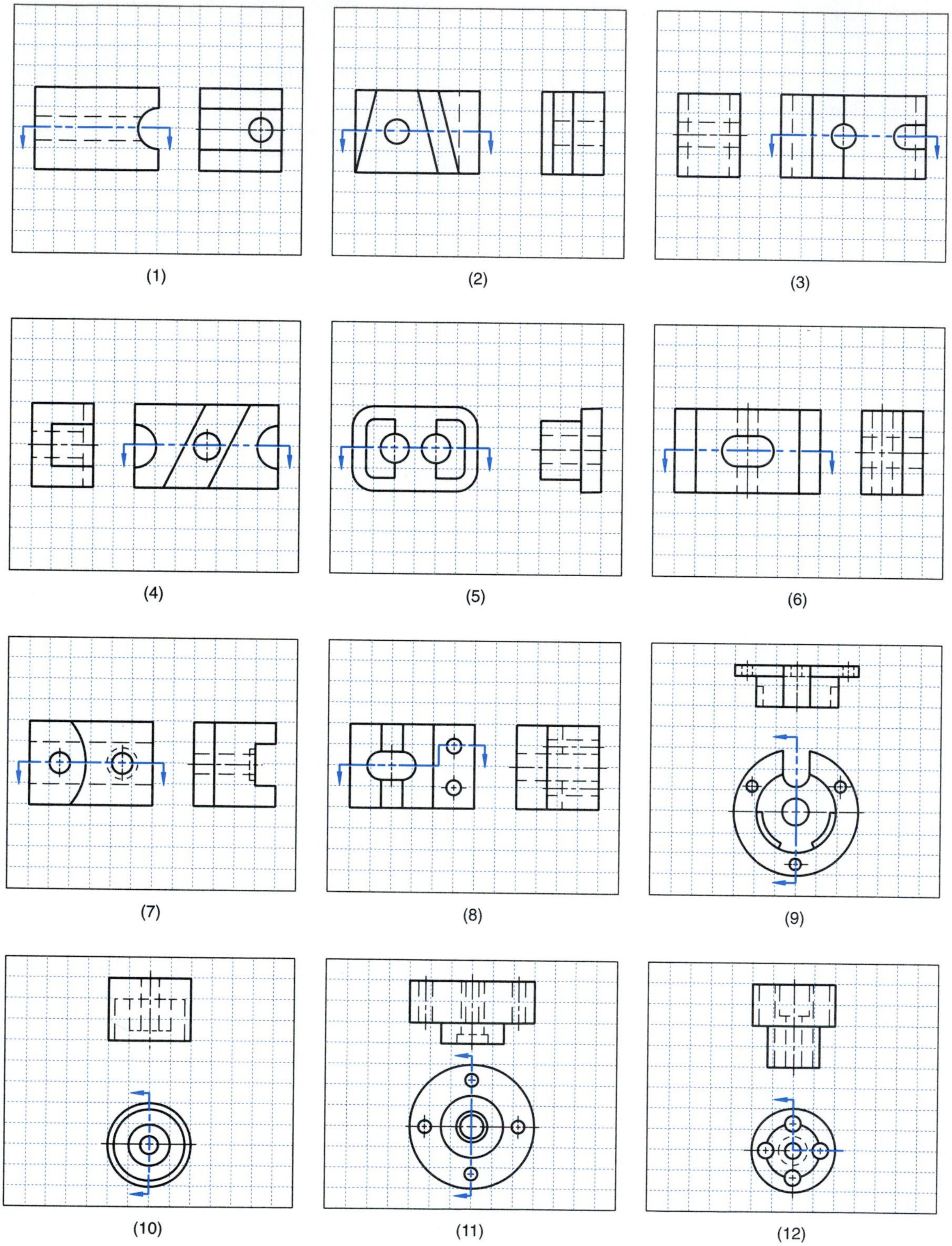

Figure 3.37 **Section View Problems**

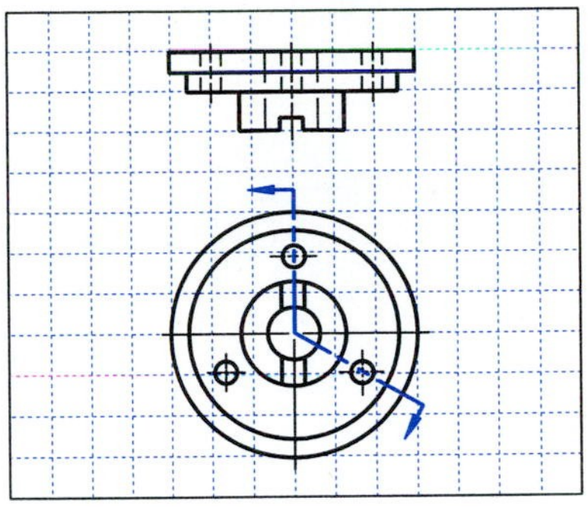

(13)

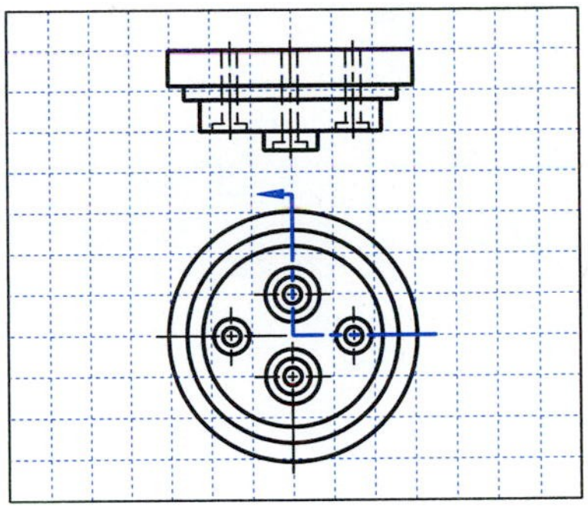

(14)

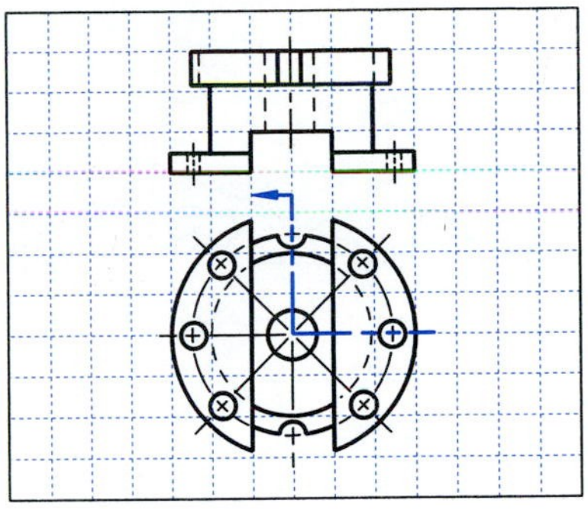

(15)

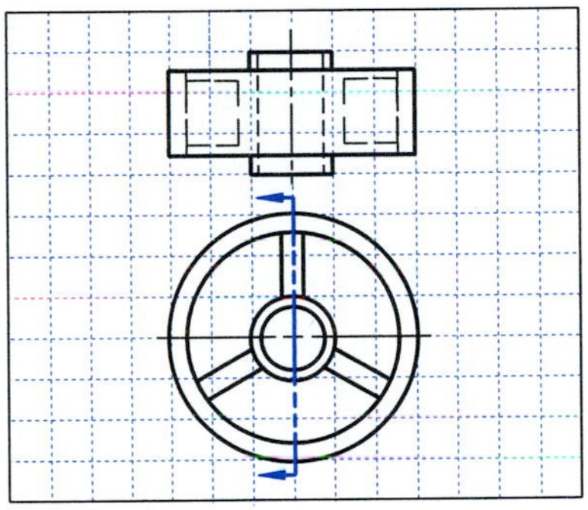

(16)

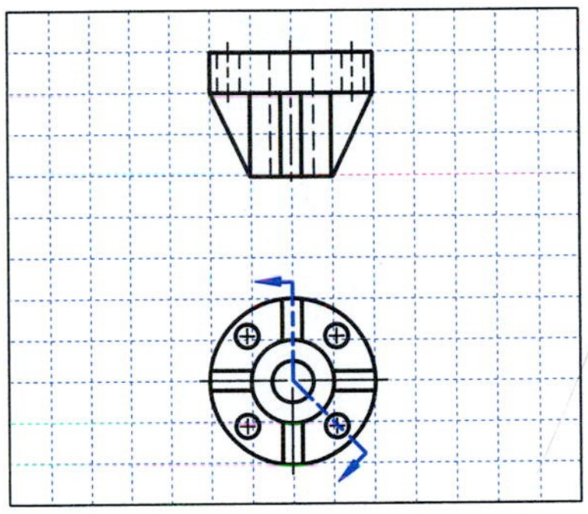

(17)

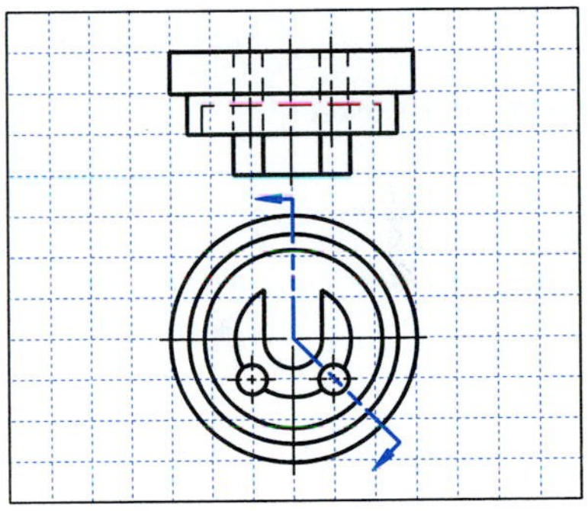

(18)

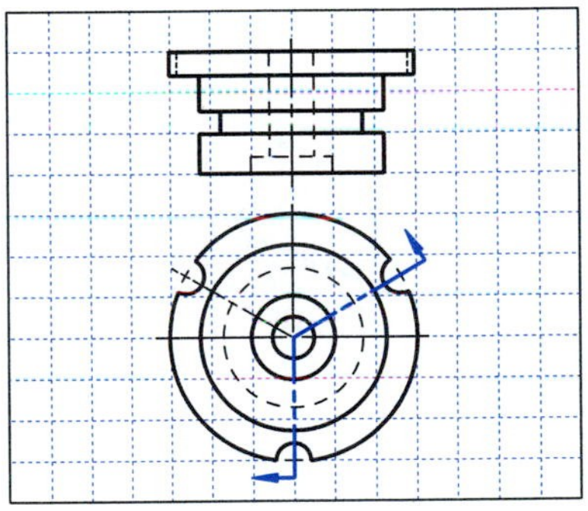

(19)

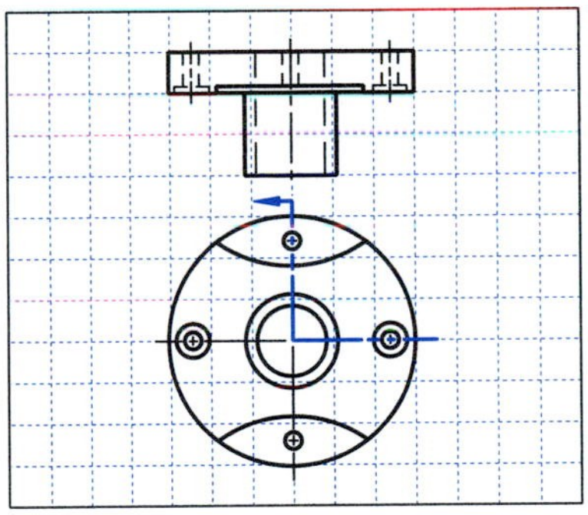

(20)

Figure 3.37 **Continued**

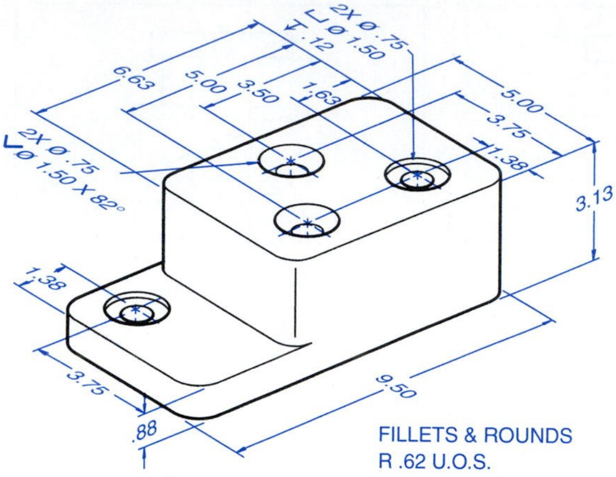

Figure 3.38 Bracket

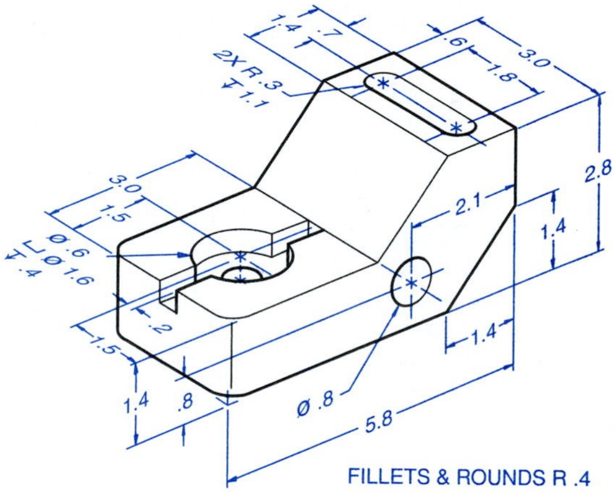

Figure 3.39 Counter Block

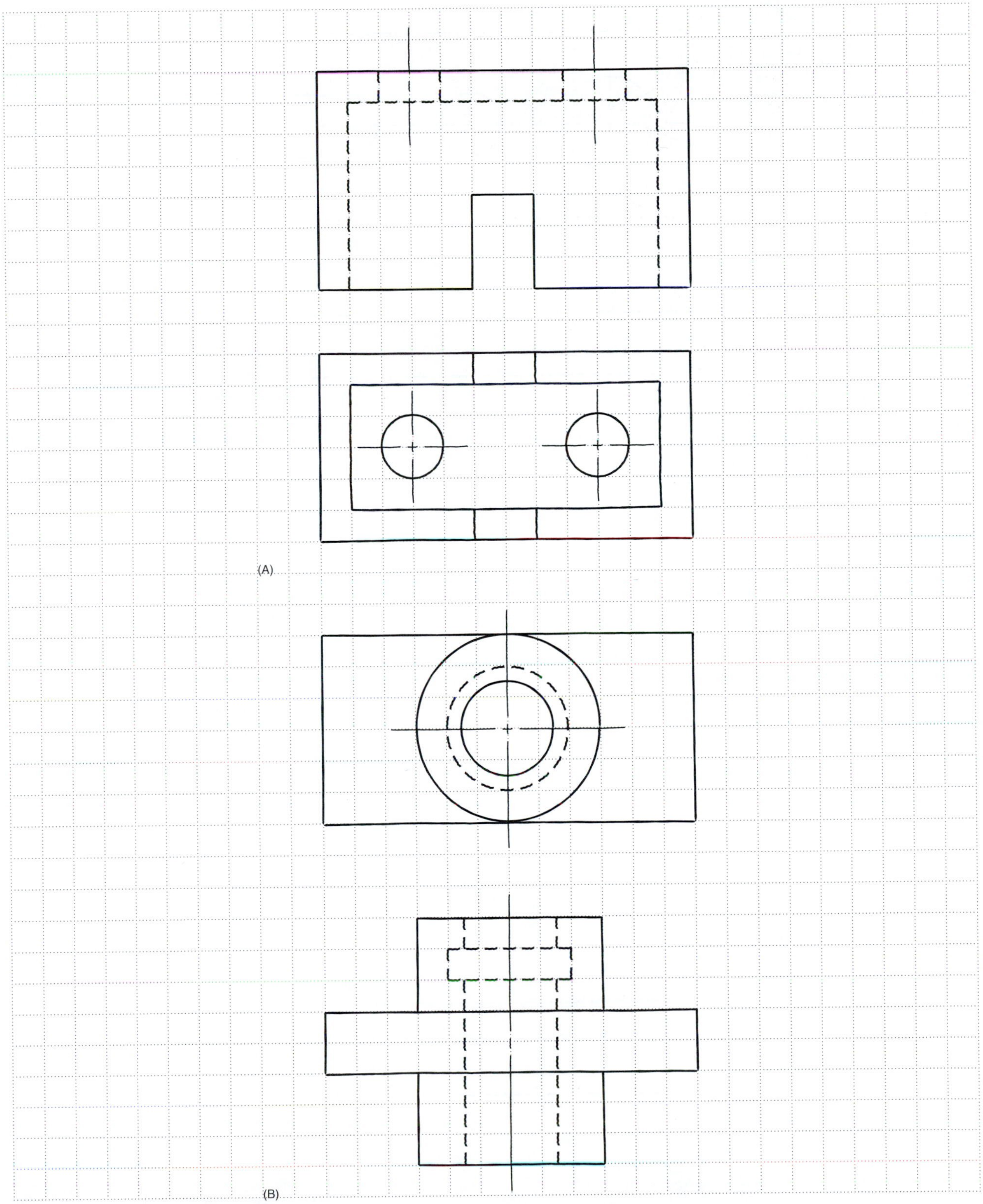

Figure 3.40 **Create Full-Section Views.**

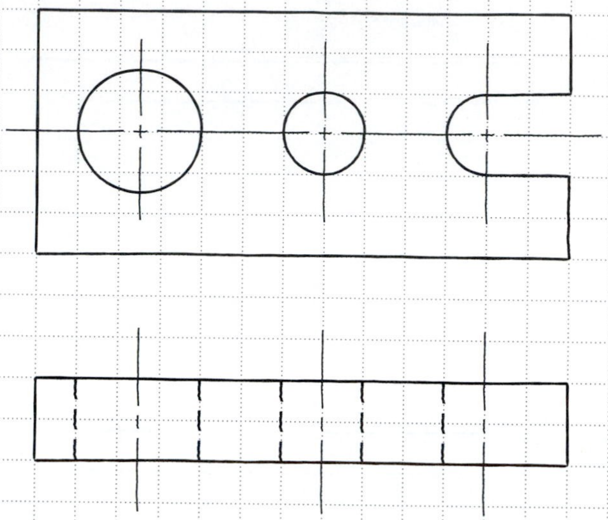
(A)

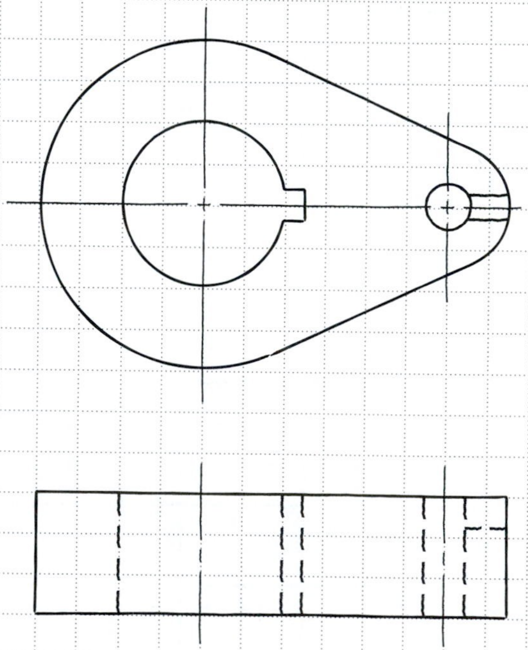
(B)

Figure 3.41 Create Full-Section Views.

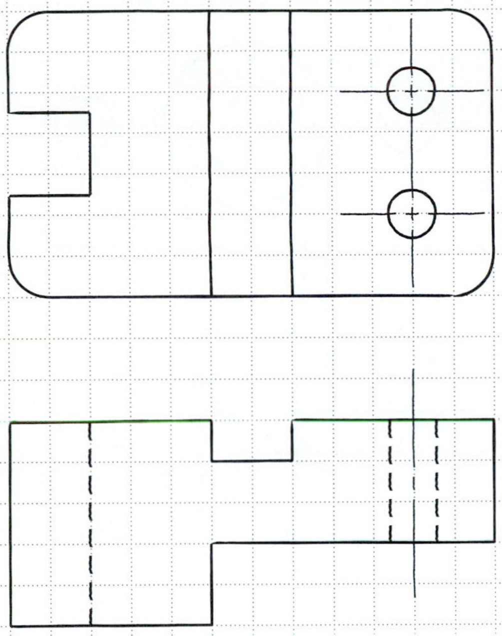

(A)

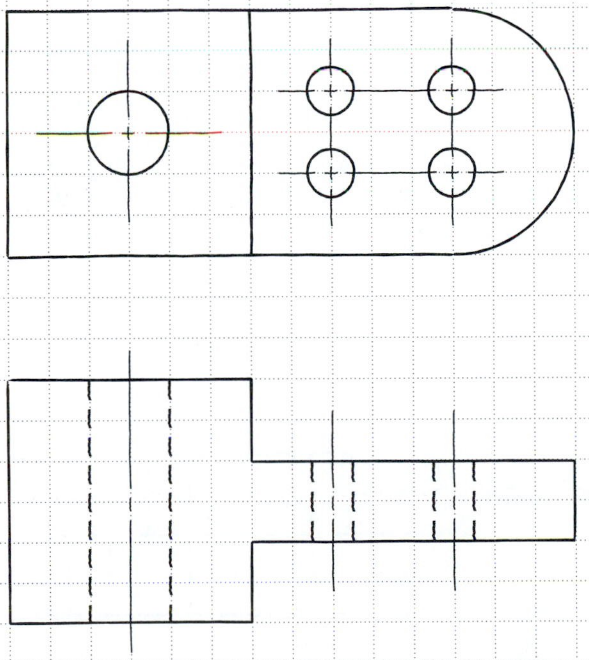

(B)

Figure 3.42 Create Offset Section Veiws.

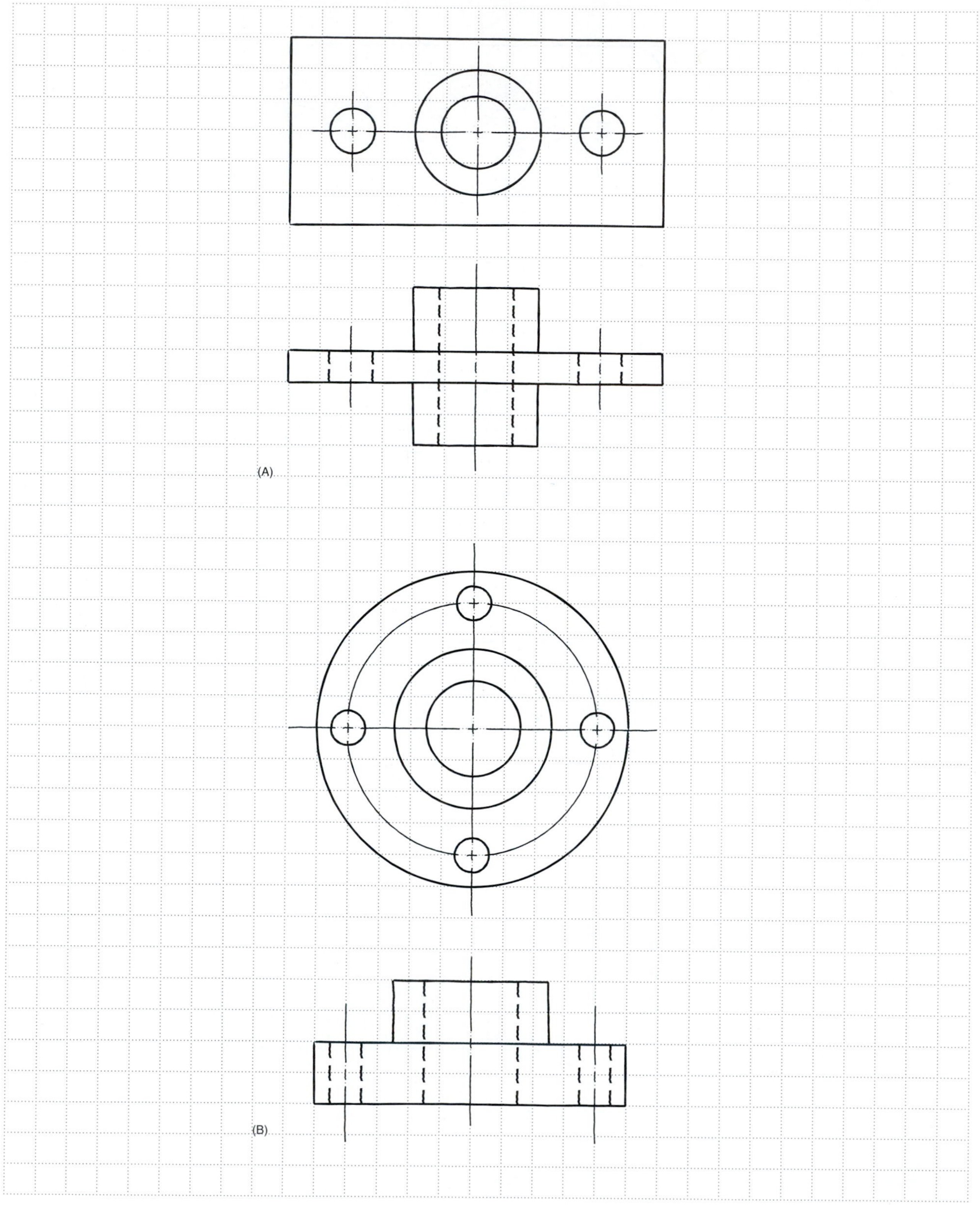

Figure 3.43 Create Half-Section Views.

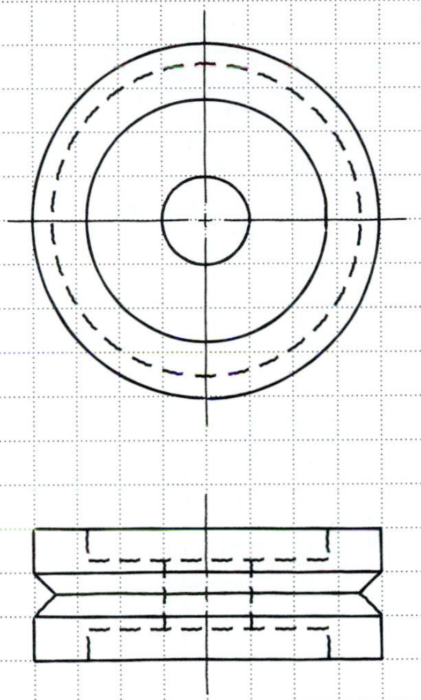
(A)

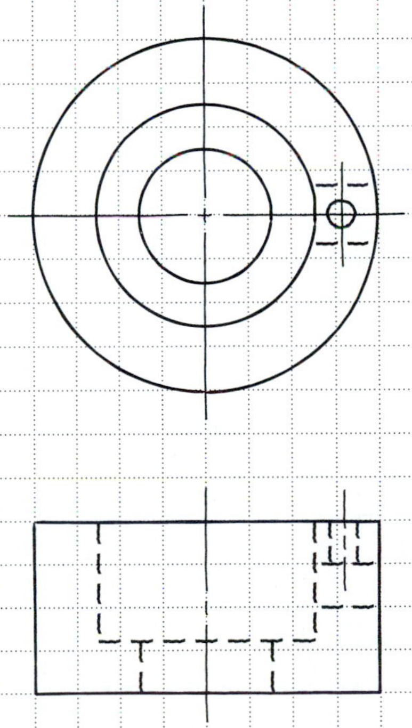
(B)

Figure 3.44 Create Half-Section Views.

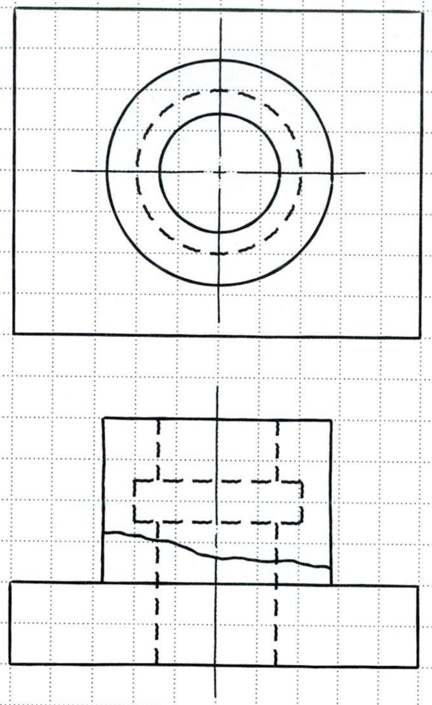

(A)

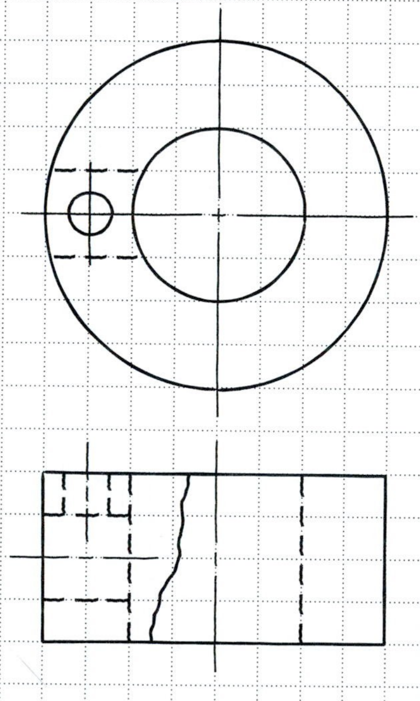

(B)

Figure 3.45 Create Broken-Out Section Views

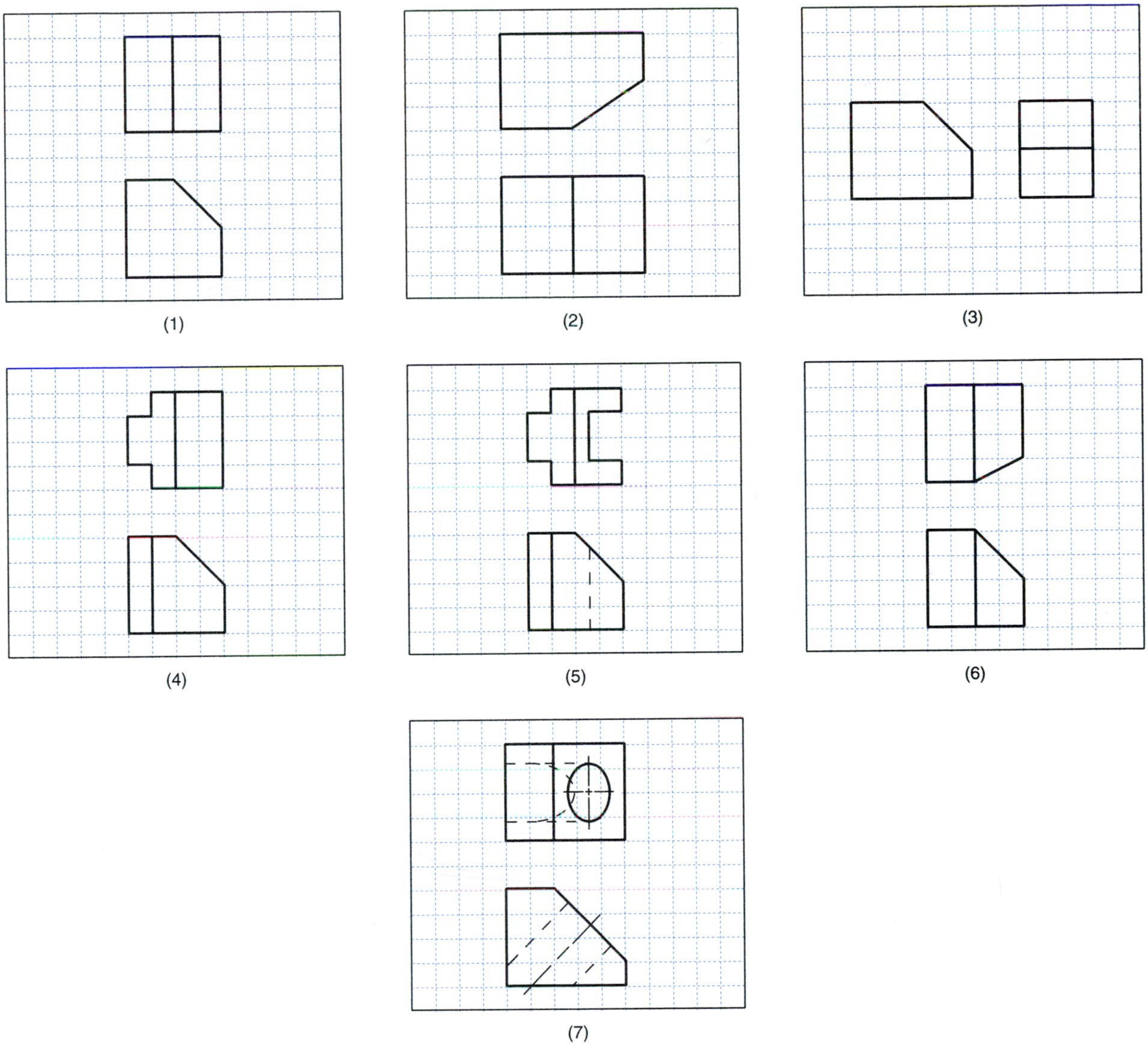

Figure 3.46 **Objects with Inclined Surfaces for Problem 3.3**

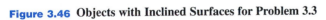

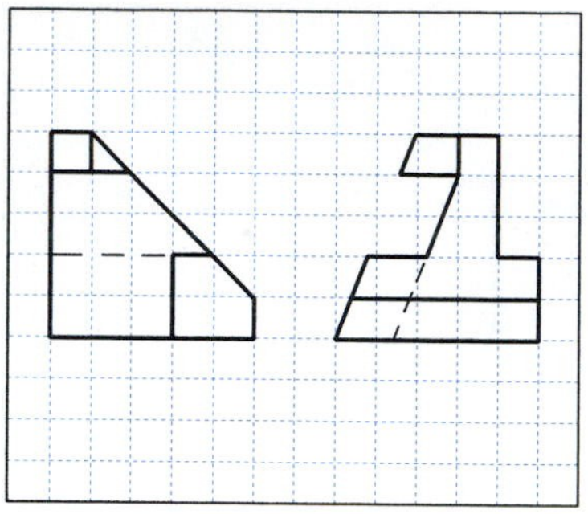

(1)

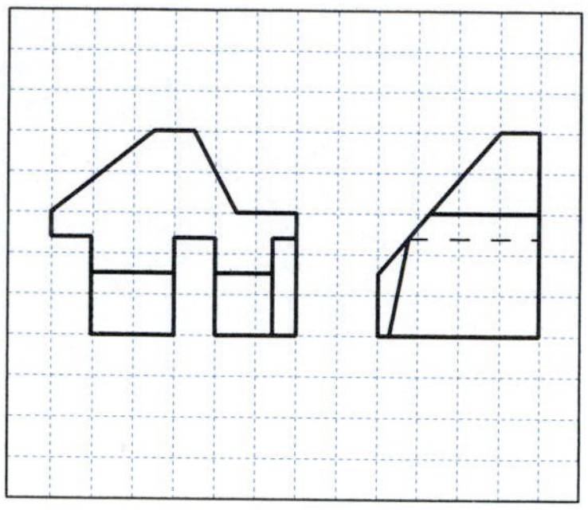

(2)

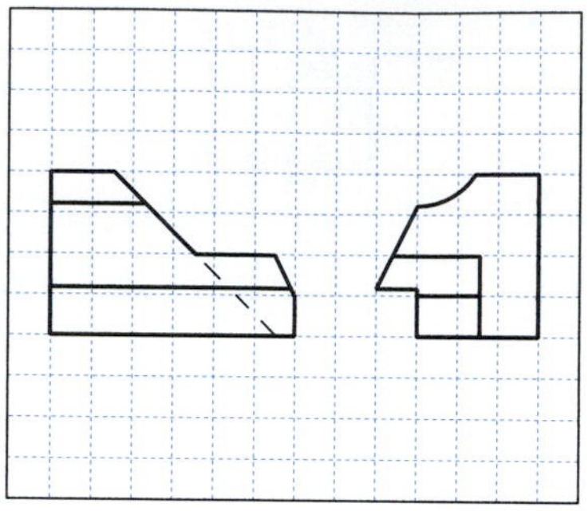

(3)

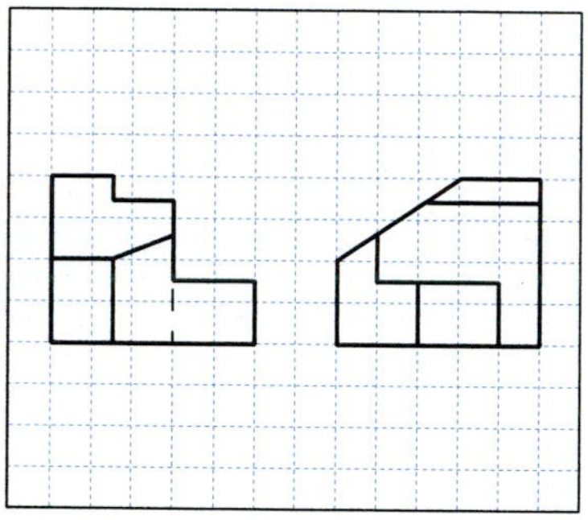

(4)

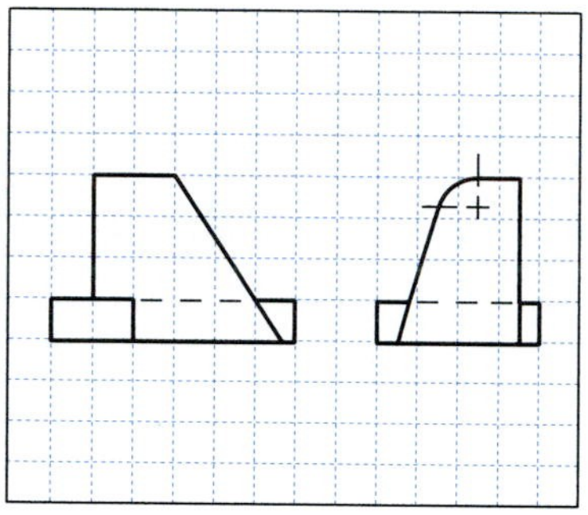

(5)

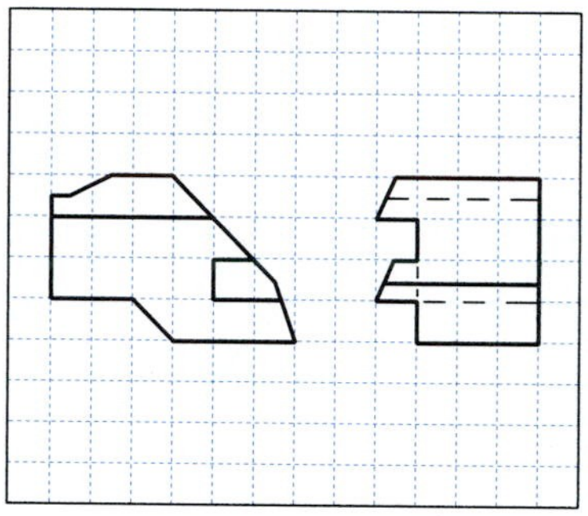

(6)

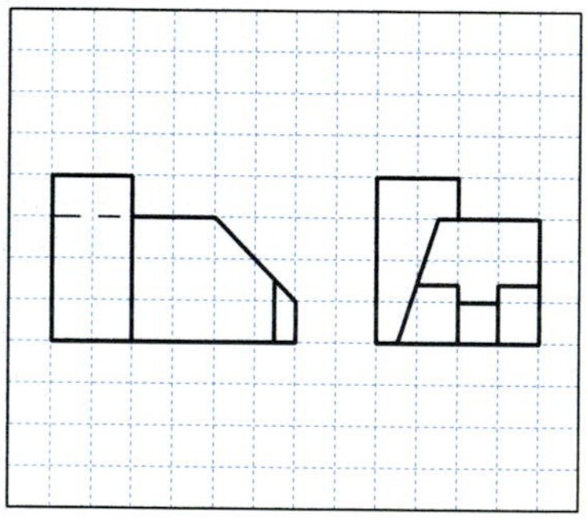

(7)

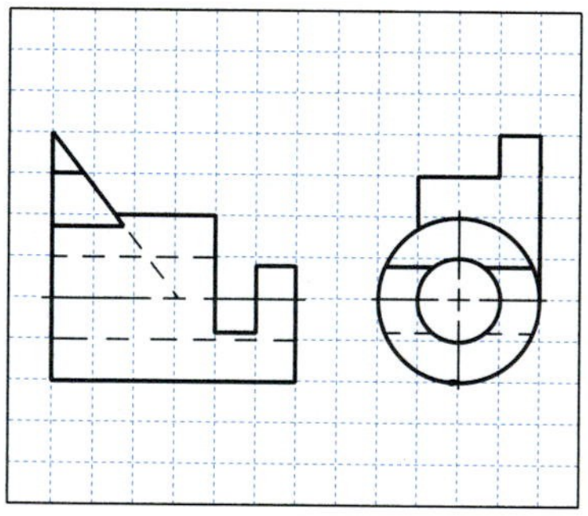

(8)

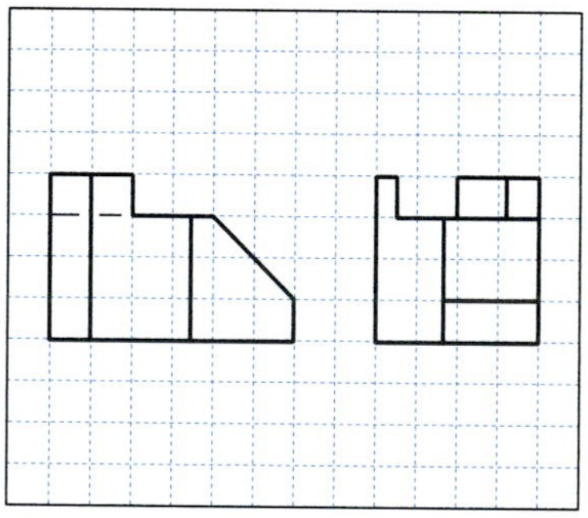

(9)

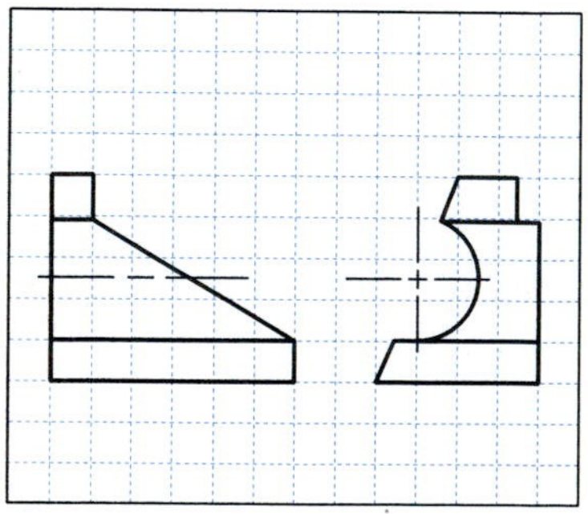

(10)

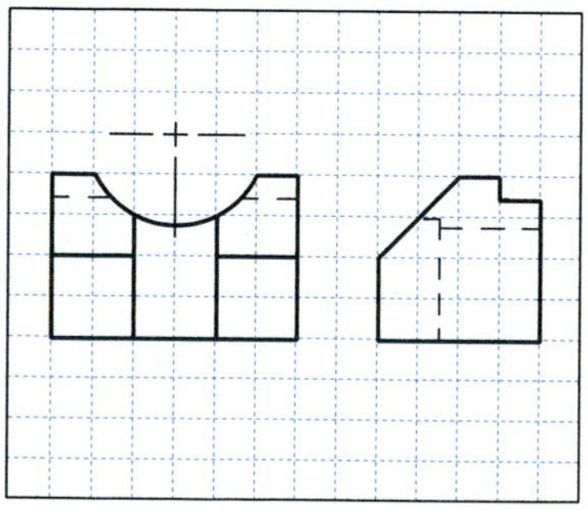

(11)

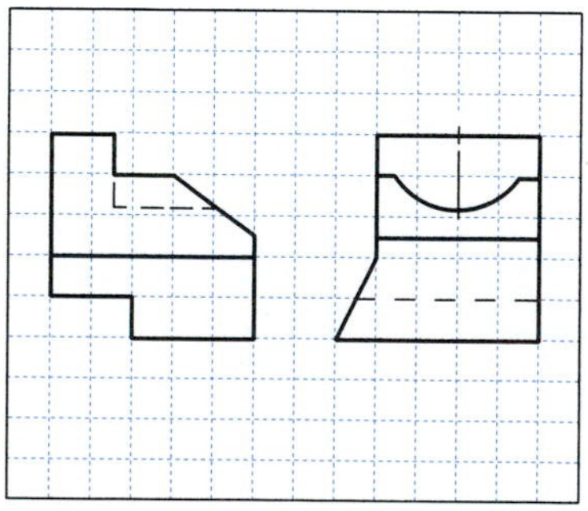

(12)

Figure 3.47 Multiviews for Problem 3.4

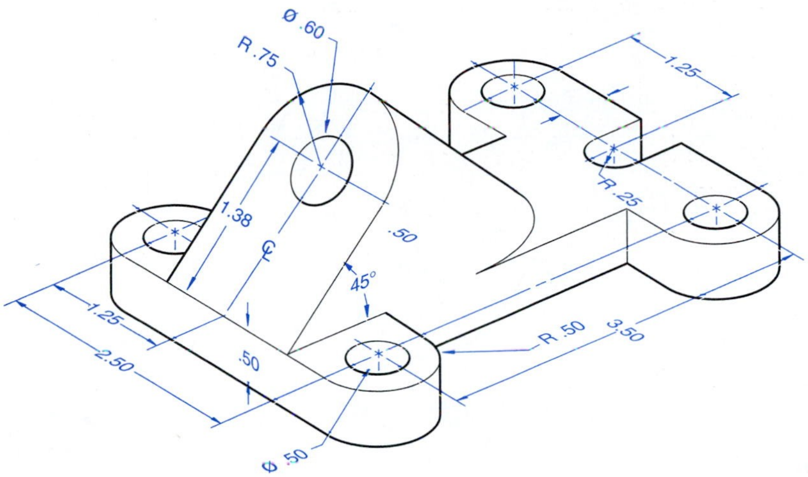

Figure 3.48 Automatic Stop

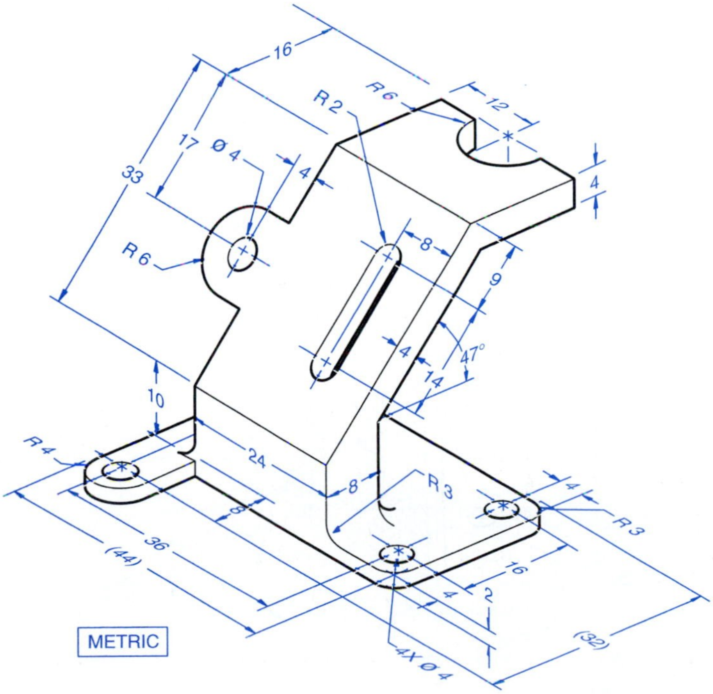

Figure 3.49 Slotted Guide

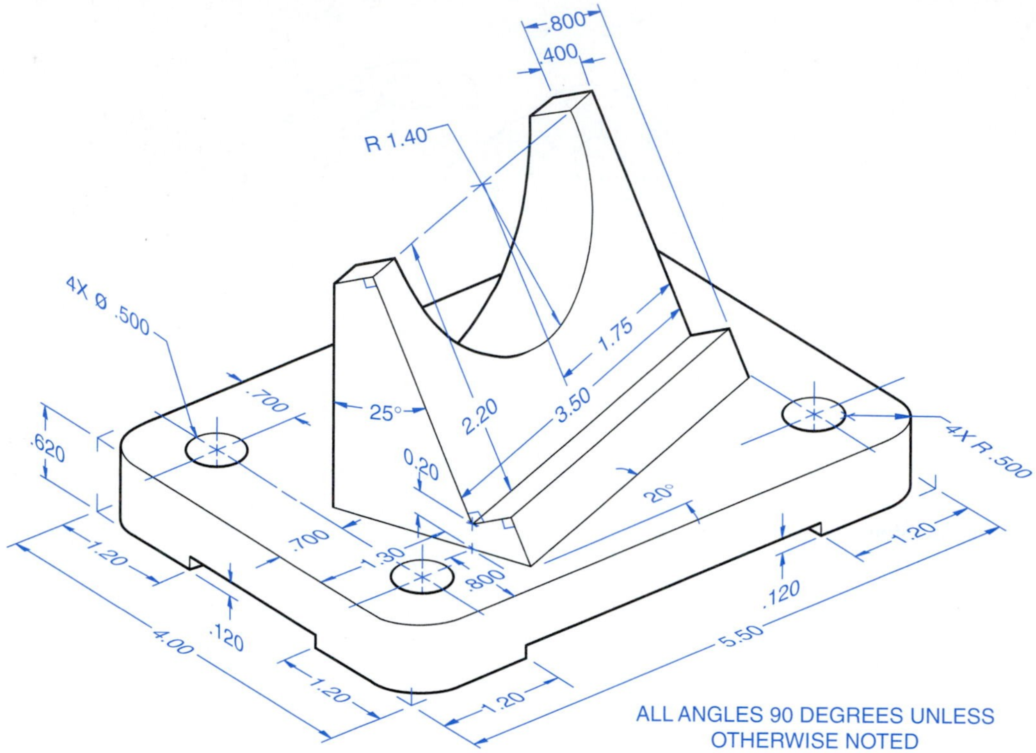

Figure 3.50 Oblique Support

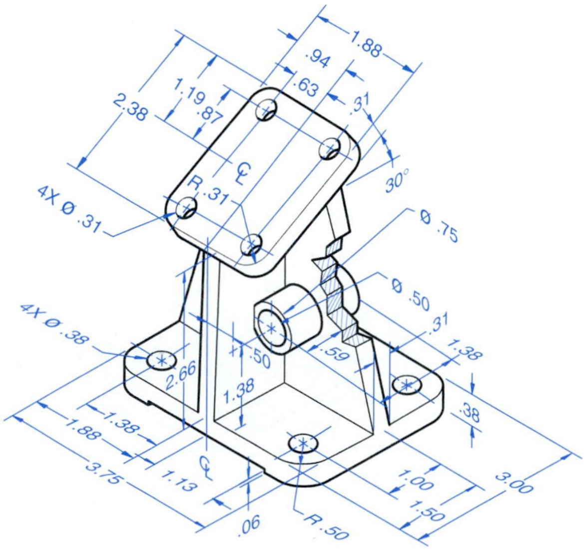

Figure 3.51 Harness Base

Problem 3.1

Refer to Figure 3.37. Sketch the necessary section views. Divide the long dimension of this problem sheet with a thick, dark line. A section view drawing can fit on either half of the paper.

Sketch Number: ____________________

Name: ____________________

Div/Sec: ____________________

Date: ____________________

Problem 3.1

Refer to Figure 3.37. Sketch the necessary section views. Divide the long dimension of this problem sheet with a thick, dark line. A section view drawing can fit on either half of the paper.

Sketch Number: ____________________

Name: ____________________

Div/Sec: ____________________

Date: ____________________

Problem 3.1

Refer to Figure 3.37. Sketch the necessary section views. Divide the long dimension of this problem sheet with a thick, dark line. A section view drawing can fit on either half of the paper.

Sketch Number: ____________________

Name: ____________________

Div/Sec: ____________________

Date: ____________________

Problem 3.3

Refer to Figure 3.46. Sketch or draw the two given views and a partial auxiliary view of the inclined surface.

Sketch Number: ____________________

Name: ____________________

Div/Sec: ____________________

Date: ____________________

Problem 3.3

Refer to Figure 3.46. Sketch or draw the two given views and a partial auxiliary view of the inclined surface.

Sketch Number: ______

Name: ______

Div/Sec: ______

Date: ______

Problem 3.4

Refer to Figure 3.47. Sketch or draw the two given views and a complete auxiliary view of the inclined surface.

Sketch Number: ____________________

Name: ____________________

Div/Sec: ____________________

Date: ____________________

Problem 3.4

Refer to Figure 3.47. Sketch or draw the two given views and a complete auxiliary view of the inclined surface.

Sketch Number: ____________________

Name: ____________________

Div/Sec: ____________________

Date: ____________________

Problem 3.4

Refer to Figure 3.47. Sketch or draw the two given views and a complete auxiliary view of the inclined surface.

Sketch Number: ____________________

Name: ____________________

Div/Sec: ____________________

Date: ____________________

Problem 3.5

Refer to Figures 3.48 through 3.51. Sketch or draw the two given views and a partial auxiliary view of the inclined surface.

Sketch Number: __________

Name: __________

Div/Sec: __________

Date: __________

Problem 3.5

Refer to Figures 3.48 through 3.51. Sketch or draw the two given views and a partial auxiliary view of the inclined surface.

Sketch Number: ____________

Name: ____________

Div/Sec: ____________

Date: ____________

Chapter

4

Dimensioning and Tolerancing Practices

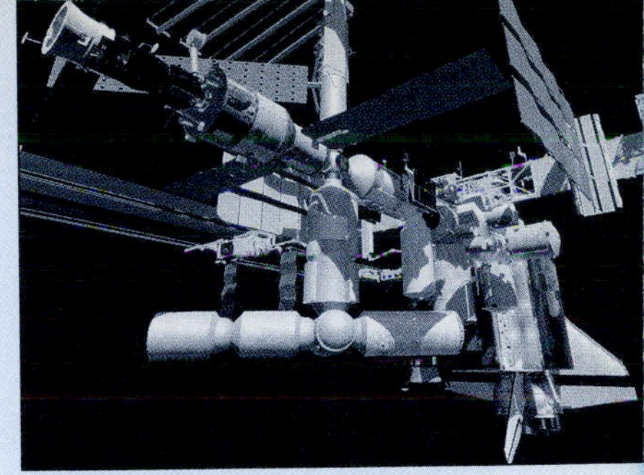

OBJECTIVES

After completing this chapter, you will be able to:

1. Apply the standard dimensioning practices for mechanical drawings.
2. Differentiate between current ASME standards and past practices for dimensioning.
3. Apply English and metric tolerances to dimensions.
4. Calculate standard tolerances for precision fits.
5. Apply tolerances using the basic shaft and basic hole systems.

4.1 DIMENSIONING

Geometrics is the science of specifying and tolerancing the shapes and locations of features on objects. In design work, it is essential that the roundness of a shaft be as clearly stated as the size. Once the shape of a part is defined with an orthographic drawing, the size information is added in the form of **dimensions.** Dimensioning a drawing also identifies the tolerance (or accuracy) required for each dimension. If a part is dimensioned properly, then the intent of the designer is clear to both the person making the part and the inspector checking the part. Everyone in this *circle of information* (design, manufacturing, quality control) must be able to speak and understand a common language. A well-dimensioned part is a component of this communications process.

4.2 SIZE AND LOCATION DIMENSIONS

A well-dimensioned part or structure will *communicate* the size and location requirements for each feature. *Communications* is the fundamental purpose of dimensions.

Designs are dimensioned based on two criteria:

1. Basic sizes and locations of features.
2. Details for construction and for manufacturing.

4.2.1 Units of Measure

The unit of measurement selected should be in accordance with the policy of the user. Construction and architecture drawings use feet and inches for dimensioning units. On a drawing for use in American industry for manufacturing, all dimensions are in inches, unless otherwise stated. Generally, if parts are more than 72 inches in length, the drawings will switch to feet and inches for the standard unit of measure. Most countries outside of the United States use the metric system of measure, or the international system of units (SI), which is based on the meter.

The SI system is being used more in the United States because of global trade and multinational company affiliations. The common metric unit of measure on engineering drawings is the *millimeter,* abbreviated as *mm.*

Angular dimensions are shown either in decimal degrees or in degrees, minutes, and seconds. The symbol used for degrees is °, for minutes ', and for seconds ". Where only minutes and seconds are specified, the number of minutes or seconds is preceded by the 0°. Figure 4.1 shows examples of angular units used to dimension angles.

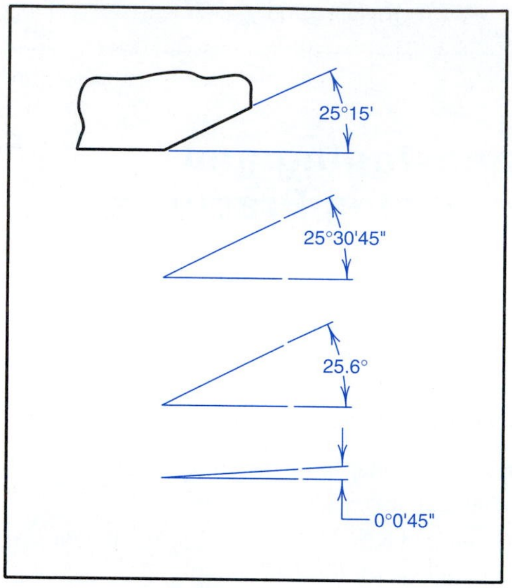

Figure 4.1 Angular Units

Angular dimensions are shown either in decimal degrees or in degrees, minutes, and seconds.

4.2.2 Terminology

There are a number of terms important to dimensioning practices. These terms are illustrated in Figure 4.2 and are defined as follows:

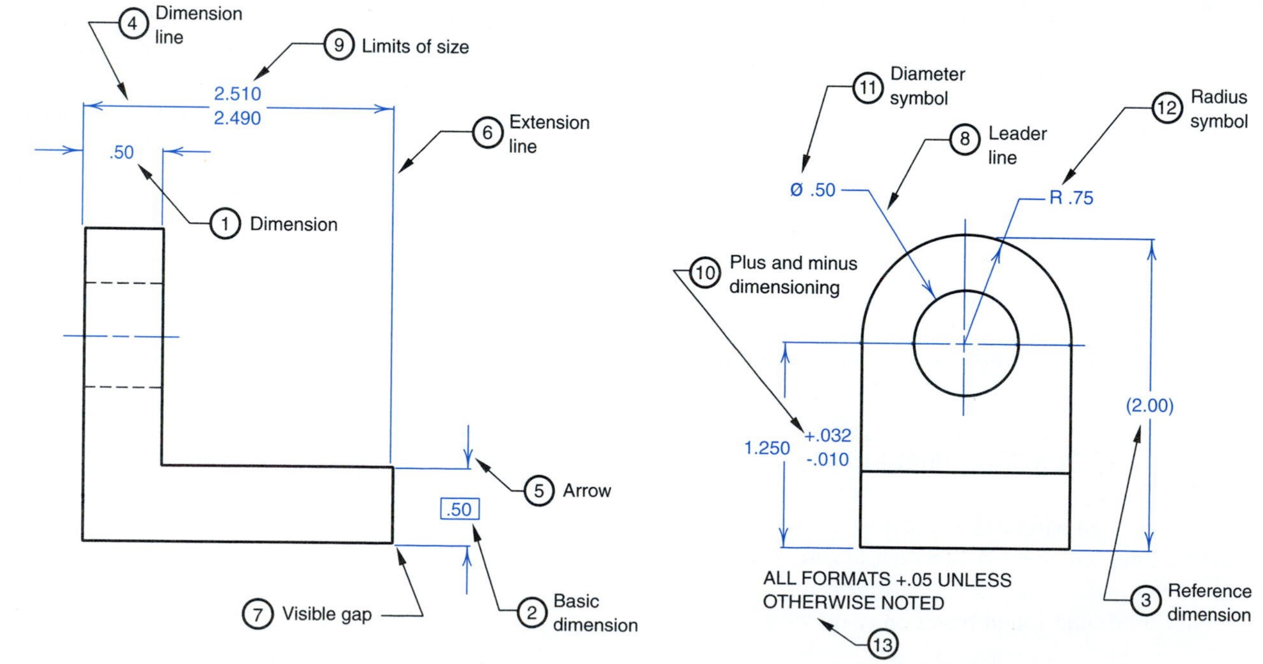

Figure 4.2 Important Elements of a Dimensioned Drawing

1. **Dimension**—the numerical value that defines the size, shape, location, surface texture, or geometric characteristic of a feature. Normally, dimension text is 3 mm or 0.125″ high, and the space between lines of text is 1.5 mm or 0.0625″ (Figure 4.3).
2. **Basic dimension**—a numerical value defining the theoretically exact size, location, profile, or orientation of a feature relative to a coordinate system established by datums. It is identified on a drawing by enclosing the dimension in a rectangular box. Basic dimensions have no tolerance. They locate the perfect geometry of a part, while the acceptable variation or geometric tolerance is described in a feature control frame.
3. **Reference dimension**—a numerical value enclosed in parentheses, provided for information only and not directly used in the fabrication of the part. A reference dimension is a calculated size without a tolerance used to show the intended design size of a part.
4. **Dimension line**—a thin, solid line that shows the extent and direction of a dimension. Dimension lines are broken for the insertion of the dimension numbers.

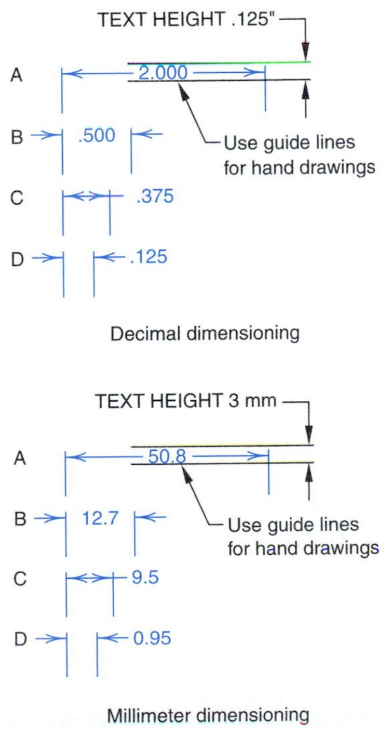

Figure 4.3 **Text Height and Standards Used for Decimal and Millimeter Dimensioning**

5. **Arrows**—symbols placed at the ends of dimension lines to show the limits of the dimension, leaders, and cutting plane lines. Arrows are uniform in size and style, regardless of the size of the drawing. Arrows are usually about 3 mm (1/8″) long and should be one-third as wide as they are long (Figure 4.4).
6. **Extension line**—a thin, solid line perpendicular to a dimension line, indicating which feature is associated with the dimension.
7. **Visible gap**—there should be a visible gap of 1 mm (1/16″) between the feature's corners and the end of the extension line.
8. **Leader line**—a thin, solid line used to indicate the feature with which a dimension, note, or symbol is associated. A leader line is generally a straight line drawn at an angle that is neither horizontal nor vertical. Leader lines are terminated with an arrow touching the part or detail. On the end opposite the arrow, the leader line will have a short, horizontal shoulder (3 mm or .125″ long). Text is extended from this shoulder such that the text height is centered with the shoulder line. Two or more adjacent leaders on a drawing should be drawn parallel to each other.
9. **Limits of size**—the largest acceptable size and the minimum acceptable size of a feature. The value for the largest acceptable size, expressed as the maximum material condition (MMC), is placed over the value for the minimum acceptable size, expressed as the least material condition (LMC), to denote the limit-dimension-based tolerance for the feature.
10. **Plus and minus dimensions**—the allowable positive and negative variance from the dimension specified. The plus and minus values may or may not be equal.

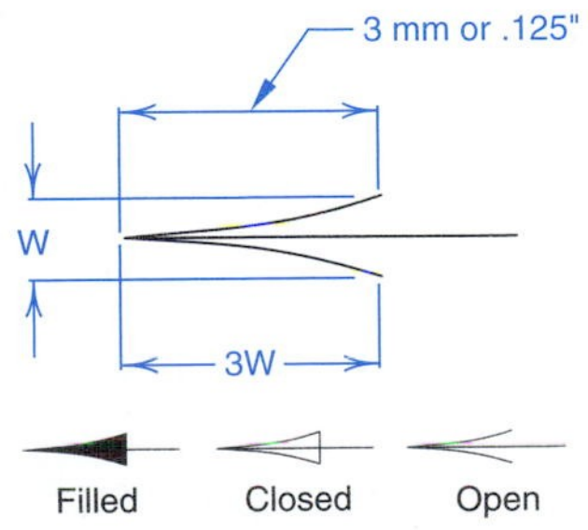

Figure 4.4 **Dimensions Used to Draw an Arrowhead**
Arrows are three times as long as they are wide.

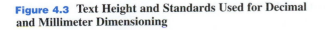

11. **Diameter symbol**—a symbol that precedes a numerical value to indicate that the dimension shows the diameter of a circle. The symbol used is the Greek letter phi (ø).
12. **Radius symbol**—a symbol that precedes a numerical value to indicate that the associated dimension shows the radius of a circle. The radius symbol used is the capital letter R.
13. **Tolerance**—the amount that a particular dimension is allowed to vary. All dimensions (except reference dimensions) have an associated tolerance. A tolerance may be expressed either through limit dimensioning, plus and minus dimensioning, or a general note. The tolerance is the difference between the maximum and minimum limits.

4.2.3 Basic Concepts

A size dimension might be the overall width of a part or structure, or the diameter of a drilled hole (Figure 4.5). A location dimension might be the length from the edge of an object to the center of a feature. The basic criterion is, "What information is necessary to manufacture or construct the object?" For example, to drill a hole, the manufacturer would need to know the diameter of the hole, the location of the center of the hole, and the depth to which the hole is to be drilled. These three dimensions describe the hole in sufficient detail for the feature to be made using machine tools.

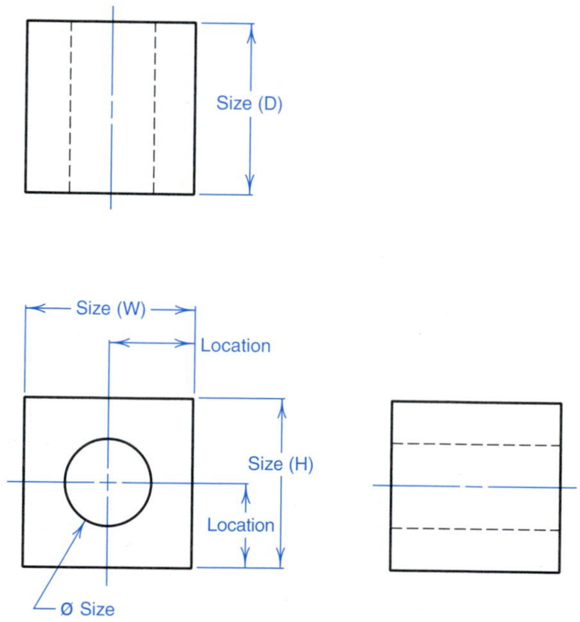

Figure 4.5 **Size and location dimensions are used to describe parts for manufacture**

4.2.4 Size Dimensions

Horizontal—the left-to-right distance relative to the drawing sheet. In Figure 4.6, the *width* is the only horizontal size dimension.

Vertical—the up and down distance relative to the drawing sheet. In Figure 4.6, the *height* and the *depth* are both vertical dimensions, even though they are in two different directions on the part.

Diameter—the full distance across a circle, measured through the center. This dimension is usually used only on full circles or on arcs that are more than half of a full circle.

Radius—the distance from the center of an arc to any point on the arc, usually used on arcs less than half circles. In Figure 4.6, the radius points to the outside of the arc, even though the distance measured is to the center, which is inside. The endpoints of the arc are tangent to the horizontal and vertical lines, making a quarter of a circle. This is assumed, and there is no need to note it. If the radius is not positioned in this manner, then the actual center of the radius must be located.

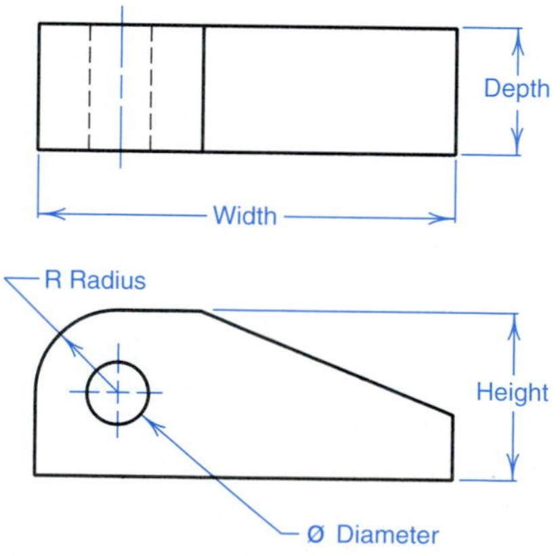

Figure 4.6 **Dimensions Showing the Sizes of Features, Such as the Height and Depth of the Part and the Diameter of the Hole**

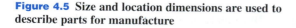

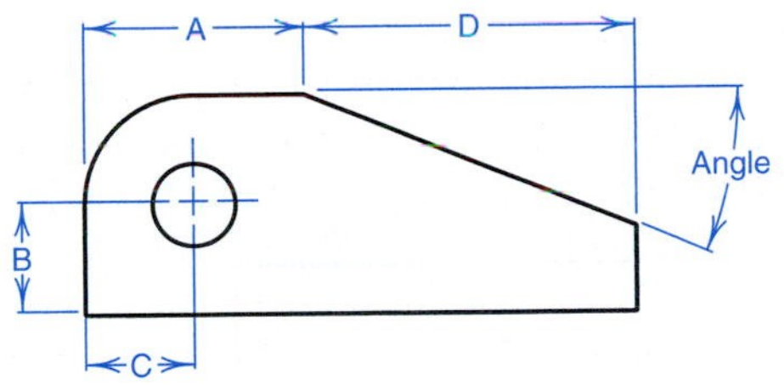

Figure 4.7 **Dimensions Showing the Location and Orientations of Features, Such as the Location of the Center of the Hole**

4.2.5 Location and Orientation Dimensions

- *Horizontal position*—In Figure 4.7, dimensions A and D are horizontal position dimensions that locate the beginnings of the angle. Dimension A measures more than one feature—the sum of the arc's radius and the straight line. The measurement for dimension A is taken parallel to the dimension line. Dimension D is the measurement of a single feature—the sloping line—but it is not the true length of the line. Rather, it is the left-to-right distance that the line displaces. This is called the "delta X value," or the change in the X direction. The C dimension measures the horizontal location of the center of the hole and arc.
- *Vertical position*—The B dimension in Figure 4.7 measures the vertical position of the center of the hole. For locating the hole, the dimensions are given to the center, rather than the edges of the hole. All circular features are located from their centers.
- *Angle*—The angle dimension in Figure 4.7 gives the angle between the horizontal plane and the sloping surface. The angle dimension can be taken from several other directions, measuring from any measurable surface.

4.2.6 Standard Practices

The guiding principle for dimensioning a drawing is *clarity*. To promote clarity, ANSI developed standard practices for showing dimensions on drawings.

Placement Dimension placement depends on the space available between extension lines. When space permits, dimensions and arrows are placed *between* the extension lines, as shown in Figures 4.8A and E.

When there is room for the numerical value but not the arrows as well, the value is placed between the extension lines and the arrows are placed outside the extension lines, as shown in Figures 4.8B and F.

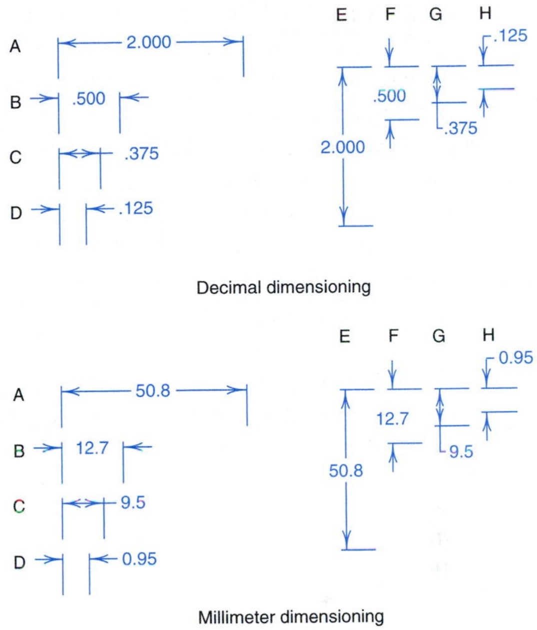

Figure 4.8 **Dimension Text Placement**

Standard practice for the placement of dimensions depends on the space available.

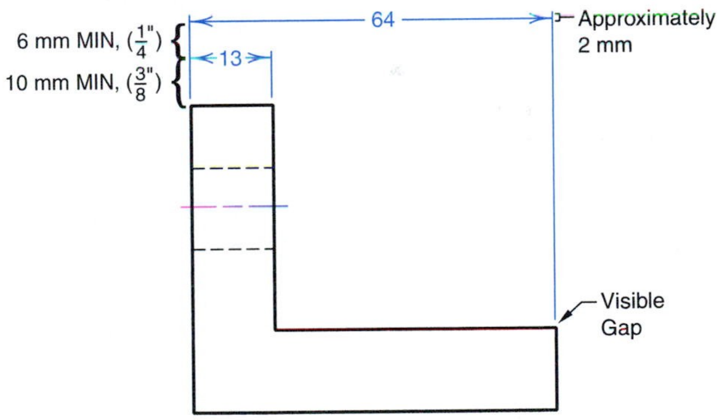

Figure 4.9 **Minimum Dimension Line Spacing**

Standard practice for the spacing of dimensions is 10 mm from the view and 6 mm between dimension lines.

When there is room for the arrows but not the numerical value, the arrows are placed between the extension lines, and the value is placed outside the extension lines and adjacent to a leader, as shown in Figures 4.8C and G.

When the space is too small for either the arrows or the numerical value, both are placed outside the extension lines, as shown in Figures 4.8D and H.

Spacing The minimum distance from the object to the first dimension is 10 mm (3/8″), as shown in Figure 4.9.

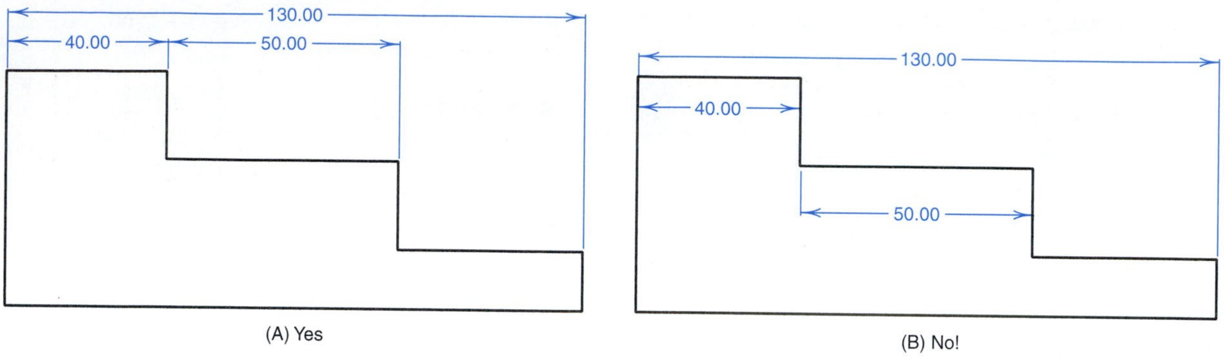

Figure 4.10 Group Dimensions
In standard practice, dimensions are grouped on a drawing. Do not use object lines as extension lines for a dimension.

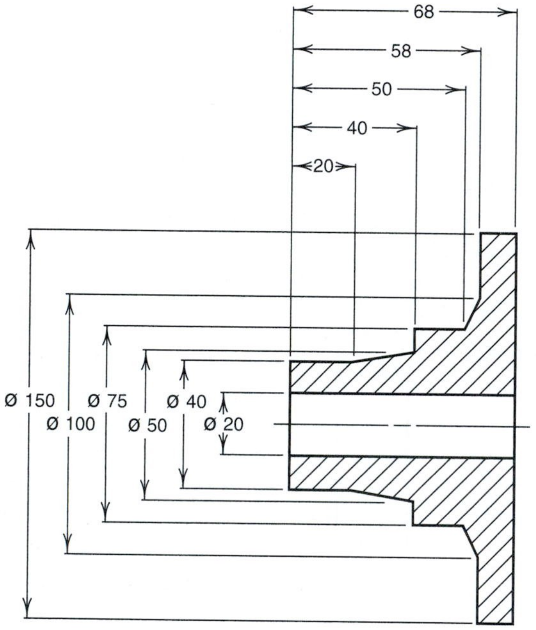

Figure 4.11 Stagger Dimension Text
The general practice is to stagger the dimension text on parallel dimensions.

The minimum spacing between dimensions is 6 mm (¼″). These values may be increased where appropriate.

There should be a visible gap between an extension line and the feature to which it refers. Extension lines should extend about 2–3 mm (⅛″) beyond the last dimension line.

Grouping and Staggering Dimensions should be grouped for uniform appearance, as shown in Figure 4.10, which also shows the continuous line dimensioning style. As a general rule, do not use object lines as part of your dimension (Figure 4.10B). Where there are several parallel dimensions, the values should be staggered, as shown in Figure 4.11, which also shows the baseline dimensioning style.

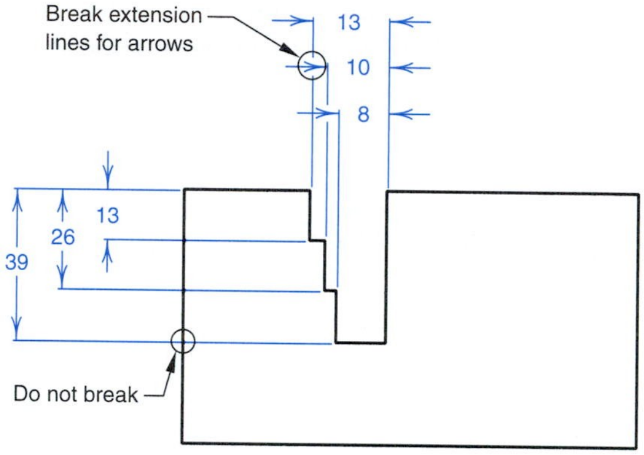

Figure 4.12 Extension Line Practice
Extension lines should not cross dimension lines, are not broken when crossing object or other extension lines, and are broken when crossing arrows.

Extension Lines **Extension lines** are used to relate a dimension to one or more features and are usually drawn perpendicular to the associated dimension line. Where angled extension lines are used, they must be parallel, and the associated dimension lines must be drawn in the direction to which they apply.

Extension lines should not cross dimension lines, and they should avoid crossing other extension lines whenever possible. When extension lines cross object lines or other extension lines, they should not be broken. When extension lines cross or are close to arrowheads, they should be broken for the arrowhead (Figure 4.12).

When the center of a feature is being dimensioned, the center line of the feature is used as an extension line (Figure 4.13A). When a point is being located by extension lines only, the extension lines must pass through the point (Figure 4.13B).

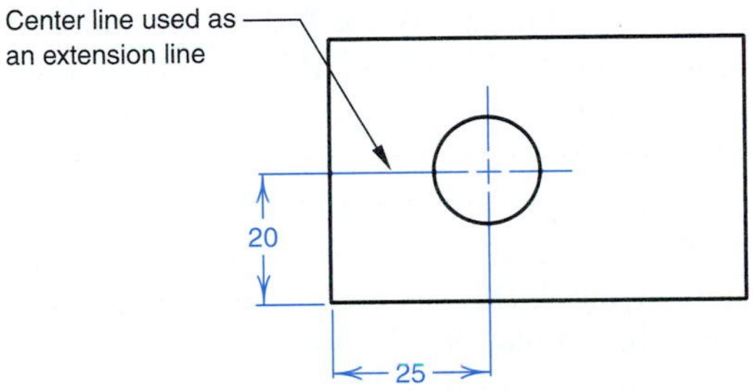

(A)

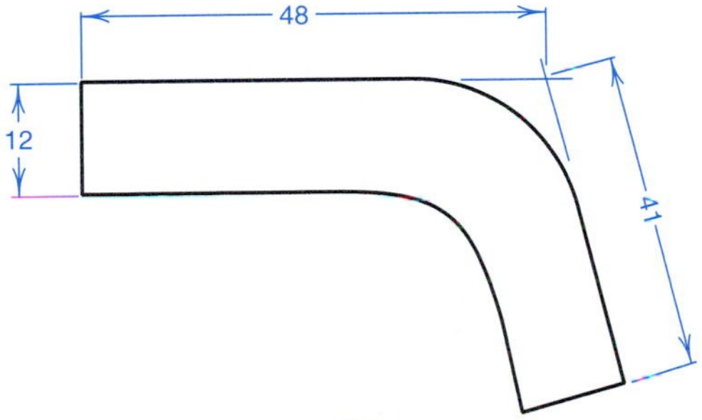

(B)

Figure 4.13
The center of a feature, such as a hole, is located by making the center lines extension lines for the dimension. Extension lines can also cross to mark a theoretical point.

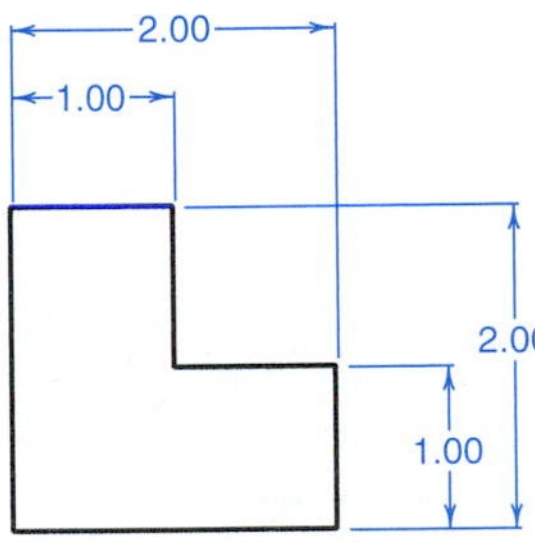

Unidirectional
Current standard

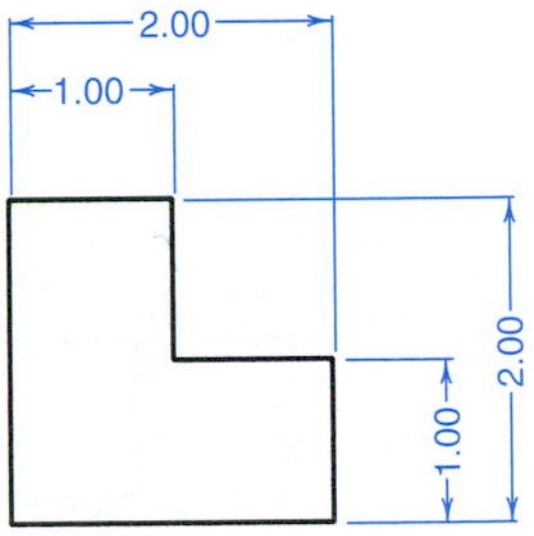

Aligned
Old standard

Figure 4.14 Unidirectional and Aligned Methods
The unidirectional method of placing text is the current standard practice.

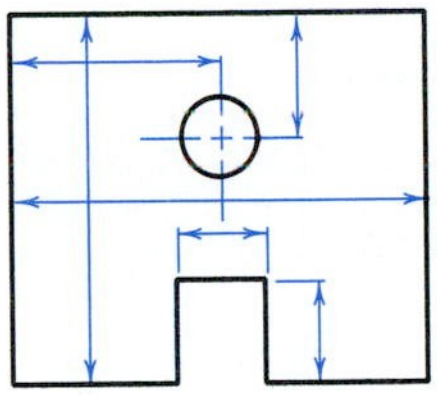

(A) No!

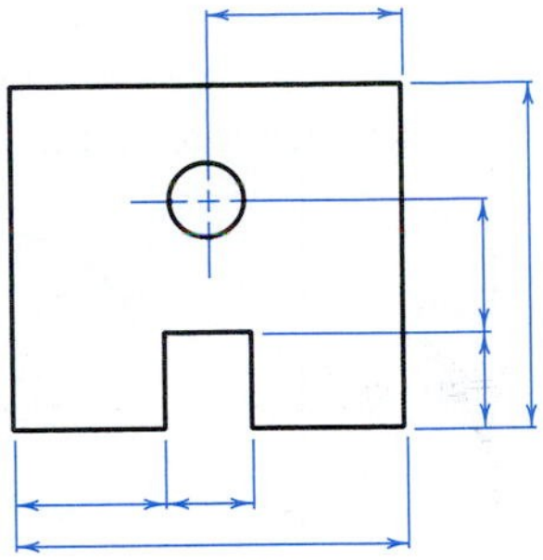

(B) Yes

Figure 4.15 Dimension Outside of View
Dimensions are kept off the view, unless necessary for clarity.

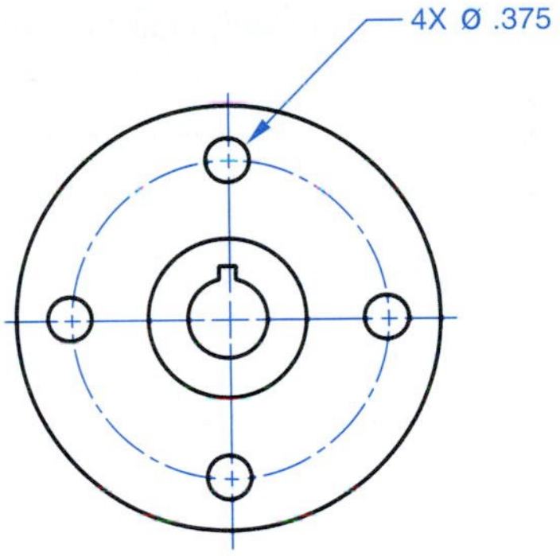

Figure 4.16 Using the Symbol × to Dimension Repetitive Features

Reading Direction All dimensions and note text must be oriented to read from the bottom of the drawing (relative to the drawing format). This is called **unidirectional dimensioning** (Figure 4.14). The *aligned method* of dimensioning may be seen on older drawings or architectural drawings but is not approved by the current ANSI standards. **Aligned dimensions** have text placed parallel to the dimension line, with vertical dimensions read from the right of the drawing sheet.

View Dimensioning Dimensions are to be kept outside the boundaries of views, wherever practical (Figure 4.15B). Dimensions may be placed within the boundaries where extension or leader l ines would be too long or where clarity would be improved.

Repetitive Features The symbol × is used to indicate the number of times a feature is to be repeated. The number of repetitions, followed by the symbol × and a space, precedes the dimension text. For example, in Figure 4.16, 4 × ø .375 means that there are 4 holes with a diameter of .375″.

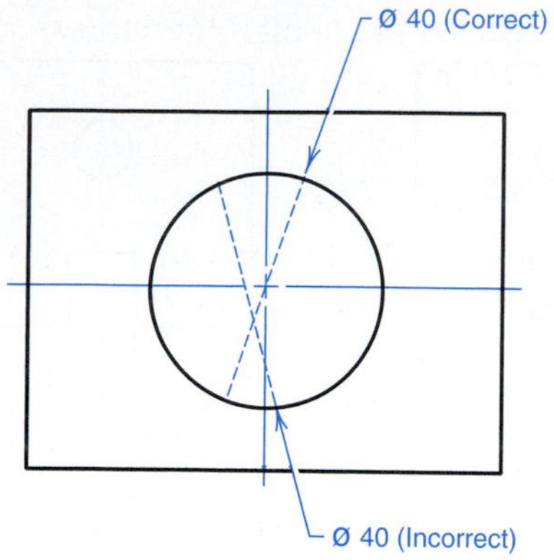

Figure 4.17 Radial Leader Lines

Leader lines used to dimension holes must be radial.

4.3 DETAIL DIMENSIONING

Holes are typically dimensioned in a view that best describes the shape of the hole. For diameters, the diameter symbol must precede the numerical value. When holes are dimensioned with a leader line, the leader line must be radial (Figure 4.17). A **radial line** is one that passes through the center of a circle or arc if extended. If it is not clear whether a hole extends completely through a part, the word THRU can follow the numerical value.

Symbols are used for spotface, counterbored, and countersunk holes. These symbols must always precede the diameter symbol. (Figure 4.18) The depth symbol is used to indicate the depth of a hole. The depth symbol precedes the numerical value. When the depth of a blind hole is specified, the depth is to the full diameter of the hole and not to the point (Figure 4.19). When a chamfer or countersink is placed in a curved surface, the diameter refers to the minimum diameter of the chamfer or countersink.

Slotted holes may be dimensioned in any of several ways, depending on which is most appropriate for the

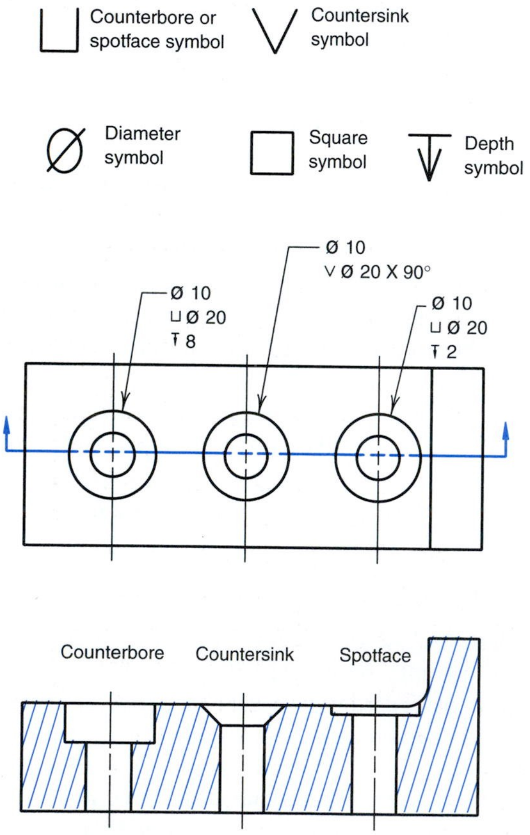

Figure 4.18 Symbols for Drilling Operations

Symbols to represent machine drilling operations always precede the diameter symbol.

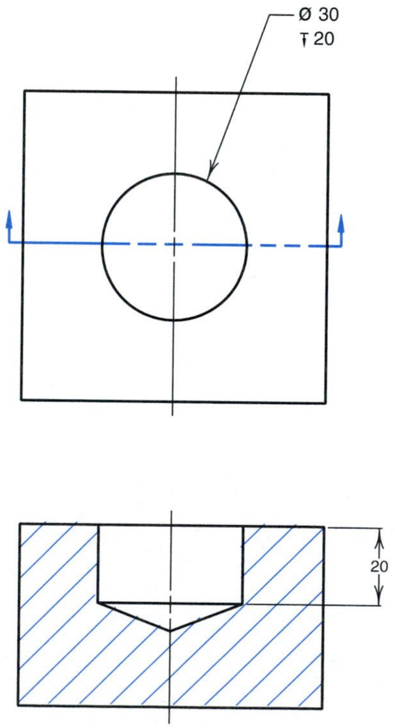

Figure 4.19 Dimensioning a Blind Hole

The depth of a blind hole reflects the depth of the full diameter.

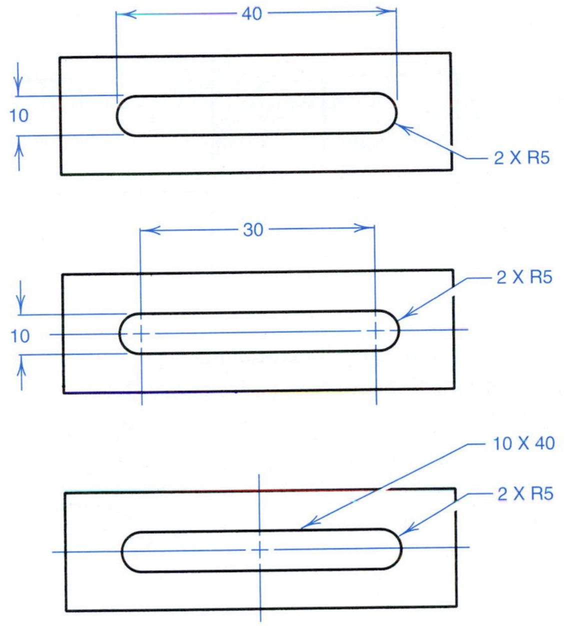

Figure 4.20 Dimensioning Slots
Several methods are appropriate for dimensioning slots.

application. The various options for dimensioning slotted holes are shown in Figure 4.20.

4.3.1 Diameter versus Radius

If a full circle or an arc of more than half a circle is being dimensioned, the diameter is specified, preceded by the diameter symbol, which is the Greek letter phi. If the arc is less than half a circle, the radius is specified, preceded by an R. Concentric circles are dimensioned in the longitudinal view whenever practical (Figure 4.21).

As previously stated, radii are dimensioned with the radius symbol preceding the numerical value. The dimension line for radii shall have a single arrowhead touching the arc. When there is adequate room, the dimension is placed between the center of the radius and the arrowhead (Figure 4.22). When space is limited, a radial leader line is used. When an arc is not clearly defined by being tangent to other dimensioned features on the object, the center of the arc is noted with a small cross (Figure 4.22). The center is not shown if the arc is tangent to other defined surfaces.

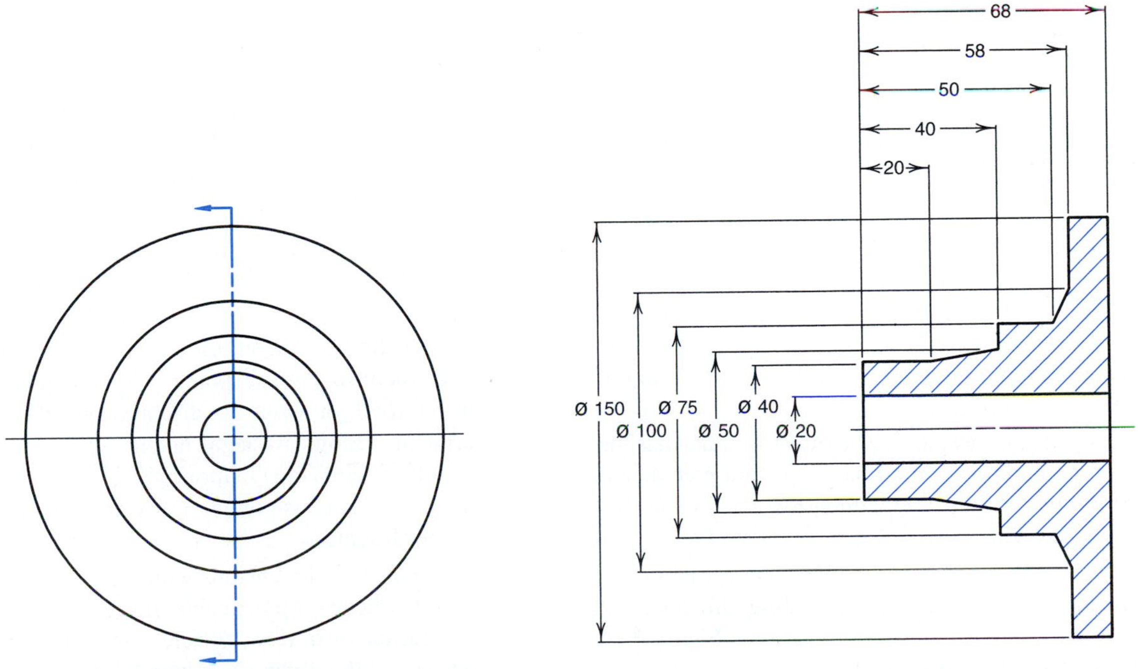

Figure 4.21 Dimensioning Concentric Circles
Concentric circles are dimensioned in the longitudinal view.

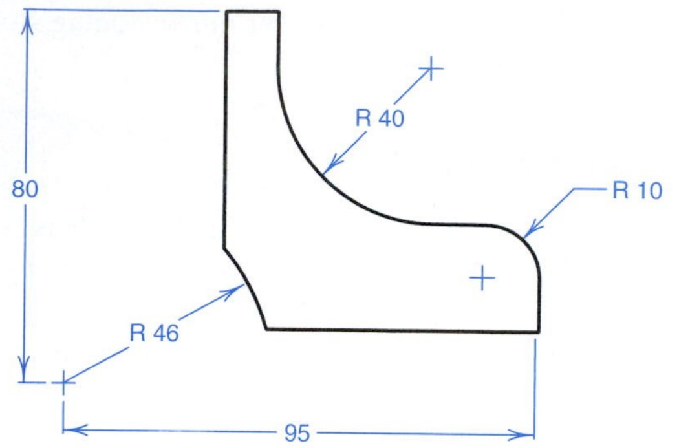

Figure 4.22 Dimensioning Arcs
Arcs of less than half a circle are dimensioned as radii, with the R symbol preceding the dimension value.

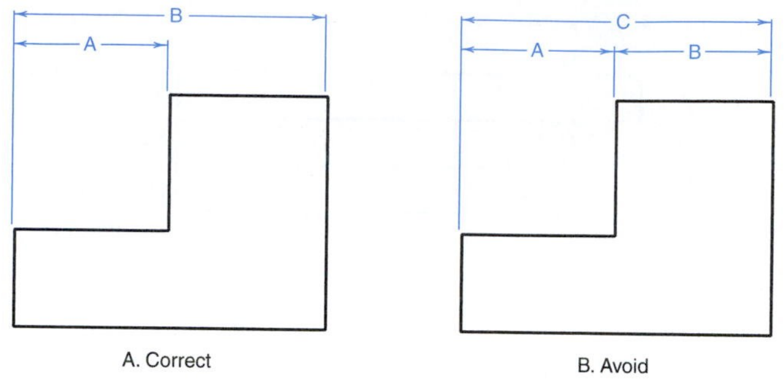

Figure 4.23 Avoid Overdimensioning
Double dimensioning can cause problems because of tolerancing.

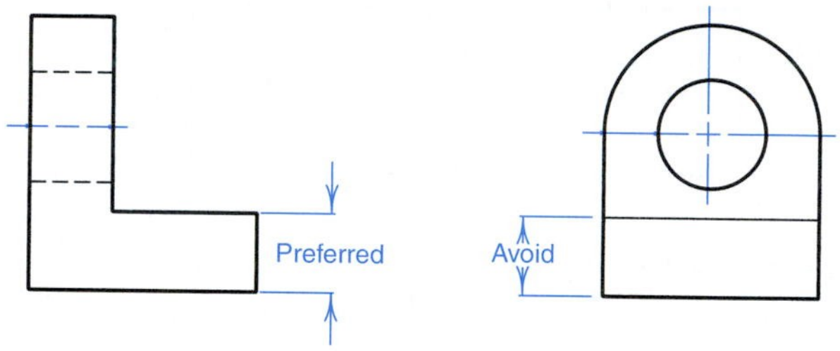

Figure 4.24 Dimension the Most Descriptive View
Dimensions are placed in the most descriptive or contour view.

4.3.2 Dimensioning Guidelines

The importance of accurate, unambiguous dimensioning cannot be overemphasized. There are many cases where an incorrect or unclear dimension has added considerable expense to the fabrication of a product, caused premature failure, or, in some cases, caused loss of life. *The primary guideline is clarity: Whenever two guidelines appear to conflict, the method that most clearly communicates the size information shall prevail.* Use the following dimensioning guidelines:

1. Every dimension must have an associated tolerance, and that tolerance must be clearly shown on the drawing.
2. Double dimensioning of a feature is not permitted. For example, in Figure 4.23, there are two ways to arrive at the overall length of the object: by adding dimensions A and B or by directly measuring the dimension C. Since each dimension must have a tolerance, it is not clear which tolerance would apply to the overall length: the tolerance for dimension C or the sum of the tolerances for dimensions A and B. This ambiguity can be eliminated by removing one of the three dimensions. The dimension that has the least importance to the function of the part should be left out. In this case, dimension A would probably be deleted.
3. Dimensions should be placed in the view that most clearly describes the feature being dimensioned (contour dimensioning). For example, Figure 4.24 illustrates a situation in which the height of a step is being dimensioned. In this case, the front view more clearly describes the step feature.
4. Maintain a minimum spacing between the object and the dimension and between multiple dimensions. This spacing is shown in Figure 4.9. If the spacing is reduced, the drawing will be more difficult to read and a lack of clarity will result.
5. A visible gap shall be placed between the ends of extension lines and the feature to which they refer.
6. Manufacturing methods should not be specified as part of the dimension, unless no other method of manufacturing is acceptable. The old practice of using *drill* or *bore* is discouraged. Because a drawing becomes a legal document for manufacturing, specifying inappropriate manufacturing methods can cause unnecessary expense and may trigger litigation.
7. Avoid placing dimensions within the boundaries of a view, whenever practicable. If several dimensions are placed in a view, differentiation between the object and the dimensions may become difficult.
8. Dimensions for materials typically manufactured to gages or code numbers shall be specified by

Table 4.1 Basics of Dimensioning

Parts of a Dimension

Dimension—A dimension is a numerical value shown on a drawing to define the size of an object or a part of an object. Dimensions may be expressed in either U.S. or metric units.

Dimension line—A dimension line is a thin, solid line used to show the extent and direction of a dimension.

Arrowheads—Arrowheads are placed at the ends of dimension lines to show the limits of the dimension.

Extension line—Extension lines are thin lines drawn perpendicular to dimension lines, and they indicate the feature of the object to which the dimension refers.

Leader line—A leader line is a thin, solid line used to direct dimensions or notes to the appropriate feature.

Tolerance—Tolerances are the amount a dimension is allowed to vary. The tolerance is the difference between the maximum and minimum permitted sizes.

Principles of Good Dimensioning

The overriding principle of dimensioning is clarity.

1. Each feature of an object is dimensioned once and only once.
2. Dimensions should be selected to suit the function of the object.
3. Dimensions should be placed in the most descriptive view of the feature being dimensioned.
4. Dimensions should specify only the size of a feature. The manufacturing method should only be specified if it is a mandatory design requirement.
5. Angles shown on drawings as right angles are assumed to be 90 degrees unless otherwise specified, and they need not be dimensioned.
6. Dimensions should be located outside the boundaries of the object whenever possible.
7. Dimension lines should be aligned and grouped where possible to promote clarity and uniform appearance.
8. Crossed dimension lines should be avoided whenever possible. When dimension lines must cross, they should be unbroken.
9. The space between the first dimension line and the object should be at least 3/8 inch (10 mm). The space between dimension lines should be at least 1/4 inch (6 mm).
10. There should be a visible gap between the object and the origin of an extension line.
11. Extension lines should extend 1/8 inch (3 mm) beyond the last dimension line.
12. Extension lines should be broken if they cross or are close to arrowheads.
13. Leader lines used to dimension circles or arcs should be radial.
14. Dimensions should be oriented to be read from the bottom of the drawing.
15. Diameters are dimensioned with a numerical value preceded by the diameter symbol.
16. Concentric circles should be dimensioned in a longitudinal view whenever possible.
17. Radii are dimensioned with a numerical value preceded by the radius symbol.
18. When a dimension is given to the center of an arc or radius, a small cross is shown at the center.
19. The depth of a blind hole may be specified in a note. The depth is measured from the surface of the object to the deepest point where the hole still measures a full diameter in width.
20. Counterbored, spotfaced, or countersunk holes should be specified in a note.

numerical values. The gages or code numbers may be shown in parentheses following the numerical values.

9. Unless otherwise specified, angles shown in drawings are assumed to be 90 degrees.
10. Avoid dimensioning hidden lines. Hidden lines are less clear than visible lines.
11. The depth and diameter of blind, counterbored, or countersunk holes may be specified in a note (Figures 4.18 and 4.19).
12. Diameters, radii, squares, counterbores, spotfaces, countersinks, and depths should be specified with the appropriate symbol preceding the numerical value (Figure 4.18).
13. Leader lines for diameters and radii should be radial lines (Figure 4.17).

Table 4.1 summarizes the basics of dimensioning.

▼ Practice Problem 4.1

Sketch dimensions in decimal inches for the object shown in the multiview drawing.

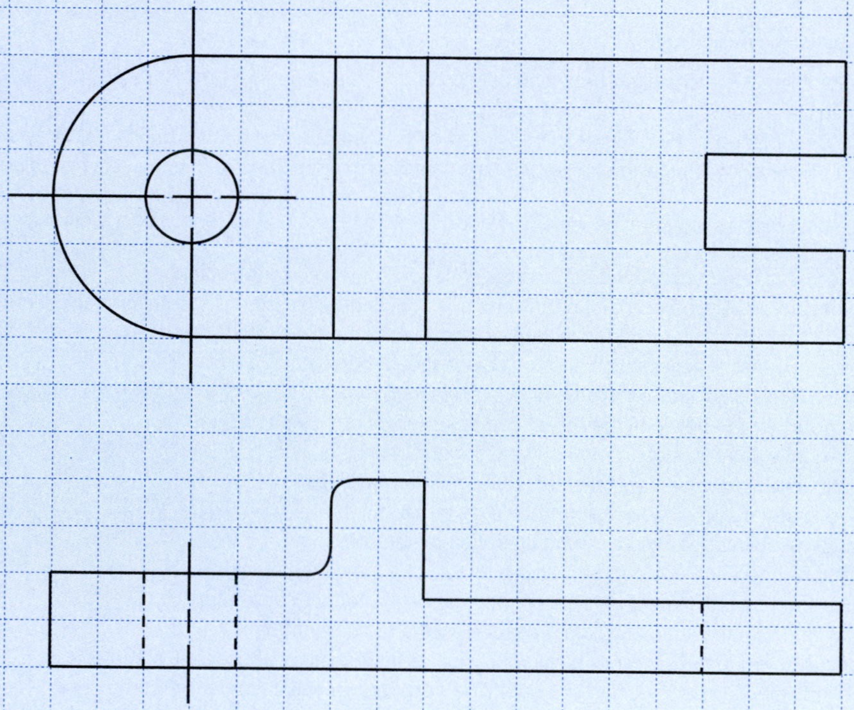

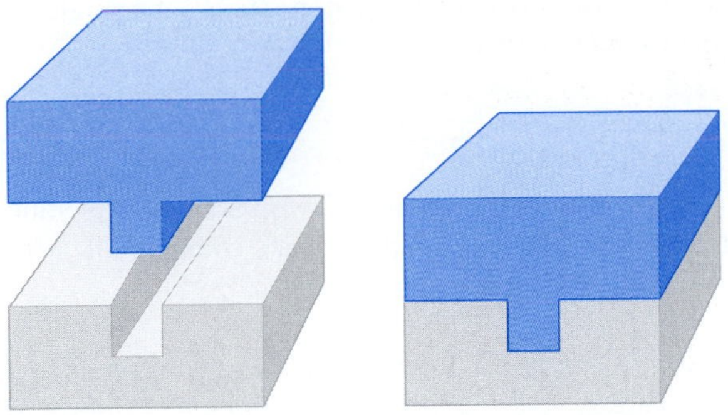

Figure 4.25
A system is two or more mating parts.

4.4 | TOLERANCING

Tolerances are used to control the variation that exists on all manufactured parts. Toleranced dimensions control the amount of variation on each part of an assembly. The amount each part is allowed to vary depends on the function of the part and of the assembly. For example, the tolerances placed on electric hand-drill parts are not as stringent as those placed on jet engine parts. The more accuracy needed in the machined part, the higher the manufacturing cost. Therefore, tolerances must be specified in such a way that a product functions as it should at a cost that is reasonable.

A tolerance of 4.650 ± .003 means that the final measurement of the machined part can be anywhere from 4.653 to 4.647 and the part would still be acceptable. The lower and upper allowable sizes are referred to as the *limit dimensions,* and the *tolerance* is the difference between the limits. In the example, the **upper limit** (largest value) for the part is 4.653, the **lower limit** (smallest value) is 4.647, and the tolerance is .006.

Tolerances are assigned to mating parts. For example, the slot in the part shown in Figure 4.25 must accommodate another part. A *system* is two or more mating parts.

4.5 | TOLERANCE REPRESENTATION

Tolerance is the total amount a dimension may vary and is the difference between the maximum and minimum limits. Because it is impossible to make everything to an exact size, tolerances are used on production drawings to control the manufacturing process more accurately and to control the variation between mating parts. ASME standard Y14.5M-1994 is commonly used in U.S. industries to specify tolerances on engineering drawings.

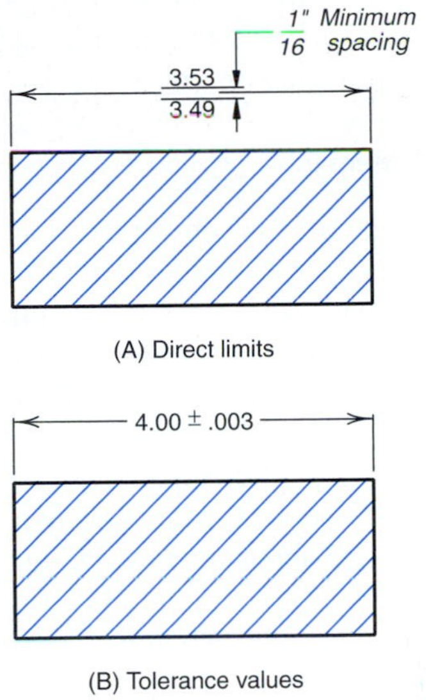

Figure 4.26 **Representing Tolerance Values**
Tolerances are represented as direct limits or as tolerance values.

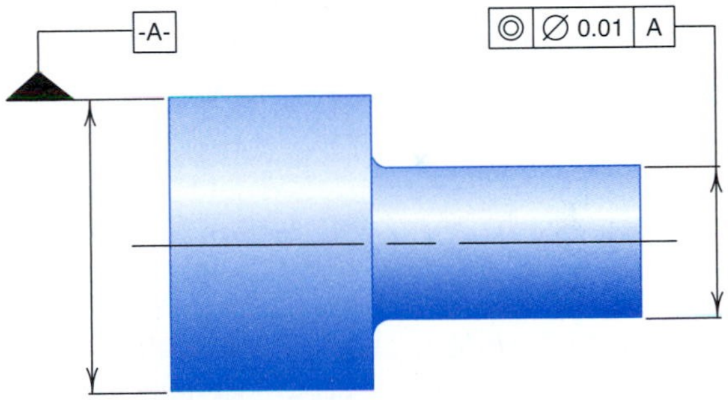

Figure 4.27 **Geometric Tolerance System Used to Dimension Parts**

Tolerances can be expressed in several ways:

1. Direct limits, or as tolerance values applied directly to a dimension (Figure 4.26)
2. Geometric tolerances (Figure 4.27)
3. Notes referring to specific conditions
4. A general tolerance note in the title block

4.5.1 General Tolerances

General tolerances are given in a note or as part of the title block. A general tolerance note would be similar to:

ALL DECIMAL DIMENSIONS TO BE HELD TO ± .002″

This means that a dimension such as .500 would be assigned a tolerance of ± .002, resulting in an upper limit of .502 and a lower limit of .498.

4.5.2 Limit Dimensions

Tolerances can be applied directly to dimensioned features, using limit dimensioning. This is the ASME preferred method; the maximum and minimum sizes are specified as part of the dimension. (See Figure 4.26A.) Either the upper limit is placed above the lower limit, or, when the dimension is written in a single line, the lower limit precedes the upper limit, and they are separated by a dash. A minimum spacing of 1⁄16 inch is required when the upper limit is placed above the lower limit.

4.5.3 Plus and Minus Dimensions

With this approach, the basic size is given, followed by a plus/minus sign and the tolerance value (Figure 4.28). Tolerances can be unilateral or bilateral. A **unilateral tolerance** varies in only one direction. A **bilateral tolerance** varies in both directions from the basic size. If the variation is equal in both directions, then the variation is preceded by a ± symbol. *The plus and minus approach can only be used when the two variations are equal.*

4.5.4 Single Limit Dimensions

When other elements of a feature will determine one limit dimension, MIN or MAX is placed after the other limit dimension. Items such as depth of holes, length of threads, corner radii, and chamfers can use the single limit dimension technique.

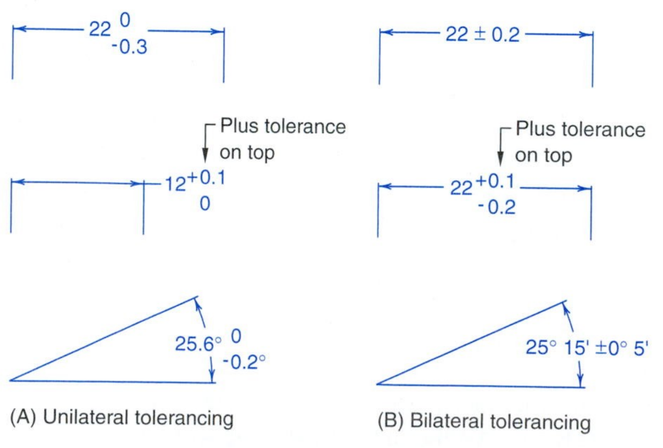

Figure 4.28 **Plus and Minus Tolerance System Applied to Various Dimensioning Conditions**

4.5.5 Important Terms

Figure 4.29 shows a system of two parts with toleranced dimensions. The two parts are an example of ASME Y14.5M-1994 important terms.

- **Nominal size**—a dimension used to describe the general size, usually expressed in common fractions. The slot in Figure 4.29 has a nominal size of ½″.
- **Basic size**—the theoretical size used as a starting point for the application of tolerances. The basic size of the slot in Figure 4.29 is .500″.
- **Actual size**—the measured size of the finished part after machining. In Figure 4.29, the actual size is .501″.
- **Limits**—the maximum and minimum sizes shown by the toleranced dimension. The slot in Figure 4.29 has limits of .502 and .498, and the mating part has limits of .495 and .497. The larger value for each part is the *upper limit,* and the smaller value is the *lower limit.*
- **Allowance**—the minimum clearance or maximum interference between parts, or the *tightest* fit between two mating parts. In Figure 4.29, the allowance is .001, meaning that the tightest fit occurs when the slot is machined to its smallest allowable size of .498 and the mating part is machined to its largest allowable size of .497. The difference between .498 and .497, or .001, is the allowance.
- **Tolerance**—the total allowable variance in a dimension; the difference between the upper and lower limits. The tolerance of the slot in Figure 4.29 is .004″ (.502 – .498 = .004) and the tolerance of the mating part is .002″(.497 – .495 = .002).
- **Maximum material condition (MMC)**—the condition of a part when it contains the greatest amount of material. The MMC of an external feature, such as a shaft, is the upper limit. The MMC of an internal feature, such as a hole, is the lower limit.
- **Least material condition (LMC)**—the condition of a part when it contains the least amount of material possible. The LMC of an external feature is the lower limit. The LMC of an internal feature is the upper limit.
- **Piece tolerance**—the difference between the upper and lower limits of a single part.
- **System tolerance**—the sum of all the piece tolerances.

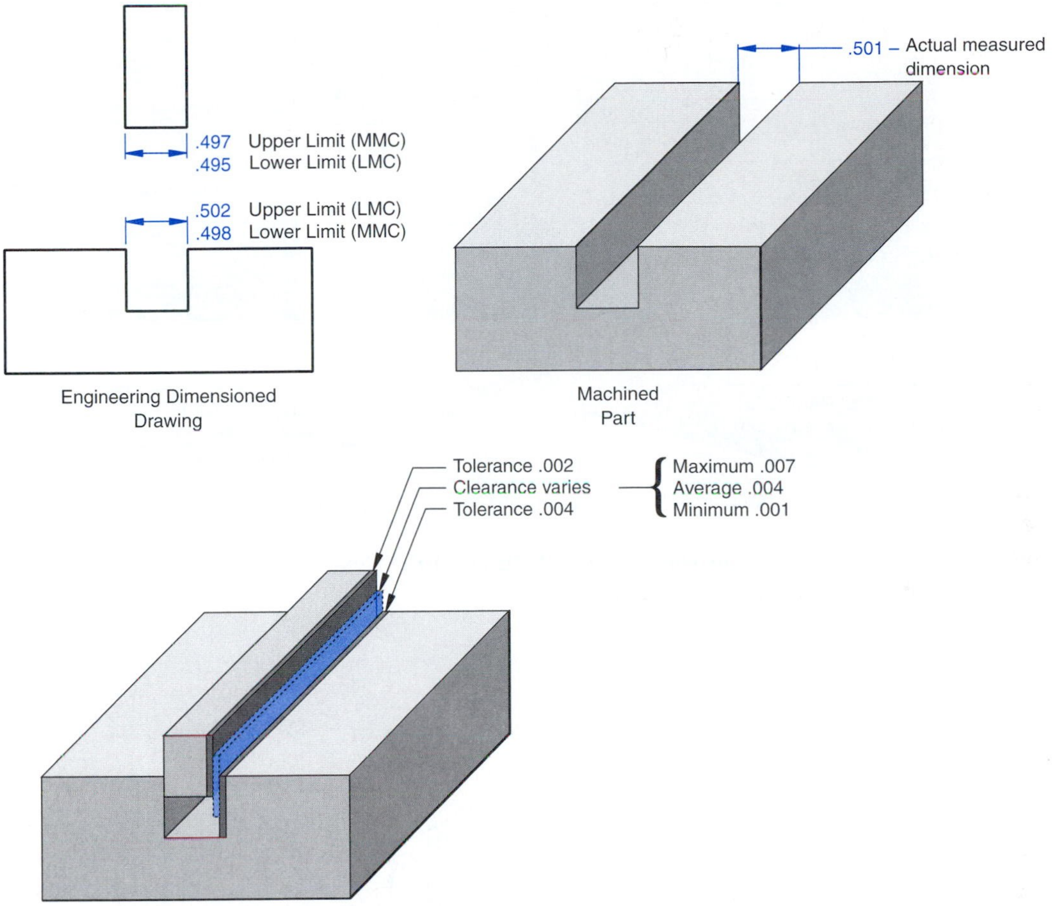

Figure 4.29 **Toleranced Parts and the Important Terms**

4.5.6 Fit Types

The degree of tightness between mating parts is called the *fit*. The basic hole and shaft system shown in Figures 4.30 and 4.31 is an example of the three most common types of fit found in industry.

Clearance fit occurs when two toleranced mating parts will always leave a space or clearance when assembled. In Figure 4.30, the largest that shaft A can be manufactured is .999, and the smallest the hole can be is 1.000. The shaft will always be smaller than the hole, resulting in a minimum clearance of +.001, also called an *allowance*. The maximum clearance occurs when the smallest shaft (.998) is mated with the largest hole (1.001), resulting in a difference of +.003.

Interference fit occurs when two toleranced mating parts will always interfere when assembled. An interference fit *fixes* or *anchors* one part into the other, as though the two parts were one. In Figure 4.30, the smallest that shaft B can be manufactured is 1.002, and the largest the hole can be manufactured is 1.001. This means that the shaft will always be larger than the hole, and the minimum interference is −.001. The maximum interference would occur when the smallest hole (1.000) is mated with the largest shaft (1.003), resulting in an interference of −.003. To assemble the parts under this condition, it would be necessary to *stretch* the hole or *shrink* the shaft or to use force to press the shaft into the hole. Having an interference is a desirable situation for some design applications. For example, it can be used to fasten two parts together without the use of mechanical fasteners or adhesive.

Transition fit occurs when two toleranced mating parts are sometimes an interference fit and sometimes a clearance

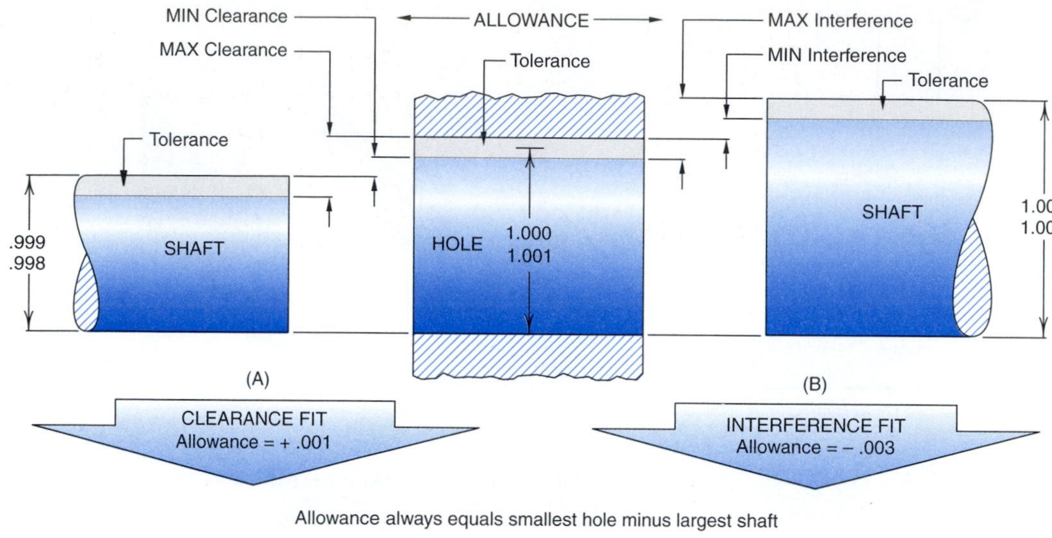

Figure 4.30 **Clearance and Interference Fits between Two Shafts and a Hole**
Shaft A is a clearance fit, and shaft B is an interference fit.

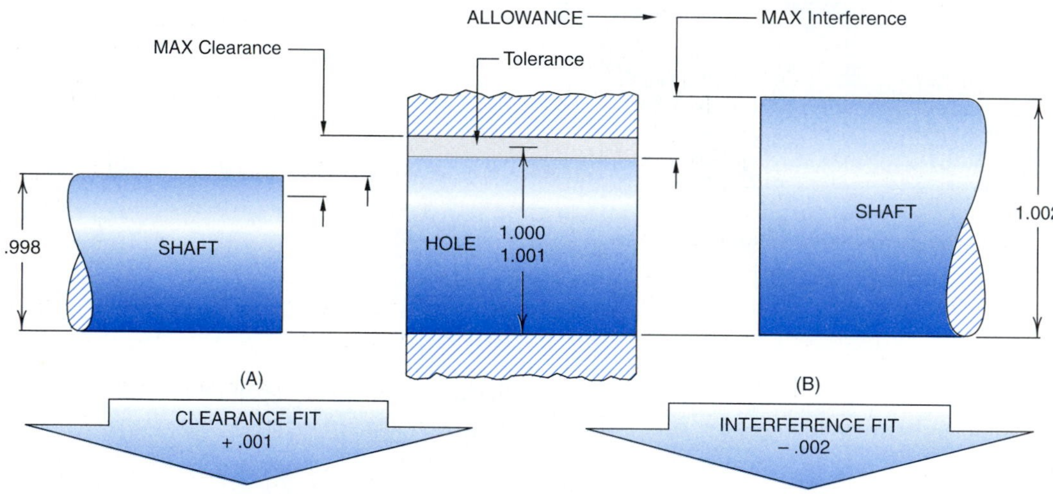

Figure 4.31 **Transition Fit between a Shaft and a Hole**
When the shaft is machined to its smallest diameter (.998), there is a clearance fit with the hole. When the shaft is machined to its largest diameter (1.002), there is an interference fit with the hole.

fit when assembled. In Figure 4.31, the smallest the shaft can be manufactured is .998, and the largest the hole can be manufactured is 1.001, resulting in a clearance of +.003. The largest the shaft can be manufactured is 1.002, and the smallest the hole can be is 1.000, resulting in an interference of −.002.

4.5.7 Basic Hole System

In the basic hole system, which is used to apply tolerances to a hole-and-shaft assembly, the smallest hole is assigned the basic diameter from which the tolerance and allowance are applied (Figure 4.32). The basic hole system is widely

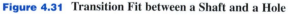

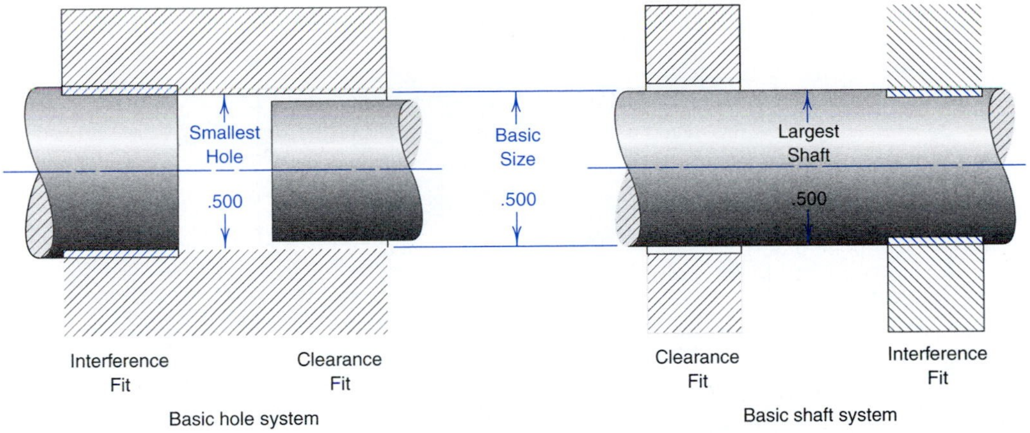

Figure 4.32 The Basic Hole and Shaft Systems for Applying English Unit Tolerances to Parts

used in industry because many of the tools used to drill holes, such as drills and reamers, are designed to produce standard-sized holes.

Step by Step: Creating a Clearance Fit Using the Basic Hole System

Step 1. Using the basic hole system, assign a value of .500″ to the smallest diameter of the hole, which is the lower limit (Figure 4.33).

Step 2. The allowance of .004″ is subtracted from the diameter of the smallest hole to determine the diameter of the largest shaft, .496″, which is the upper limit.

Step 3. The lower limit for the shaft is determined by subtracting the part tolerance from .496″. If the tolerance of the part is .003″, the lower limit of the shaft is .493″.

Step 4. The upper limit of the hole is determined by adding the tolerance of the part to .500″. If the tolerance of the part is .003″, the upper limit of the hole is .503″.

Step 5. The parts are dimensioned on the drawing, as shown in Figure 4.33.

Step 6. Using the assigned values results in a clearance fit between the shaft and the hole. This is determined by finding the difference between the smallest hole (.500″ lower limit) and the largest shaft (.496″ upper limit), which is a positive .004″. As a check, this value should equal the allowance used in Step 2.

Step 7. The difference between the largest hole (.503″ upper limit) and the smallest shaft (.493″ lower limit) equals a positive .010″. Because both the tightest and loosest fits are positive, there will always be clearance between the shaft and the hole, no matter which manufactured parts are assembled.

Step 8. Check the work by determining the *piece tolerances* for the shaft and the hole. To do so, first find the difference between the upper and lower limits for the hole. Subtract .500″ from .503″ to get .003″ as a piece tolerance. This value matches the tolerance applied in Step 4. For the shaft, subtract .493″ from .496″ to get .003″ as the piece tolerance. This value matches the tolerance applied in Step 3.

Step 9. The *system tolerance* is the sum of all the piece tolerances. To determine the system tolerance for the shaft and the hole, add the piece tolerances of .003″ and .003″ to get .006″.

An interference fit would be possible if an allowance is added to the basic size assigned to the hole (Figure 4.34).

Step by Step: Creating an Interference Fit Using the Basic Hole System

Step 1. Using the basic hole system, assign a value of .500″ to the smallest diameter of the hole, which is the lower limit.

Step 2. The allowance of .007″ is added to the smallest hole diameter to determine the diameter of the largest shaft, .507″, which is the upper limit.

Step 3. The lower limit for the shaft is determined by subtracting the part tolerance from .507″. If the part tolerance is .003″, the lower limit of the shaft is .504″.

Step 4. The upper limit of the hole is determined by adding the part tolerance to .500″. If the part tolerance is .003″, the upper limit of the hole is .503″.

Step 5. The parts are dimensioned on the drawing, as shown in Figure 4.34.

Step 6. Using the assigned values results in an interference fit between the shaft and the hole. This is determined by finding the difference between the smallest hole (.500″ lower limit) and the largest shaft (.507″ upper limit),

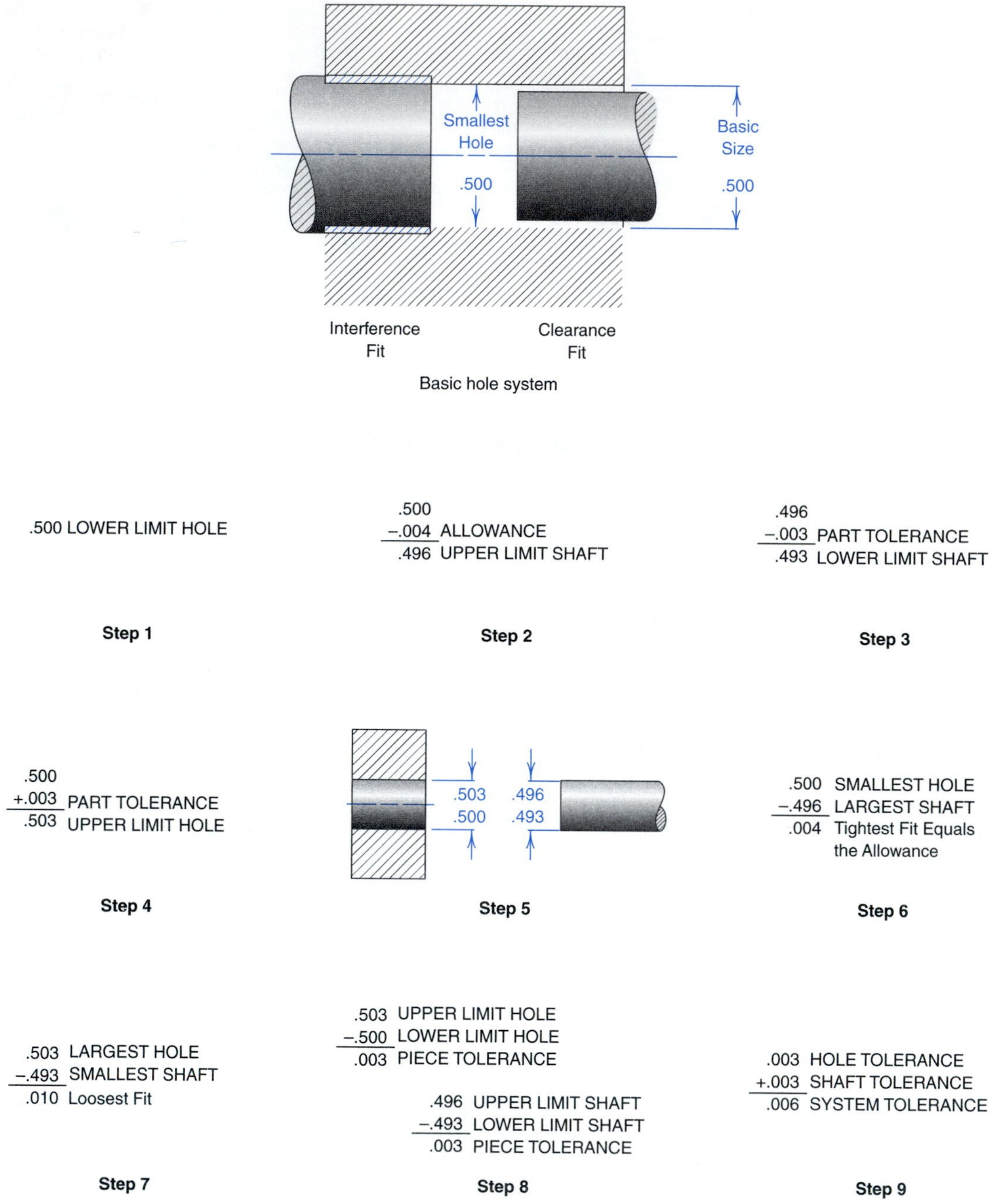

Figure 4.33
Applying tolerances for a clearance fit using the basic hole system

which is a negative .007″. This value equals the allowance used in Step 2.

Step 7. The difference between the largest hole (.503″ upper limit) and the smallest shaft (.504″ lower limit) equals a negative .001″. Because both the tightest and loosest fits are negative, there will always be an interference between the shaft and the hole, no matter which manufactured parts are assembled.

.500 LOWER LIMIT HOLE

Step 1

.500
+.007 ALLOWANCE
.507 UPPER LIMIT SHAFT

Step 2

.507
−.003 TOLERANCE
.504 LOWER LIMIT SHAFT

Step 3

.500
+.003 TOLERANCE
.503 UPPER LIMIT HOLE

Step 4

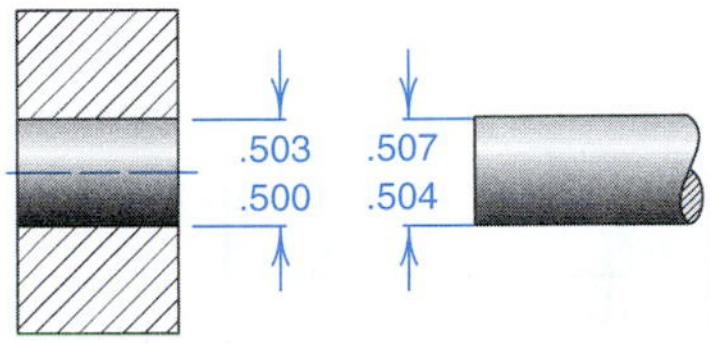

Step 5

.500 SMALLEST HOLE
−.507 LARGEST SHAFT
−.007 Tightest Fit Equals the Allowance

Step 6

.503 LARGEST HOLE
−.504 SMALLEST SHAFT
−.001 Loosest Fit

Step 7

Figure 4.34 **Applying Tolerances for an Interference Fit Using the Basic Hole System**

Step by Step: Creating a Clearance Fit Using the Basic Shaft System

Step 1. Using the basic shaft system, assign a value of .500″ to the largest diameter of the shaft.

Step 2. The allowance of .004″ is added to the largest shaft diameter to determine the diameter of the smallest hole, .504″.

Step 3. The largest limit for the hole is determined by adding the part tolerance to .504″. If the part tolerance is .003″, the upper limit of the hole is .507″.

Step 4. The smallest limit of the shaft is determined by subtracting the part tolerance from .500″. If the part tolerance is .003″, the lower limit of the shaft is .497″. This results in a clearance fit between the shaft and the hole. An interference fit would be possible by subtracting the allowance from the basic size assigned to the shaft.

4.5.8 Basic Shaft System

The basic shaft system, a less popular method of applying tolerances to a shaft and hole, can be used for shafts that are produced in standard sizes. For this system, the largest diameter of the shaft is assigned the basic diameter from which all tolerances are applied (Figure 4.35).

4.6 THREAD NOTES

Threads are only symbolically represented on drawings; therefore, thread notes are needed to provide the required information. A thread note must be included on all threaded parts, with a leader line to the external thread or to an internal thread in the circular view. The recommended thread note size is $^1/_8$″ or 3 mm lettering, whether done by hand or with CAD. The thread note must contain all the information necessary to specify the threads completely (Figure 4.36). External thread notes are given in the longitudinal view. Internal thread notes are given on the end view, with a pointer to the solid circle.

A thread note should contain the following information, in the order given:

1. *Major diameter,* in three-place decimal form, followed by a dash. Fractional sizes are permitted. If a standard number designation is used, the decimal equivalent should be given in parentheses, such as *No. 10 (.190)-32 UNF-2A*.
2. *Number of threads* per inch, followed by a space.
3. *Thread form* designation.
4. *Thread series* designation, followed by a dash.
5. *Thread class* designation (1, 2, or 3).
6. *Internal or external* symbol (A is for external threads, B is for internal threads), followed by a space.
7. *Qualifying information,* such as:

   LH for left-hand threads. If the thread is right-hand, RH is omitted.

   DOUBLE or TRIPLE for multiple threads.

   Thread length.

   Material.

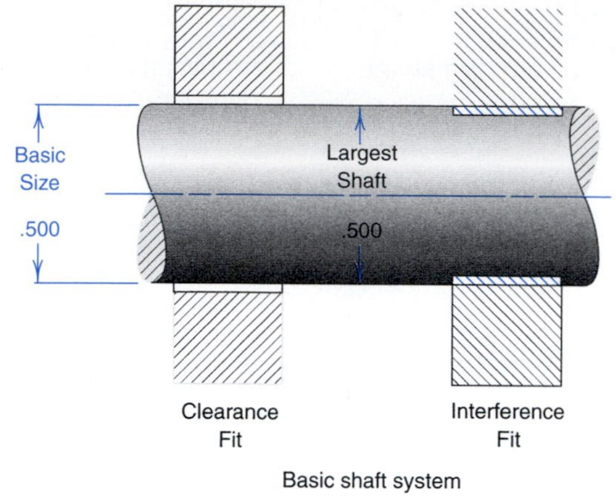

.500 UPPER LIMIT SHAFT

Step 1

.500
+.004 ALLOWANCE
.504 LOWER LIMIT HOLE

Step 2

.504
+.003 TOLERANCE
.507 UPPER LIMIT HOLE

Step 3

.500
−.003 TOLERANCE
.497 LOWER LIMIT SHAFT

Step 4

Figure 4.35 **Applying Tolerances for a Clearance Fit Using the Basic Shaft System**

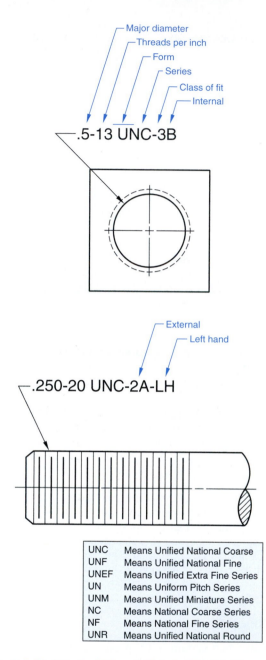

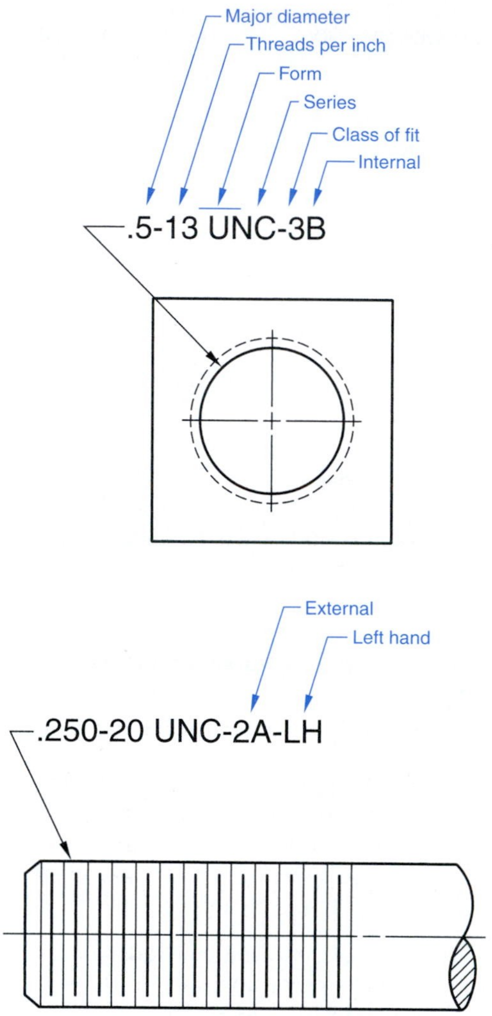

UNC	Means Unified National Coarse
UNF	Means Unified National Fine
UNEF	Means Unified Extra Fine Series
UN	Means Uniform Pitch Series
UNM	Means Unified Miniature Series
NC	Means National Coarse Series
NF	Means National Fine Series
UNR	Means Unified National Round

Figure 4.36 **Standard Thread Note for English Dimension Fasteners**

Thread notes can also provide information about tap drill depth and size, drill and thread depths, countersinking, counterboring, and number of holes to be threaded (Figure 4.37). Tap drill sizes are found in *Machinery's Handbook,* Table 3, in the section "Tapping and Thread Cutting."

Refer to ANSI Y14.6–1978 for more detailed examples of thread notes.

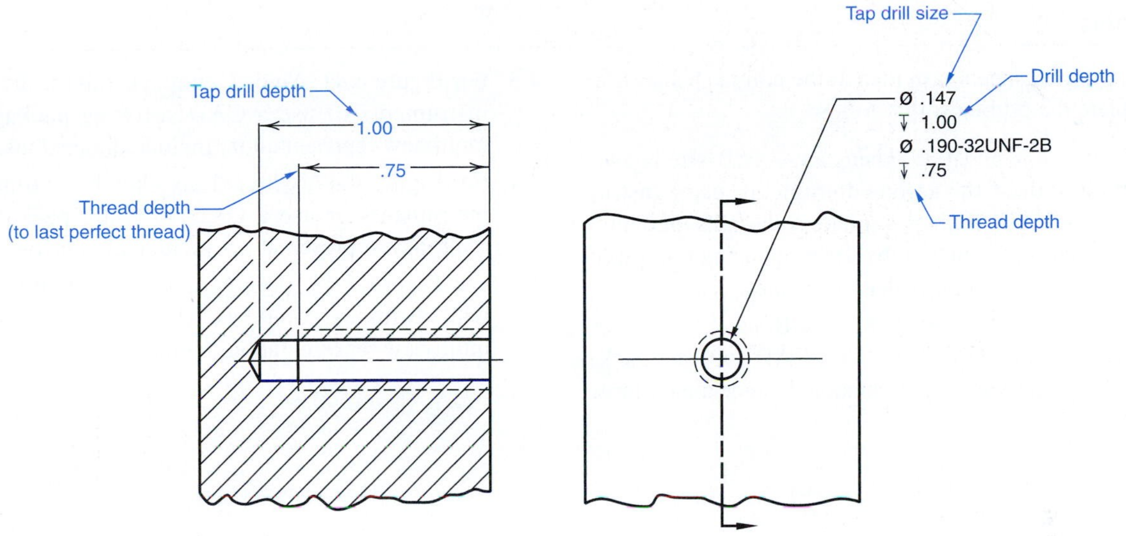

Figure 4.37 **Standard Thread Note for Specifying Tap Drill Size**

Questions for Review

1. How are concentric circles best dimensioned?
2. Sketch the symbols for diameter, radius, depth, counterbore, countersink, and square.
3. Where are these symbols placed with respect to their numerical values?
4. When is a small cross required at the center of a radius?
5. When should extension lines be broken?
6. Write a note showing that a .25-inch deep, .875-inch diameter hole is to be repeated six times.
7. When is an arc dimensioned with a diameter, and when is one dimensioned with a radius?
8. What is the proper proportion of width to length of arrowheads?
9. What is the difference between limit dimensioning and plus and minus dimensioning?
10. What is MMC?
11. Compare the thickness of dimension lines with that of object lines.
12. Compare the thickness of dimension lines with that of extension lines.

Problems

Use the gridded sheets provided at the end of this section to complete the problems that follow.

4.1 (Figures 4.38 and 4.39) Using an A- or B-size layout, sketch or draw the assigned problems, using instruments or CAD. Each grid square equals .25″ or 6 mm. Completely dimension the drawing, using one-place millimeter or two-place decimal inches.

4.2 For Figure 4.40, Gusseted Angle Bracket, sketch (or draw with instruments or use a CAD software package) the multiview representation. Include dimensions.

4.3 For Figure 4.41, Angle Clamp, sketch (or draw with instruments or use a CAD software package) the multiview representation. Include dimensions.

4.4 For Figure 4.42, Clevis Eye, sketch (or draw with instruments or use a CAD software package) the multiview representation. Include dimensions.

4.5 For Figure 4.43, Foot Mounting, sketch (or draw with instruments or use a CAD software package) the multiview representation. Include dimensions.

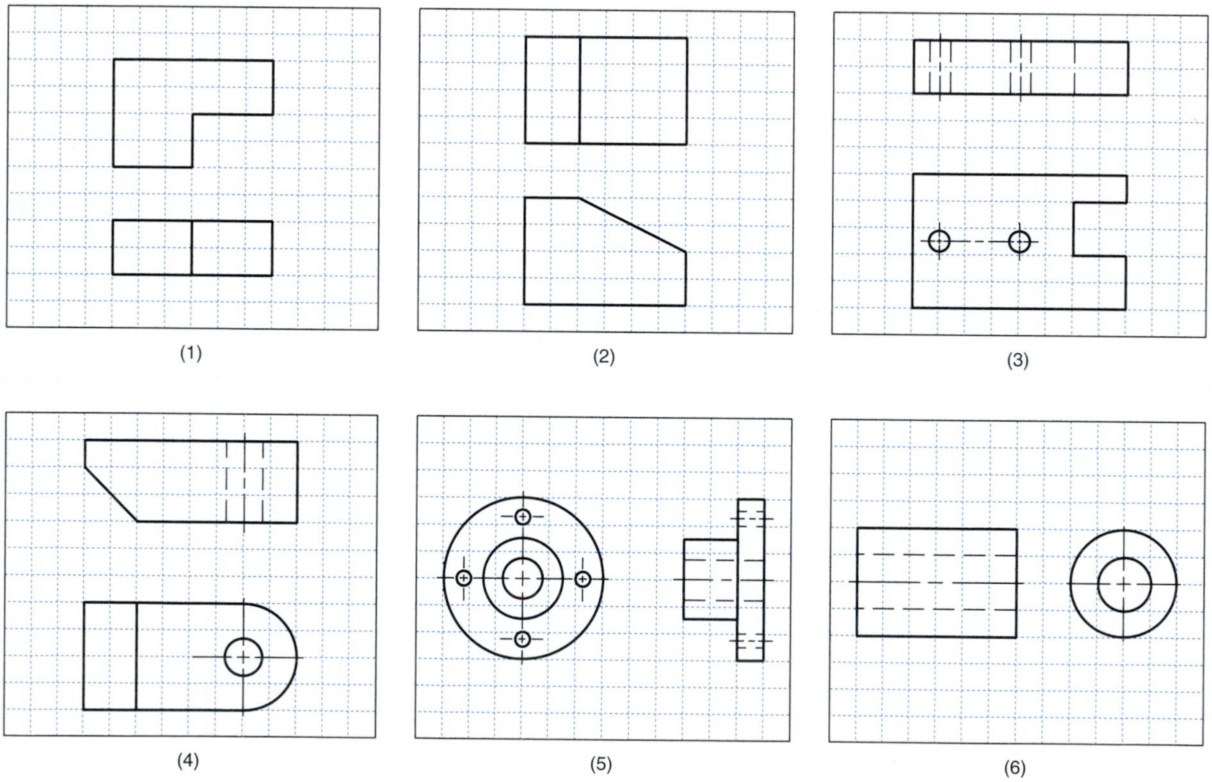

Figure 4.38 Problem 4.1

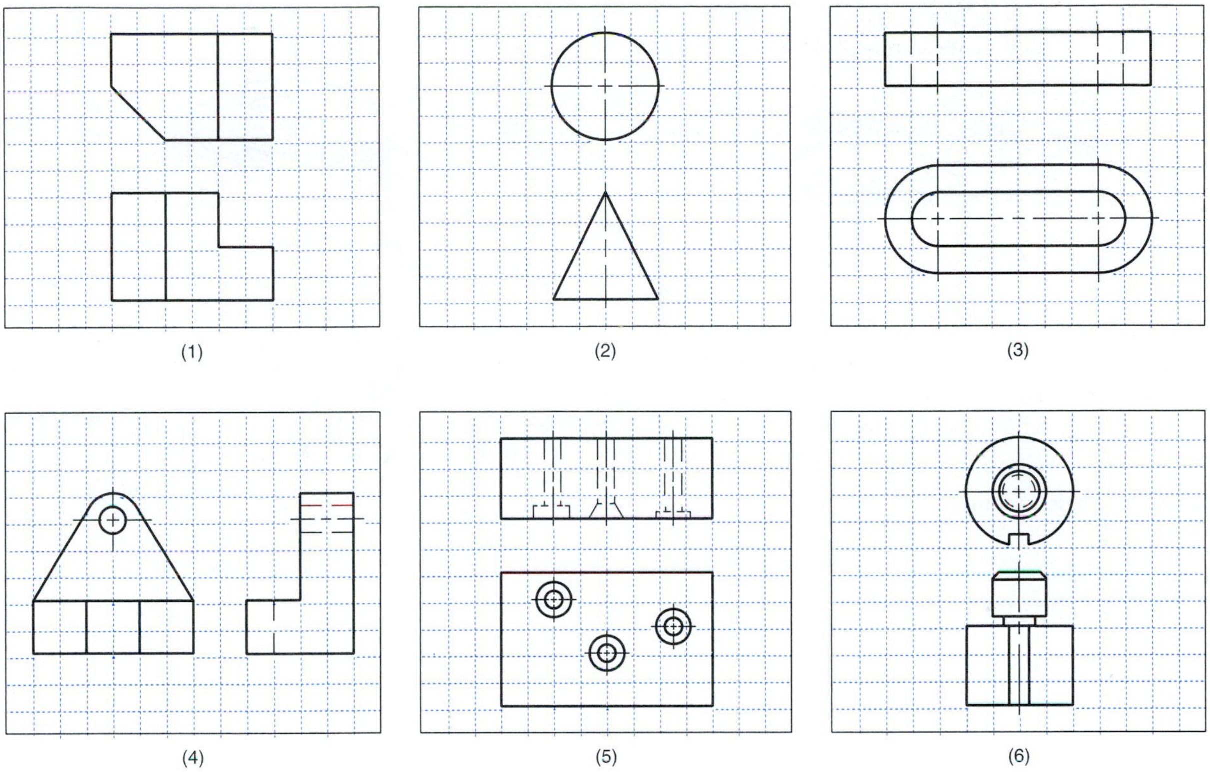

Figure 4.39 Problem 4.1

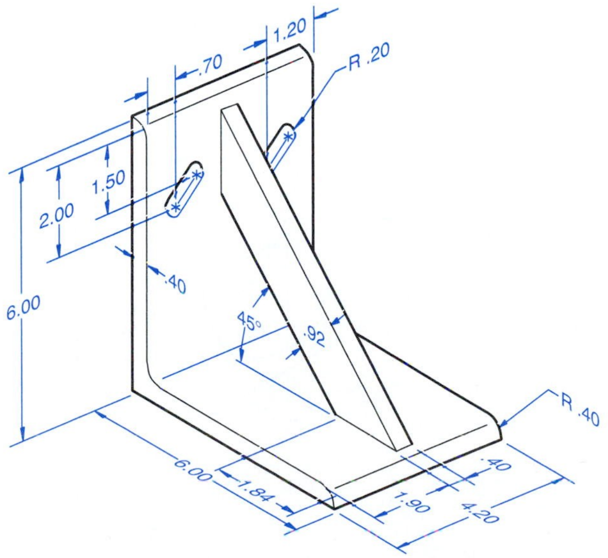

Figure 4.40 Gusseted Angle Bracket

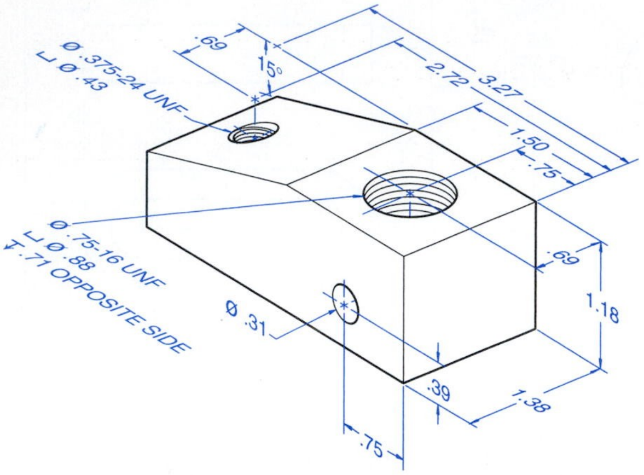

Figure 4.41 Angle Clamp

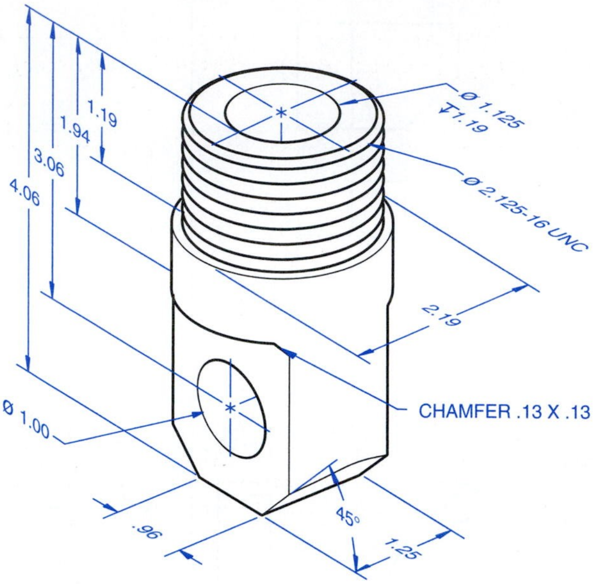

Figure 4.42 Clevis Eye

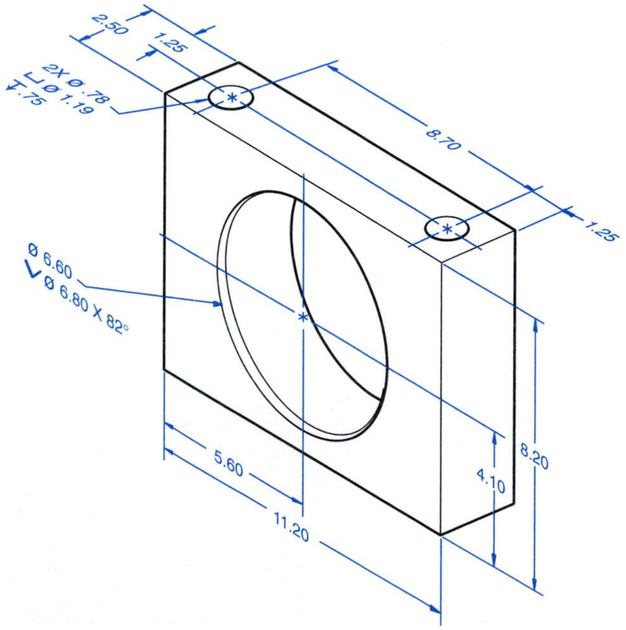

Figure 4.43 Foot Mounting

Problem 4.1

Refer to Figures 4.38 and 4.39. Sketch the necessary views, then add dimensions.

Sketch Number: __________

Name: __________

Div/Sec: __________

Date: __________

Problem 4.1

Refer to Figures 4.38 and 4.39. Sketch the necessary views, then add dimensions.

Sketch Number: ______________

Name: ______________

Div/Sec: ______________

Date: ______________

Problems 4.2 through 4.5

Refer to Figures 4.40 through 4.43. Sketch the necessary views, then add dimensions.

Sketch Number: ____________

Name: ____________

Div/Sec: ____________

Date: ____________

Problems 4.2 through 4.5

Refer to Figures 4.40 through 4.43. Sketch the necessary views, then add dimensions.

Sketch Number: ____________

Name: ____________

Div/Sec: ____________

Date: ____________

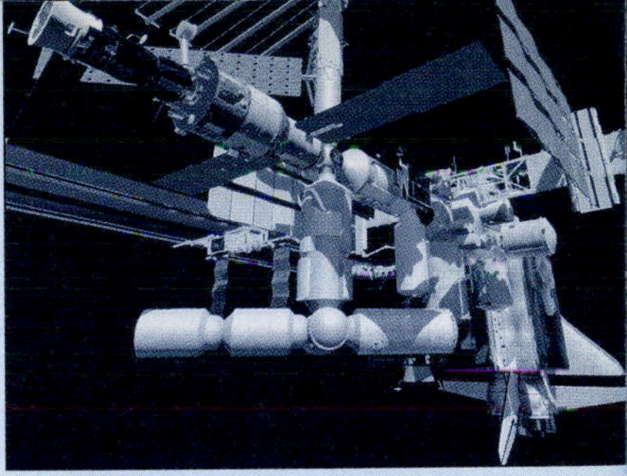

Chapter

5

Reading and Constructing Working Drawings

OBJECTIVES

After completing this chapter, you will be able to:

1. Define working drawings.
2. Describe how working drawings are used in industry.
3. List the major components of a complete set of working drawings.
4. Describe the differences between detail and assembly drawings.
5. Describe how part numbers, zoning, and tables are used on working drawings.
6. Draw standard representations of threads.
7. Specify a metric or English thread in a note.

5.1 BASIC CONCEPTS

Engineering drawings are used to communicate designs to others, document design solutions, and communicate design production information (Figure 5.1). Preceding chapters discussed the development of engineering drawings for the purpose of communicating the design to others. This chapter focuses on communication of the final design for production purposes. These types of drawings are called **working drawings** or **production drawings.**

Figure 5.2 shows the components of the production cycle. **Documenting** is the process of communicating and archiving design and manufacturing information on a product or structure. The documents created include drawings, models, change orders, memos, and reports.

Part of the documenting process includes storing, retrieving, and copying engineering drawings, a process called *reprographics. Archiving* is part of reprographics and involves the storage and retrieval aspects of that process.

Figure 5.1
Engineering drawings are used as a communications tool.

CAD has brought significant changes to this area of design documentation. With 3-D CAD and modern manufacturing techniques, the need for production drawings is minimized. Rather than creating 2-D drawings of the 3-D model, manufacturers extract production information and geometry directly from the computer model. With both 2-D and 3-D CAD, electronic file storage and distribution eliminate the need for many traditional reprographics activities. With networked computers, personnel in any phase of manufacturing can access the most current version of the production drawings or models on their computer screens.

5.2 WORKING DRAWINGS

Working drawings are the complete set of standardized drawings specifying the manufacture and assembly of a product based on its design. The complexity of the design determines the number and types of drawings. Working drawings may be on more than one sheet and may contain written instructions called **specifications.**

Working drawings are the *blueprints* used for manufacturing products. Therefore, the set of drawings must: *(a)* completely describe the parts, both visually and dimensionally; *(b)* show the parts in assembly; *(c)* identify all the parts; and *(d)* specify standard parts. The graphics and text information must be sufficiently complete and accurate to manufacture and assemble the product without error.

Generally, a complete set of working drawings for an assembly includes:

1. Detail drawings of each nonstandard part.
2. An assembly or subassembly drawing showing all the standard and nonstandard parts in a single drawing.
3. A bill of materials (BOM).
4. A title block.

5.2.1 Detail Drawings

A **detail drawing** is a dimensioned, multiview drawing of a single part (Figure 5.3), describing the part's shape, size, material, and finish, in sufficient detail for the part to be

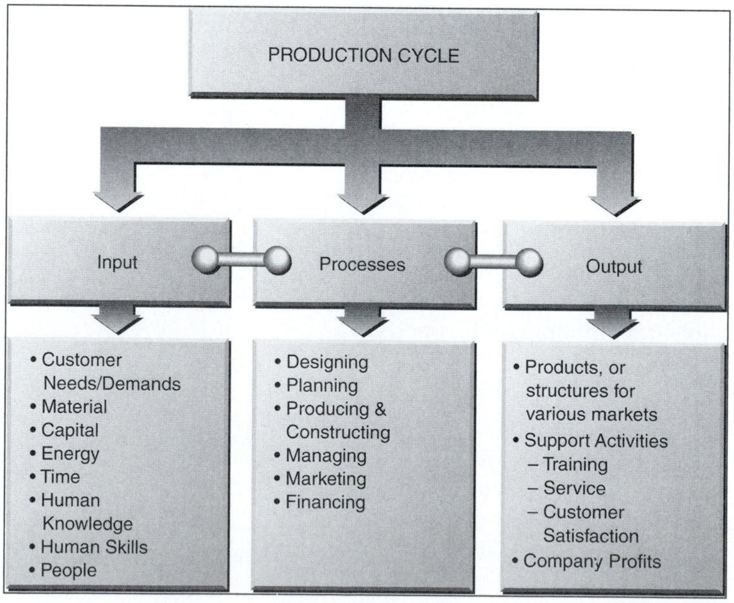

Figure 5.2 Production Cycle
The production cycle includes documenting as a major process.

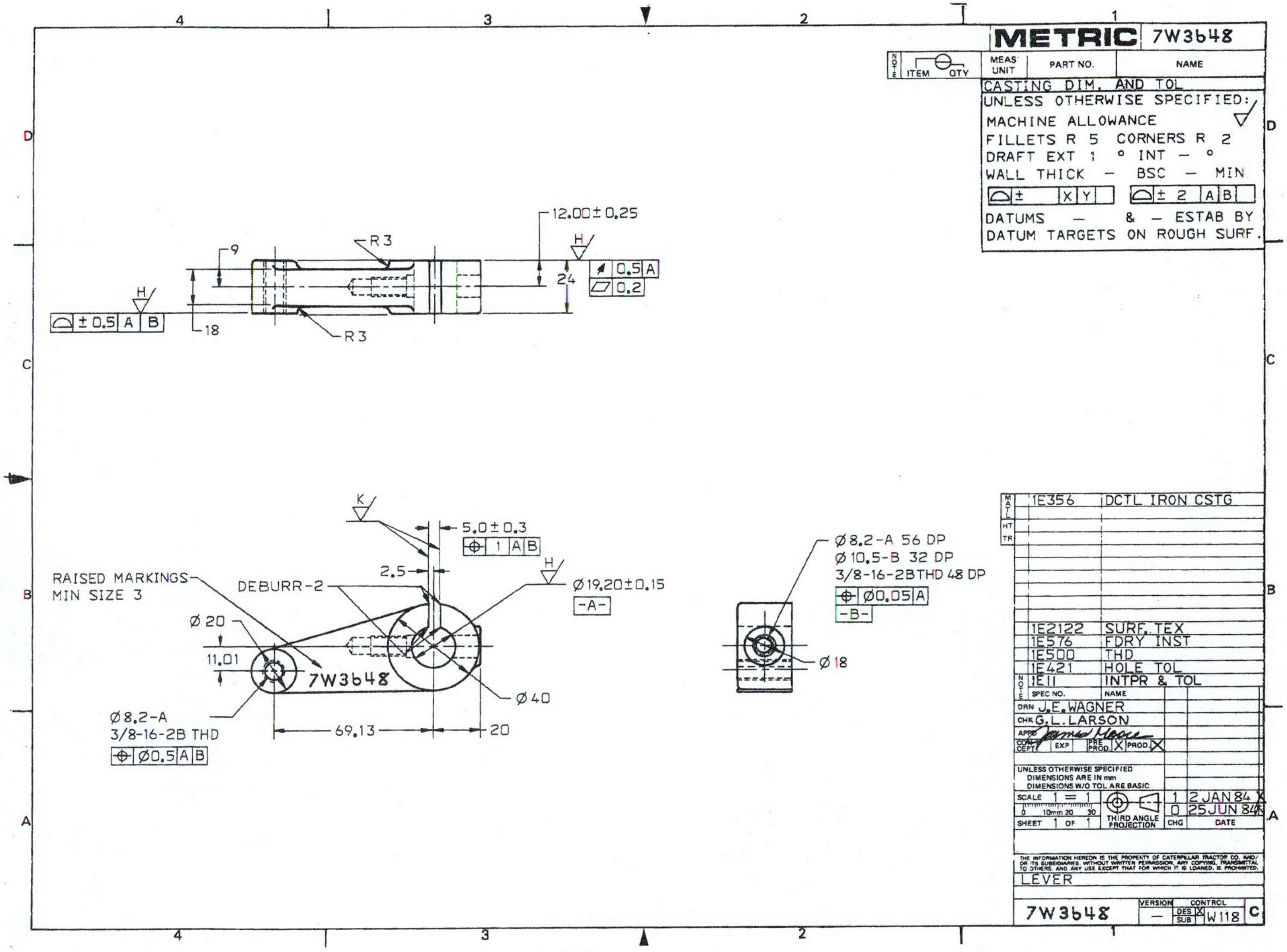

Figure 5.3 **A Detail Drawing of a Single Part Called a Lever**

This detail drawing includes three orthographic views, metric dimensions, tolerances, finish information, part number in the title block and on the front view, and material type, as well as other information.

manufactured based on the drawing alone. Detail drawings are produced from design sketches or extracted from 3-D computer models. They adhere strictly to ANSI standards, and the standards for the specific company, for lettering, dimensioning, assigning part numbers, notes, tolerances, etc.

In an assembly, standard parts such as threaded fasteners, bushings, and bearings are not drawn as details, but are shown in the assembly view. Standard parts are not drawn as details because they are normally purchased, not manufactured, for the assembly.

Multiple details on a sheet are usually drawn at the same scale. If different scales are used, they are clearly marked under each detail. Also, when more than one detail is placed on a sheet, the spacing between details is carefully planned, including leaving sufficient room for dimensions and notes (Figure 5.4). One technique involves lightly blocking in the views for each detail, using construction lines. CAD detail drawings extracted from 3-D models or created in 2-D CAD are more easily manipulated to provide adequate spacing and positioning of multiple details on a single sheet.

5.2.2 Assembly Drawings

An **assembly drawing** shows how each part of a design is put together (Figure 5.5). If the design depicted is only part of the total assembly, it is referred to as a **subassembly.**

An assembly drawing normally consists of:

1. All the parts, drawn in their operating position.
2. A **parts list** or **bill of materials (BOM)** (shown in Figure 5.5 at the bottom of the drawing sheet, to the left of the title block), showing the detail number for each part, the quantity needed for a single assembly, the description or name of the part, the catalog number if it is a standard part, and the company part number.
3. Leader lines with balloons, assigning each part a **detail number,** in sequential order and keyed to the list of parts in the parts list. For example, the assembly shown in Figure 5.5 has 21 assigned detail numbers, each with a leader and circle, and detail number 5 in the sectioned assembly view is described as an end cap in the parts list.

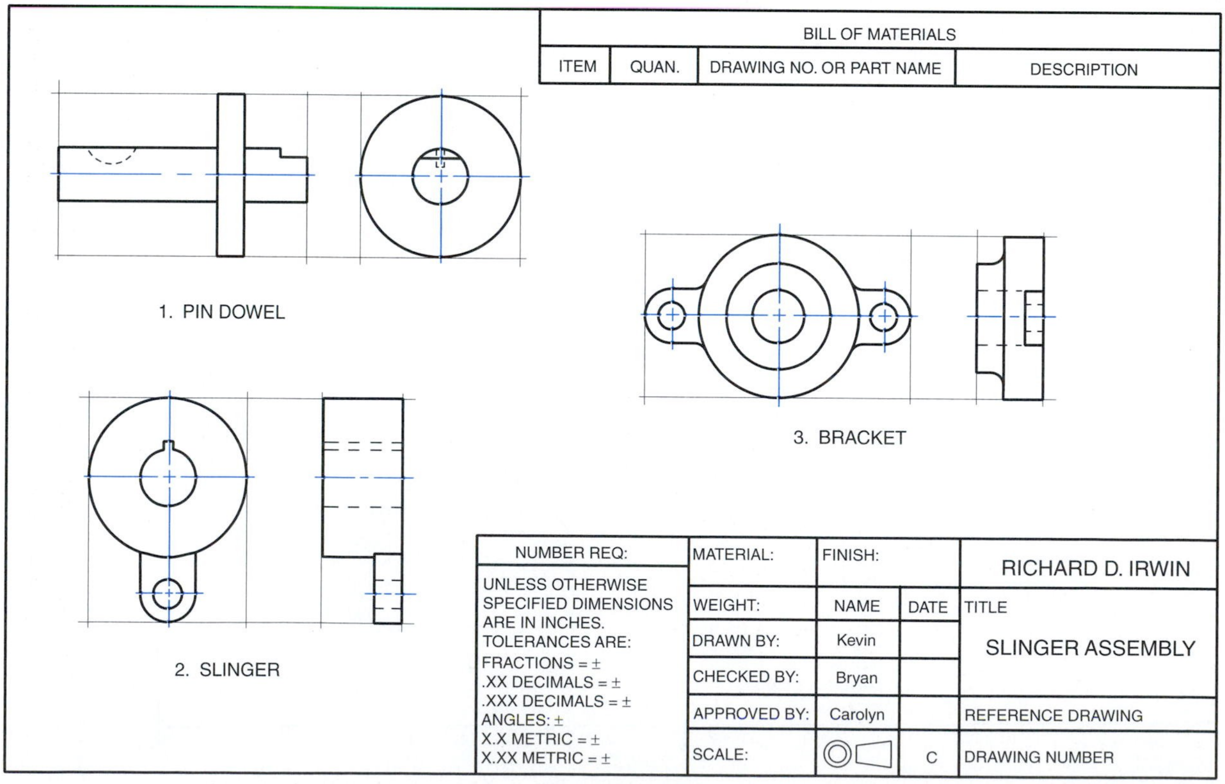

Figure 5.4 **Blocking in the Views to Determine Spacing Requirements**

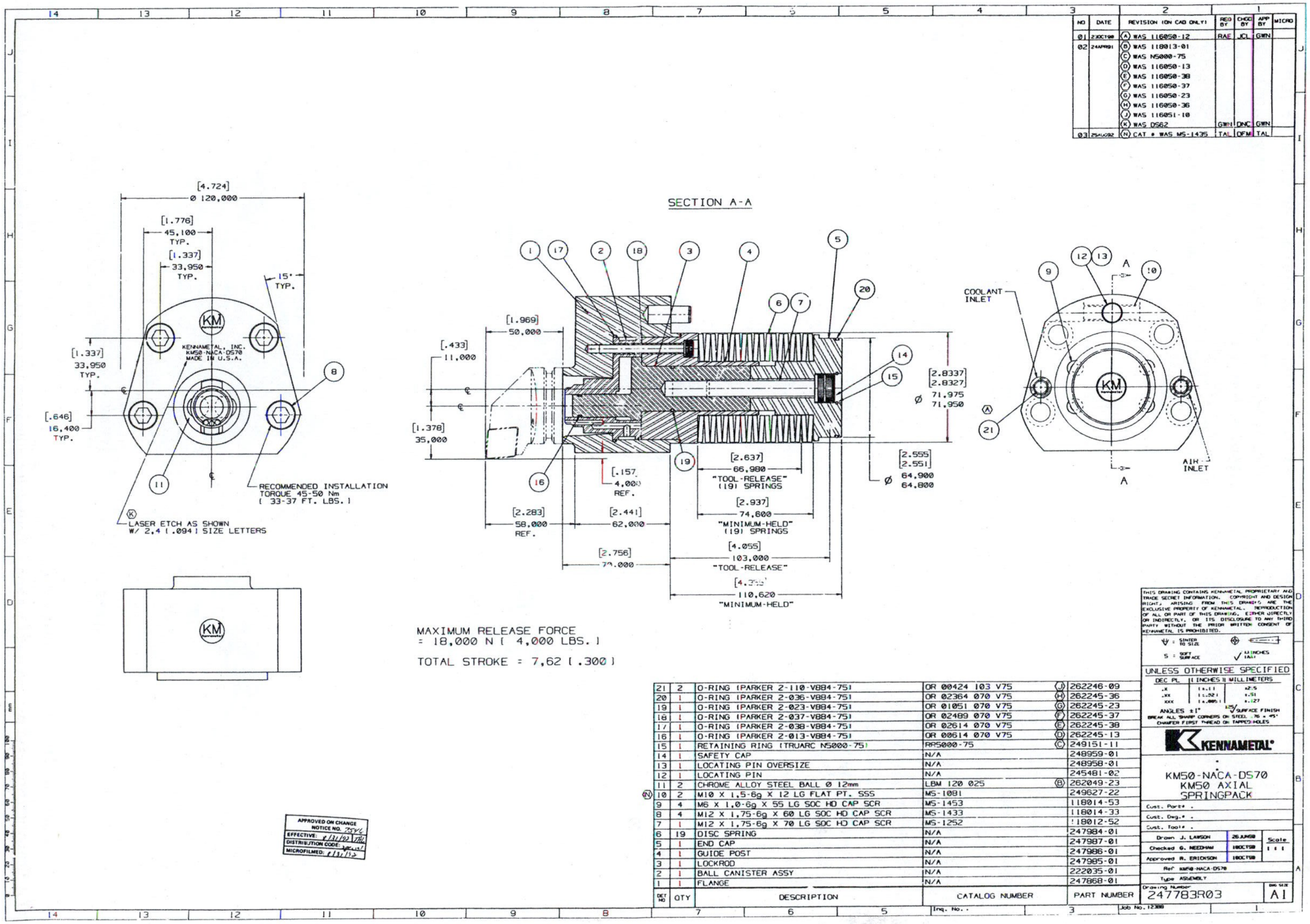

Figure 5.5 **Multiview Sectioned Assembly Drawing of a Spring Pack**

The front view uses variations of section line symbols to distinguish different parts. Courtesy of Kennametal.

4. Machining and assembly operations and critical dimensions related to these functions. In Figure 5.5, the note on the left side view, RECOMMENDED INSTALLATION TORQUE 45–50 Nm (33–37 FT. LBS.), is an assembly operation recommended by the engineer to the person tightening the socket head cap screw.

Dimensions are not shown on assembly drawings, unless necessary to provide overall assembly dimensions, or to assist machining operations necessary for assembly.

Also, hidden lines are omitted in assembly drawings, except when needed for assembly or clarity.

The three basic types of assembly drawings are outline, sectioned, and pictorial.

A **sectioned assembly** gives a general graphic description of the interior shape by passing a cutting plane through all or part of the assembly. (See Figure 5.5.) The section assembly is usually a multiview drawing of all the parts, with one view in full section. Other types of sections, such as broken-out and half sections, can also be used.

Chapter 3, "Section and Auxiliary Views," describes the important conventions that must be followed when assemblies are sectioned. These conventions are summarized as follows:

1. Standard parts, such as fasteners, dowels, pins, bearings, and gears, and nonstandard parts, such as shafts, are not sectioned; they are drawn showing all their exterior features. For example, in Figure 5.5, fasteners, such as part number 7, socket head cap screw, are not sectioned.
2. Adjacent parts in section are lined at different angles, using the cast iron or other type of symbol (Figure 5.5).
3. Thin parts, such as gaskets, are shown solid black.

Sectioned assembly drawings are used for the manufacture and assembly of complicated devices. With CAD, a sectioned assembly can be created by copying detail views and editing them. A 3-D model can also be sectioned to create a sectioned assembly (Figure 5.6).

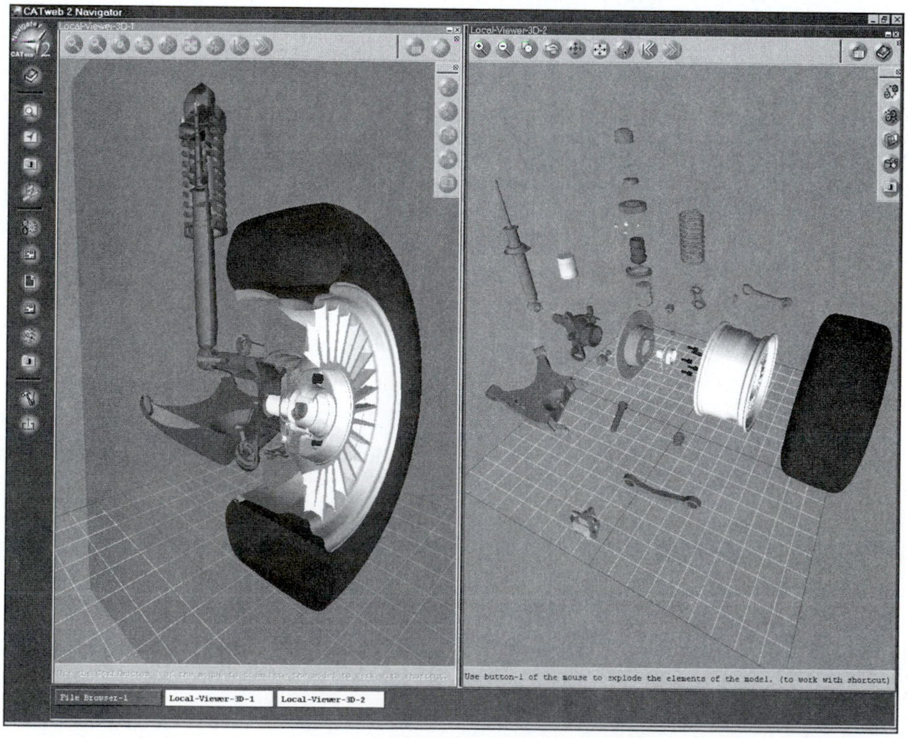

Figure 5.6 Sectioned Assembly Drawing Created with CAD
Sectioned assembly drawings are used by assembly technicians to determine how complicated devices are assembled and for design visualization.

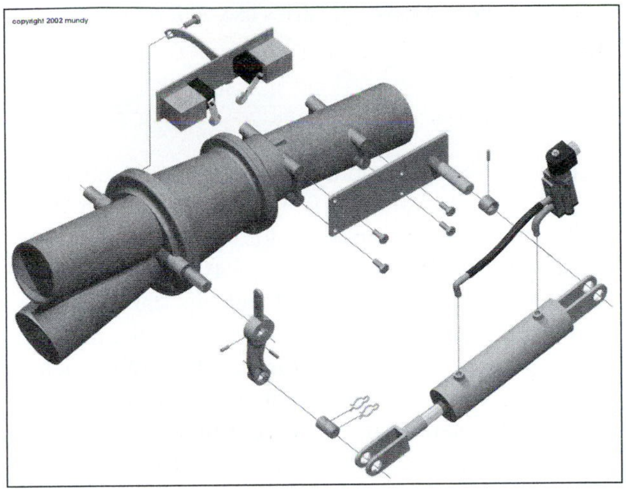

Figure 5.7 Pictorial Assembly
This model was created as an illustration for maintenance handbooks published by Northwind Engineering.

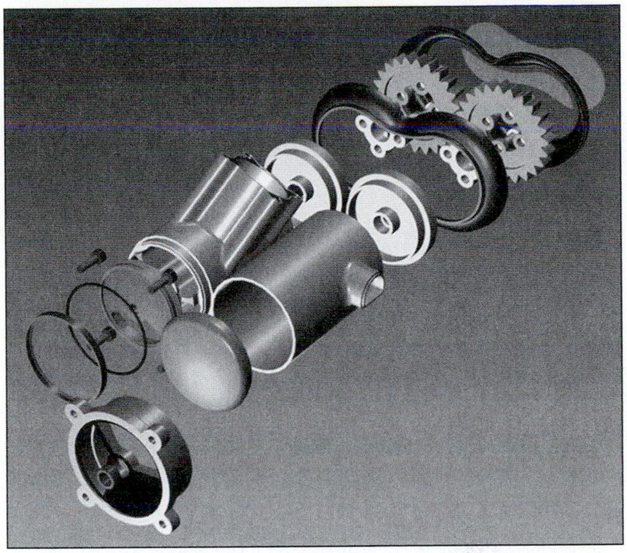

Figure 5.8 Pictorial Shaded Exploded Assembly Drawing of a Cordless Screwdriver Made from a 3-D CAD Model
Courtesy of SDRC.

A **pictorial assembly** gives a general graphic description of each part and uses center lines to show how the parts are assembled (Figure 5.7). The pictorial assembly is normally an isometric view and is used in installation and maintenance manuals.

With 2-D CAD, pictorial assembly drawings can be created using traditional techniques. A 3-D CAD model can also be used to create pictorial assemblies by positioning each part in a pictorial view (Figure 5.8). Center lines and a parts list are added to complete the drawing. With more sophisticated CAD systems, the part models are referenced from a central CAD database. When specifications on an individual part change, this change is automatically reflected in the assembly model.

5.2.3 Part Numbers

Every part in an assembly is assigned a **part number,** which is usually a string of numbers coded in such a way that a company can keep accurate records of its products. For example, in Figure 5.5, the part with detail number 5 in the sectioned assembly drawing has a company part number of 247987-01, as shown in the parts list. Some assemblies are extremely complicated, with thousands of parts. For example, large aircraft have thousands of parts, and considerable documentation is necessary to design, manufacture, assemble, and maintain the aircraft.

5.2.4 Drawing Numbers

Every drawing used in industry is assigned a number. Each company develops its own standard numbering system, based on various criteria, such as sequential numbers, combinations of numbers and letters, sheet sizes, number of parts in the assembly, model numbers, function, etc. Going back to Figure 5.5, the **drawing number** assigned is 247783R03, which can be found at the lower right-hand corner of the drawing sheet in the bottom of the title block.

5.2.5 Title Blocks

Title blocks are used to record all the important information necessary for the working drawings. The title block is normally located in the lower right corner of the drawing sheet. Figure 5.9 shows the ANSI standard formats for full and continuation title blocks. A continuation title block is used for multiple-sheet drawings and does not contain as much detail as the title block found

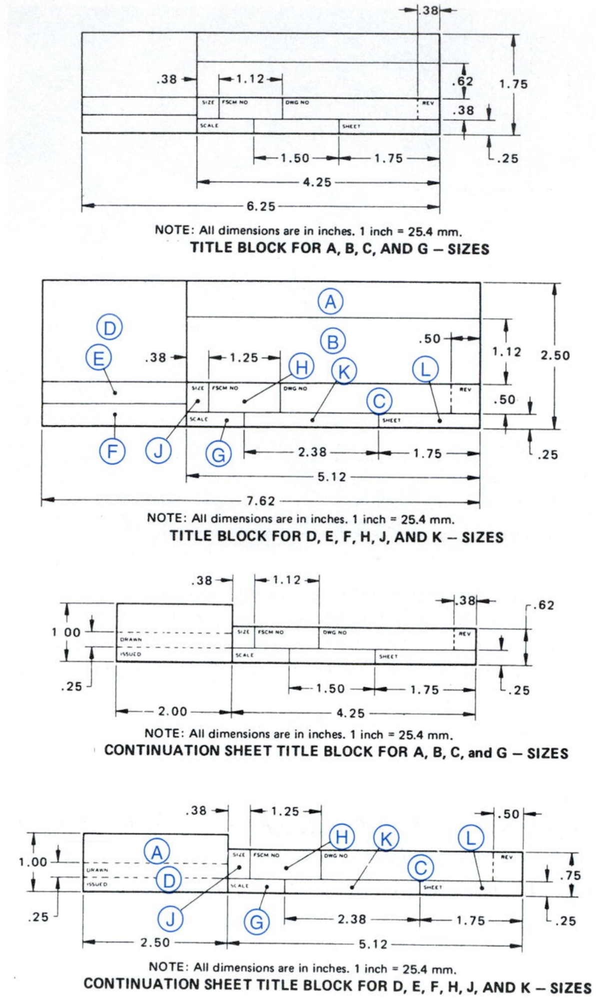

Figure 5.9 ANSI Standard Title Blocks
ANSI Y14.1–1980.

on the first sheet. Many industries use their own title block formats.

5.2.6 Parts Lists

A complete set of working drawings must include a detailed *parts list* or *bill of material.* Based on ANSI standards, a parts list should be located in the lower right corner above the title block (Figure 5.10). Additional parts lists may be located at the left of and adjacent to the original block (ANSI Y14.1–1980). As an example, in Figure 5.5, the parts list is located to the left of the title block. A parts list must include a minimum amount of information necessary to manufacture and assemble the part. The information normally included in a parts list is as follows:

1. Name of the part.
2. A detail number for the part in the assembly.
3. The part material, such as cast iron or bronze.
4. The number of times that part is used in the assembly.
5. The company-assigned part number.
6. Other information, such as weight, stock size, etc.

Information on standard parts, such as threaded fasteners, includes the part name and size or catalog number. For example, in Figure 5.11, information on flat washers

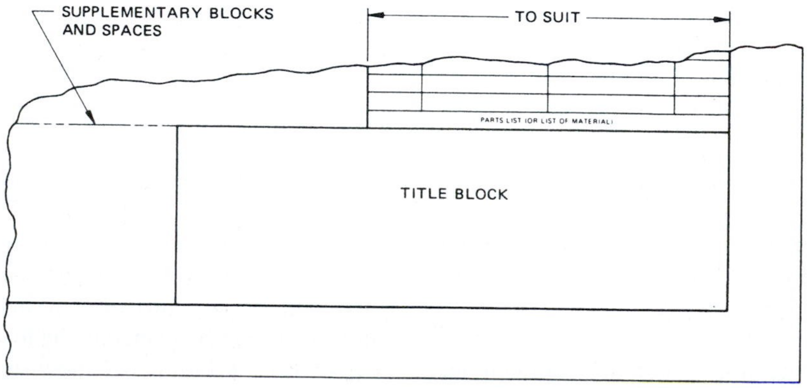

Figure 5.10 Standard Parts List
The parts list runs vertically for as many rows as are needed to list the parts. ANSI Y14.1–1980.

FLAT WASHERS

USS FLAT WASHERS
Standard washers for general purpose application.
Available in plain or zinc plated steel.
SOLD IN PACKAGE QUANTITIES OF 5 LBS. ONLY

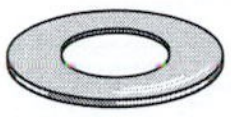

Bolt Size	Diameter I.D.	Diameter O. D.	Thk.	Pcs./ Lb.	Plain Steel No.	Plain Steel NET/LB.	Zinc Plated Steel No.	Zinc Plated Steel NET/LB.
3/16"	1/4"	9/16"	3/64"	303	91081A027	$1.79	91081A027	$1.98
1/4"	5/16"	3/4"	1/16"	154	91081A029	1.48	91081A029	1.67
5/16"	3/8"	7/8"	5/64"	92	91081A030	1.43	91081A030	1.62
3/8"	7/16"	1"	5/64"	79	91081A031	1.43	91081A031	1.56
7/16"	1/2"	1 1/4"	5/64"	41	91081A032	1.37	91081A032	1.54
1/2"	9/16"	1 3/8"	7/64"	30	91081A033	1.35	91081A033	1.48
9/16"	5/8"	1 1/2"	7/64"	22	91081A034	1.33	91081A034	1.54
5/8"	11/16"	1 3/4"	9/64"	13	91081A035	1.35	91081A035	1.46
3/4"	13/16"	2"	5/32"	11	91081A036	1.31	91081A036	1.46
7/8"	15/16"	2 1/4"	11/64"	5	91081A037	1.31	91081A037	1.46
1"	1 1/16"	2 1/2"	11/64"	4	91081A038	1.31	91081A038	1.46
1 1/8"	1 1/4"	2 3/4"	11/64"	5	91081A039	1.31	91081A039	1.61
1 1/4"	1 3/8"	3"	11/64"	4	91081A040	1.45	91081A040	1.61
1 3/8"	1 1/2"	3 1/4"	3/16"	3	91081A042	1.45	91081A042	1.65
1 1/2"	1 5/8"	3 1/2"	3/16"	2	91081A041	1.45	91081A041	1.65
1 5/8"	1 3/4"	3 3/4"	3/16"	2	91081A043	1.45	91081A043	1.65
1 3/4"	1 7/8"	4"	3/16"	2	91081A044	1.63	91081A044	1.65
2"	2 1/8"	4 1/2"	3/16"	1	91081A046	1.45	91081A046	1.65
2 1/4"	2 3/8"	4 3/4"	7/32"	1	91081A047	1.93	91081A047	2.15
2 1/2"	2 5/8"	5"	15/64"	1	91081A048	1.93	91081A048	2.15

Figure 5.11 Typical Page from a Parts Catalog

Part numbers for standard parts are listed in parts catalogs.

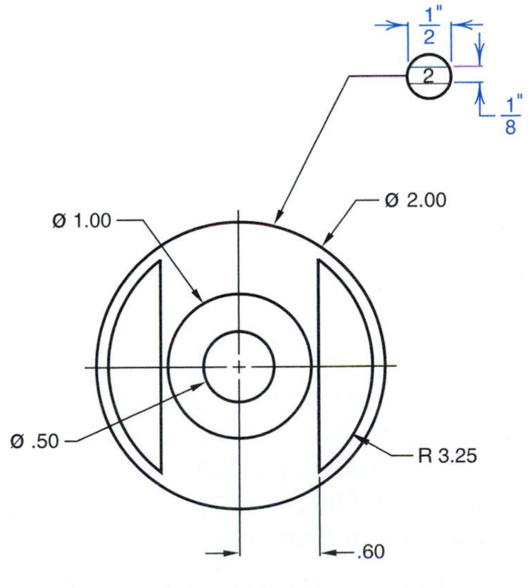

Figure 5.12 Balloons

Balloons are used to identify parts by their assigned number in the assembly.

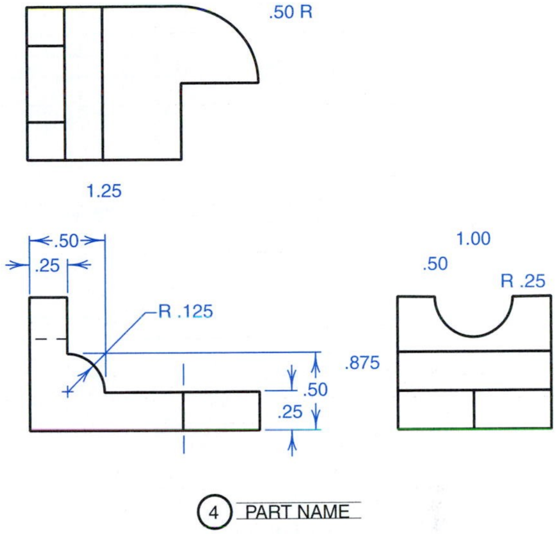

Figure 5.13 Part Name

In detail drawings of an assembly, the part name and detail number are located near one of the views or in the title block.

would be obtained from the tabular drawing from an engineering catalog. Such information includes critical dimensions and model number, such as 91081A07.

5.2.7 Part Identification

Parts are identified in assembly drawings by a leader line with an arrow that points to the part. The other end of the leader has a **balloon** showing the part detail number (Figure 5.12). The balloon is approximately four times the height of the number. Part names are given that are placed as close to the part as possible (Figure 5.13).

5.2.8 Revision Block

Drawing revisions occur because of design changes, tooling changes, customer requests, errors, etc. If a drawing is changed, an accurate record of the change must be created and should contain the date, name of the person making the change, description of the change, the change number, and approval. This information is placed in a **revision block** (Figure 5.14), which is normally in the upper right corner of the drawing, with sufficient space reserved for the block to be expanded downward.

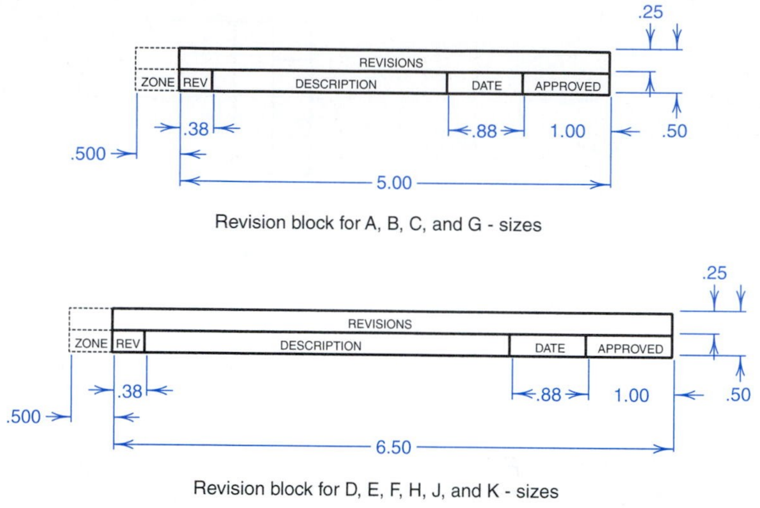

Figure 5.14 **Standard Revision Block**
ANSI Y14.1–1980.

5.2.9 Scale Specifications

The scale used on a set of working drawings is placed in the title block. If more than one scale is used, the scale is shown near the detail drawing. Scales are indicated in metric drawings by a colon, such as 1:2; for the English system, an equal sign, such as 1=2, is used.

Common English scales found on engineering drawings are:

1=1	Full
1=2	Half
1=4	Quarter
1=8	Eighth
1=10	Tenth
2=1	Double

Common metric scales include:

1:1	Full
1:2	Half
1:5	Fifth
1:10	Tenth
1:20	Twentieth
1:50	Fiftieth
1:100	Hundredth

The designations *METRIC* or *SI* appear in or near the title block to show that metric dimensions and scale are used on the drawing.

5.2.10 Tolerance Specifications

Tolerances are specified in a drawing using toleranced dimensions. For those dimensions that are not specifically toleranced, a **general tolerance note** is used. This note is placed in the lower right corner, near the title block, and usually contains a tolerance and a general statement, such as UNLESS OTHERWISE SPECIFIED.

For example, in Figure 5.15, the general tolerance note specifies the following tolerances for untoleranced dimensions:

One-place decimal inches ±.1
Two-place decimal inches ±.02
Three-place decimal inches ±.005
One-place decimal millimeters ±2.5
Two-place decimal millimeters ±0.51
Three-place decimal millimeters ±0.127

Refer to Chapter 4, "Dimensioning and Tolerancing Practices," for a more detailed explanation of tolerances.

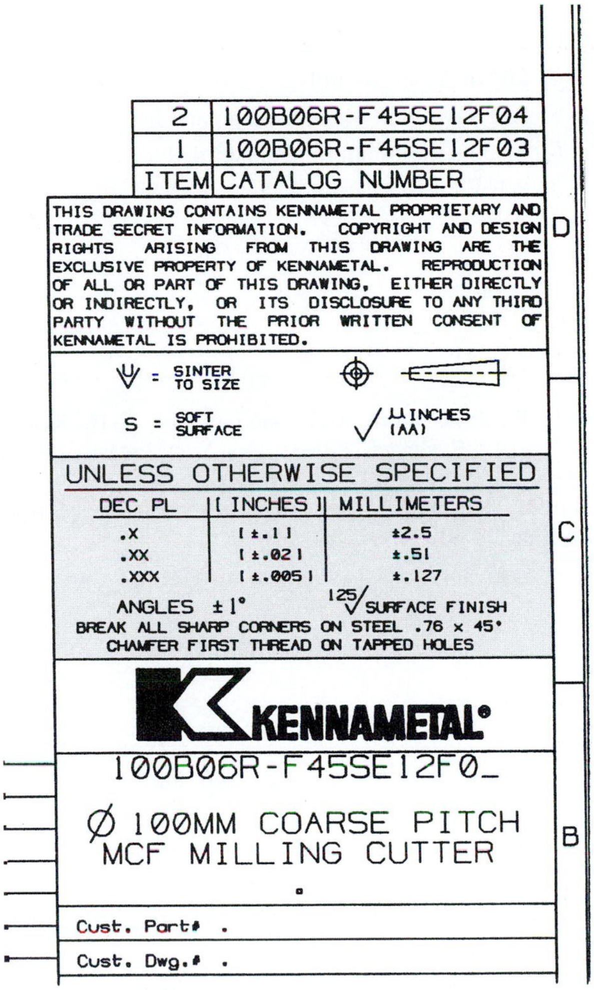

Figure 5.15 General Tolerance Note for Inch and Millimeter Dimensions

Courtesy of Kennametal.

5.2.11 Engineering Change Orders (ECOs)

Making engineering changes after the design has been finalized is very expensive and should be avoided but is necessary if there is a design error, a change request by the customer, or a change in the material or manufacturing process. To make a change in a design, most industries require the use of an Engineering Change Notice (ECN) or Engineering Change Order. The form used to request a change varies by industry, but they all have common features such as:

Identification of what has to be changed in the form of part numbers, part names, and drawing numbers.

An explanation of the need for the requested change.

A list of all documents and departments within an organization that is affected by change.

A description of the change, including drawings of the part(s) before and after the change.

Approval of the changes by project managers.

Instructions describing when the changes are to be implemented.

Questions for Review

1. Define working drawings.
2. List the types of drawings commonly found in a complete set of working drawings.
3. List the types of assembly drawings.
4. Define a subassembly.
5. List the important items in a parts list.
6. List the important items in a title block.
7. What is a revision block?

Further Reading

COADE Mechanical Engineering News. COADE Engineering Software, 12777 Jones Rd., Suite 480, Houston, TX 77070, Phone 713–890–4566.

Graves, F. E. "Nuts and Bolts." *Scientific American,* June 1984, pp. 136–144.

Oberg, E., F. D. Jones, H. L. Horton, and H. H. Ryffell. *Machinery's Handbook.* 24th ed. New York: Industrial Press, 1992.

Young, B. "Streamlining the Design Process." *Cadence,* April 1993, pp. 78–80.

Problems

Use the gridded sheets at the end of this section to complete the problems that follow. Figures 5.16 through 5.22 range from simple to complex mechanical and electromechanical assemblies. For each figure, create a complete set of working drawings. Some of the figures require tolerances to be assigned and missing information to be developed. Some of the assemblies may also be redesigned to improve on the existing design. For each problem, do the following:

1. Sketch orthographic views of each part, with dimensions. If dimensions are missing, determine what they should be by their relationship to other parts.
2. Determine tolerances as noted or assigned.
3. Determine the scale and size of sheet to use if using traditional tools, or the plot scale and sheet size if using CAD.
4. Determine finished surfaces, and mark them on the sketch.
5. Determine which parts should be represented as section views on the detail drawings.
6. Determine the specifications for all standard parts in the assembly.
7. Create dimensioned detail drawings of each nonstandard part in the assembly.
8. Create an orthographic or exploded pictorial assembly drawing in section.
9. Label all parts in the assembly drawing, using numbers and balloons.
10. Create an ANSI standard parts list with all relevant information for the parts in the assembly.

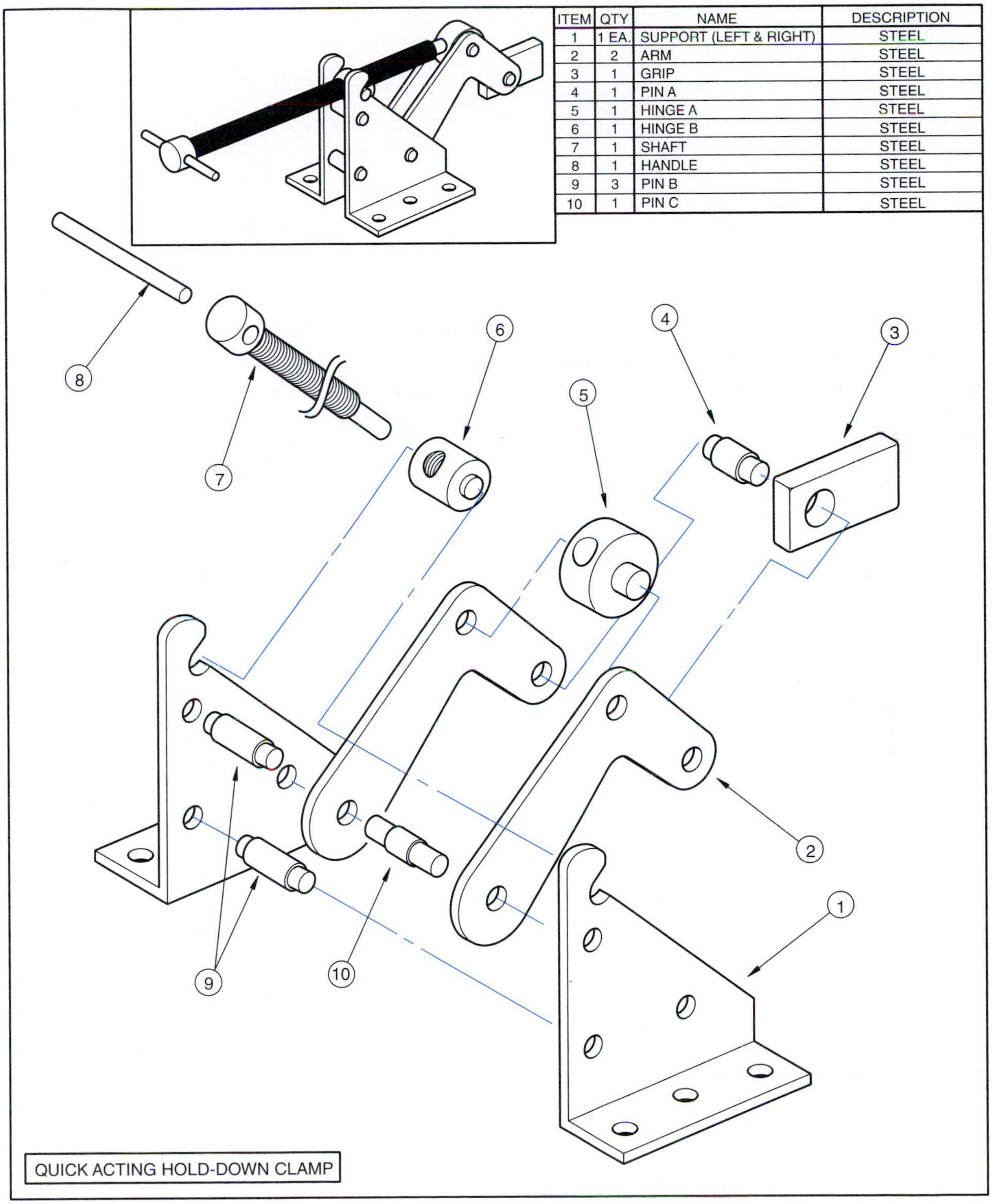

ITEM	QTY	NAME	DESCRIPTION
1	1 EA.	SUPPORT (LEFT & RIGHT)	STEEL
2	2	ARM	STEEL
3	1	GRIP	STEEL
4	1	PIN A	STEEL
5	1	HINGE A	STEEL
6	1	HINGE B	STEEL
7	1	SHAFT	STEEL
8	1	HANDLE	STEEL
9	3	PIN B	STEEL
10	1	PIN C	STEEL

Figure 5.16 Quick Acting Hold-Down Clamp

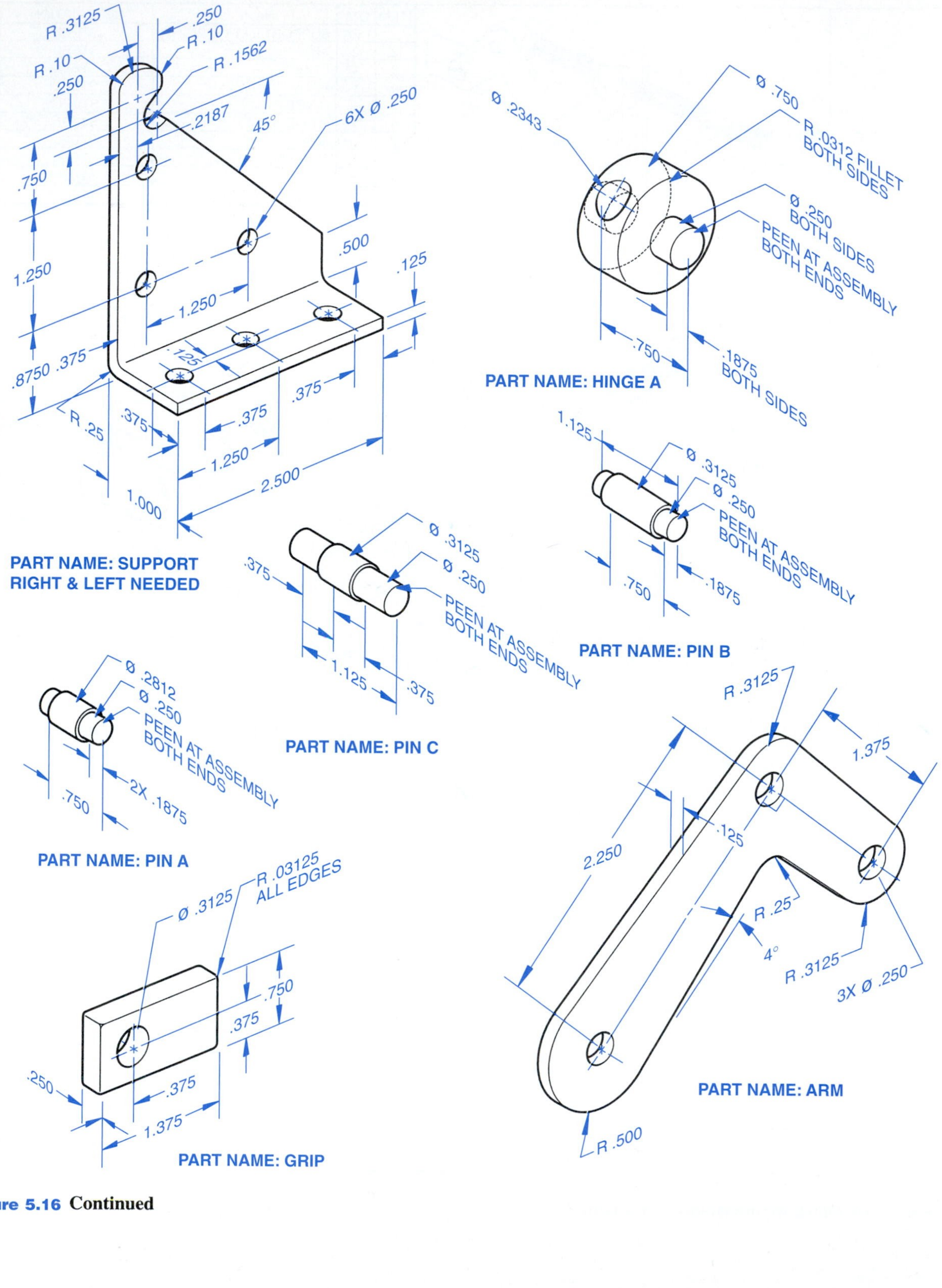

Figure 5.16 Continued

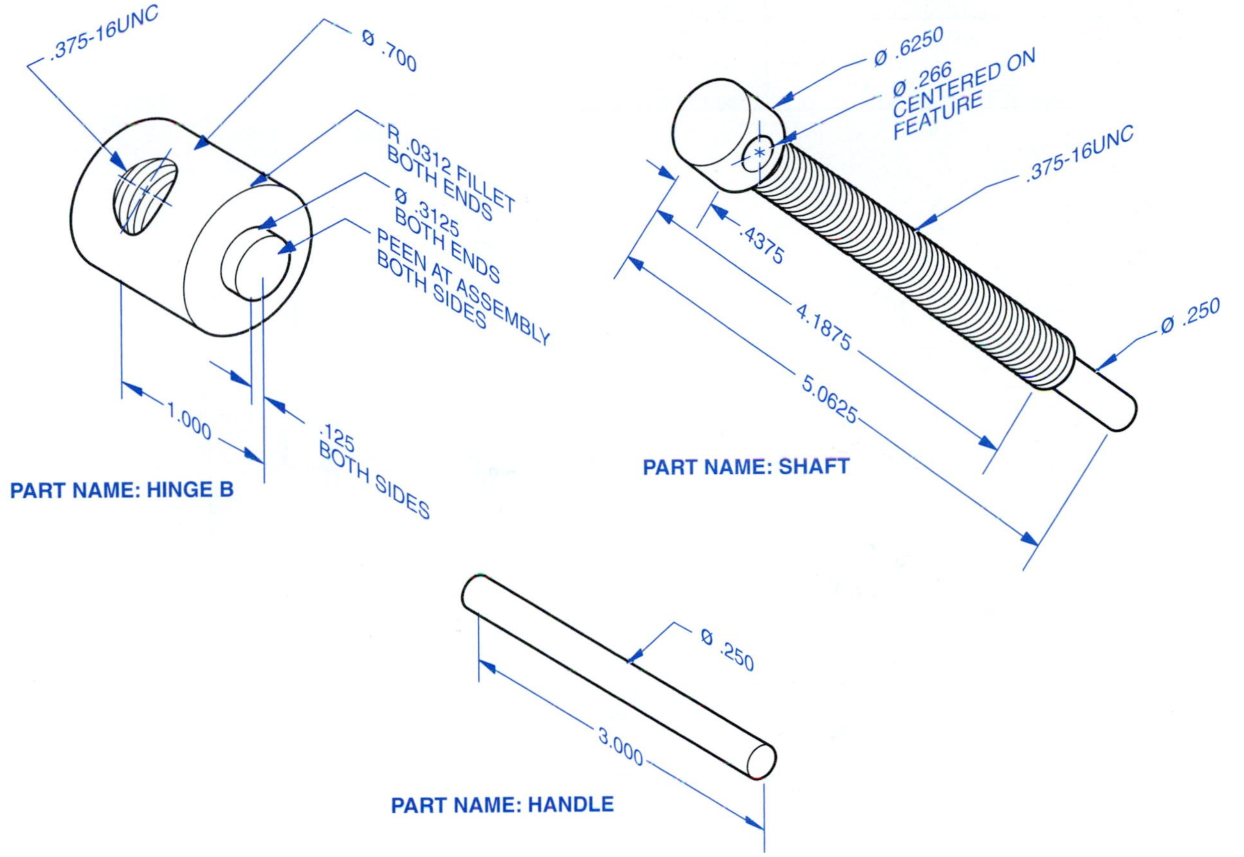
.375-16UNC
Ø .700
R .0312 FILLET BOTH ENDS
Ø .3125 BOTH ENDS
PEEN AT ASSEMBLY BOTH SIDES
1.000
.125 BOTH SIDES
PART NAME: HINGE B
Ø .6250
Ø .266 CENTERED ON FEATURE
.375-16UNC
.4375
4.1875
5.0625
Ø .250
PART NAME: SHAFT
Ø .250
3.000
PART NAME: HANDLE

Figure 5.16 Continued

ITEM	QTY	NAME	DESCRIPTION
1	2	BUSHING	BRONZE
2	1	BASE	STEEL
3	1	CAP	STEEL
4	6	HEX HEAD BOLT	.50-13UNC X 2.00

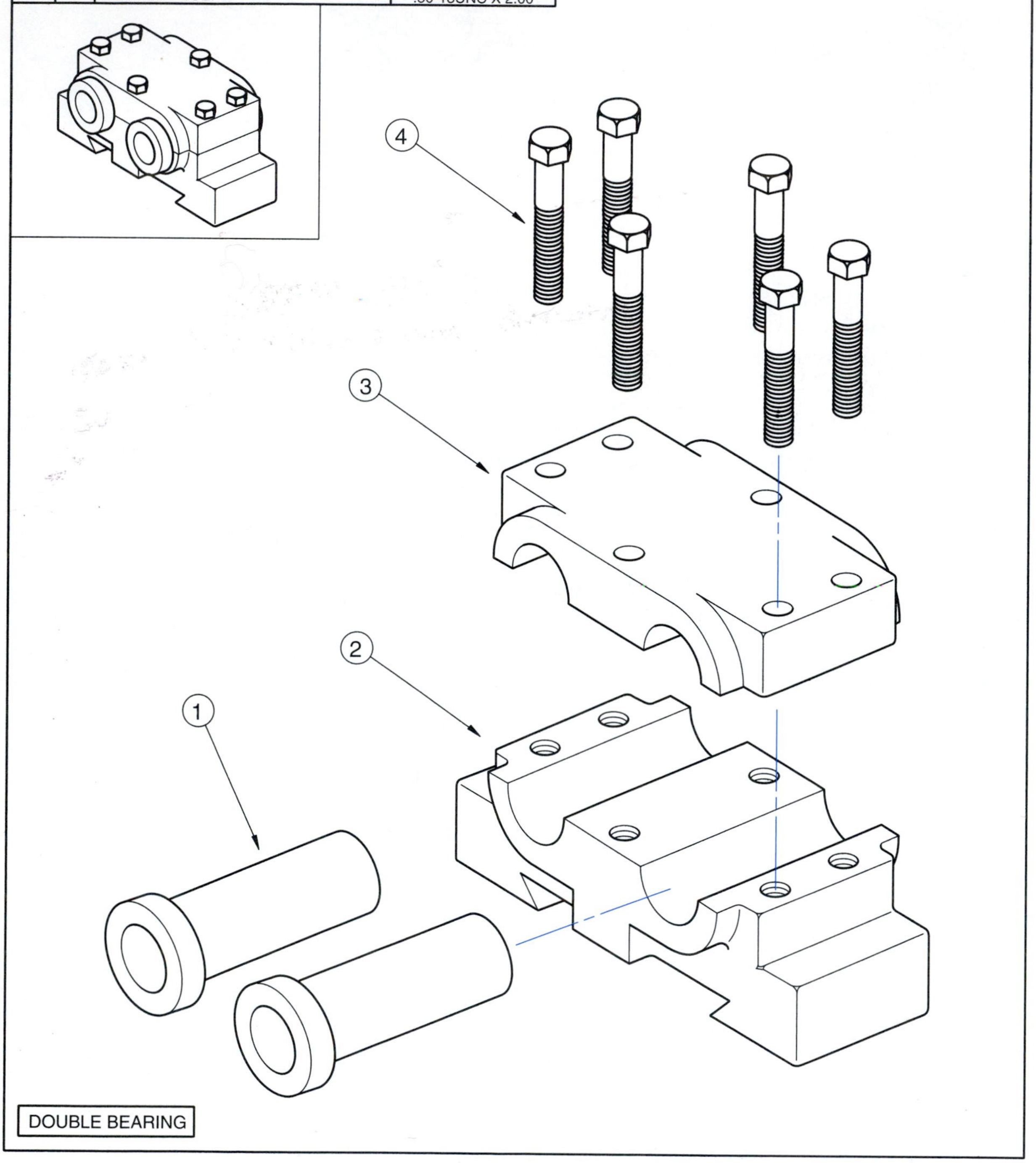

Figure 5.17 Double Bearing

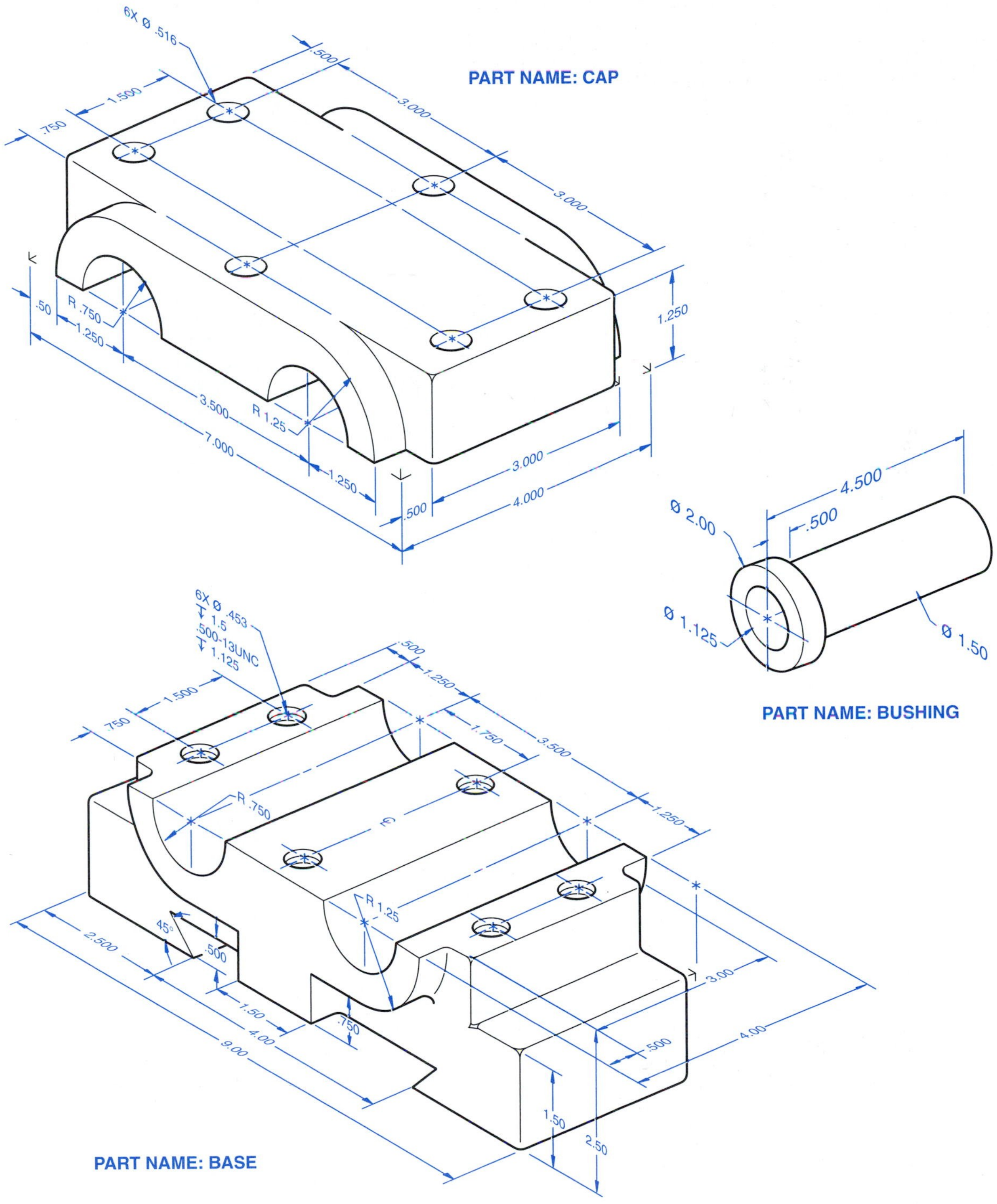

Figure 5.17 Continued

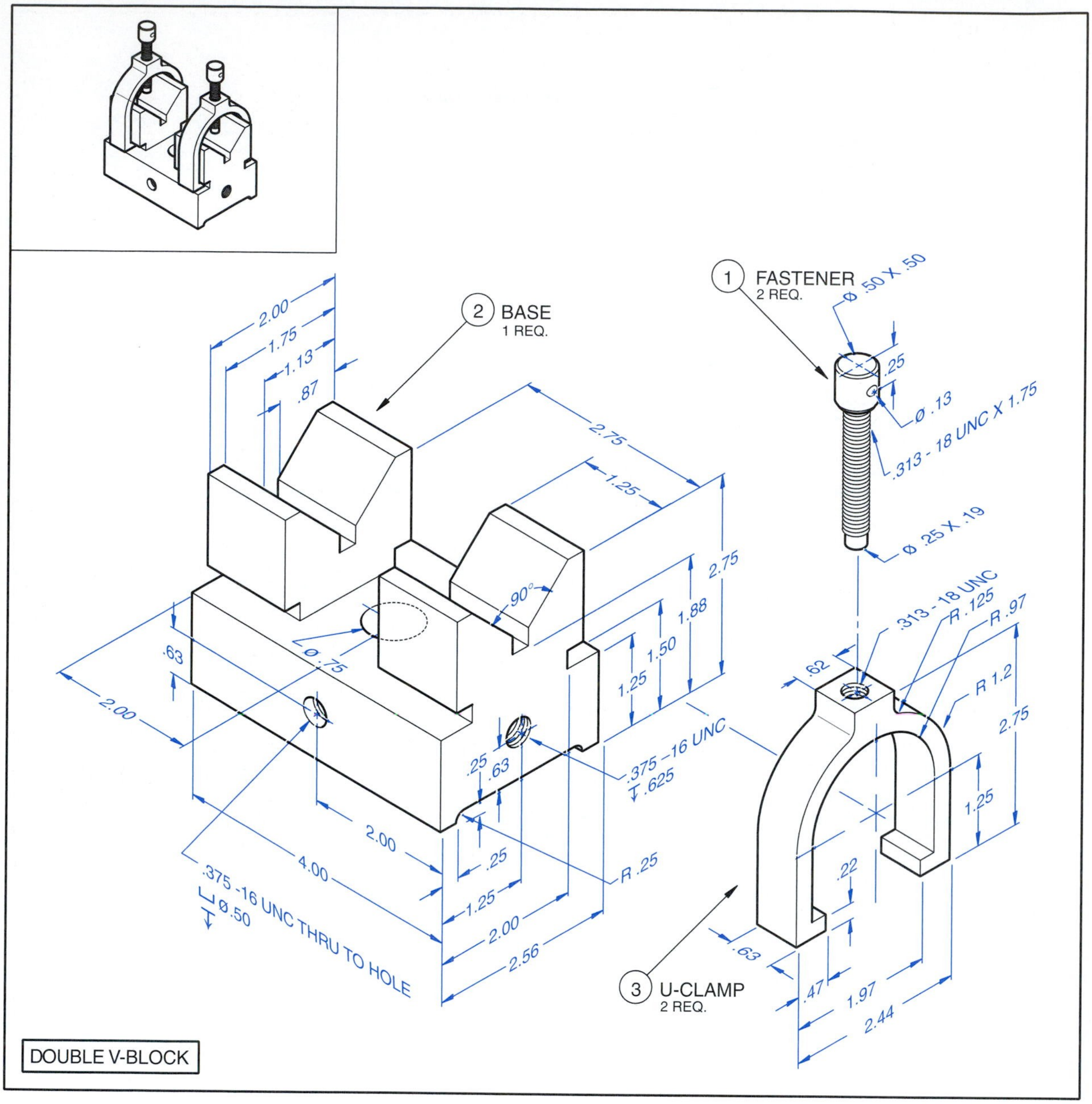

Figure 5.18 Double V-Block

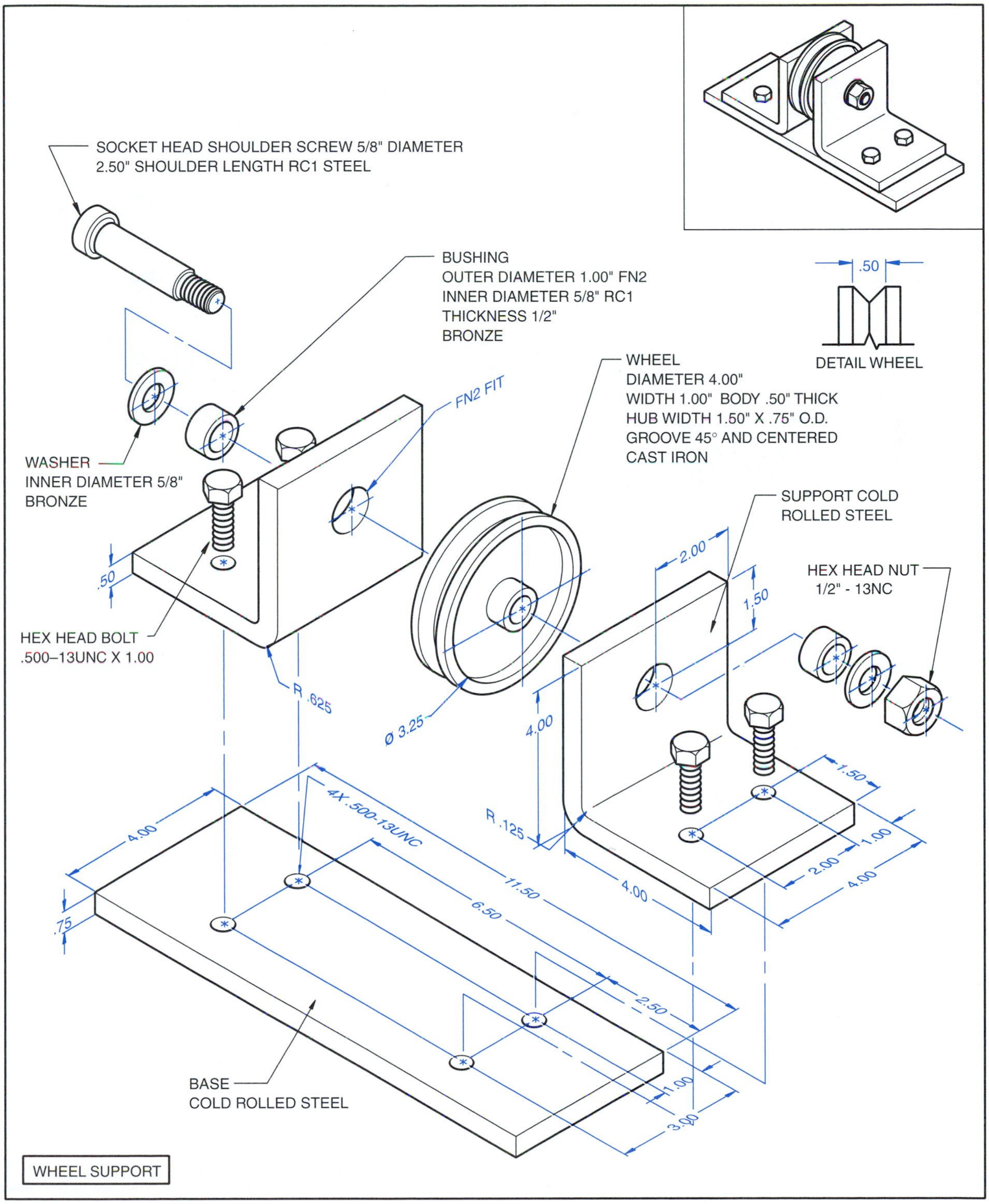
SOCKET HEAD SHOULDER SCREW 5/8" DIAMETER
2.50" SHOULDER LENGTH RC1 STEEL
BUSHING
OUTER DIAMETER 1.00" FN2
INNER DIAMETER 5/8" RC1
THICKNESS 1/2"
BRONZE
.50
DETAIL WHEEL
WHEEL
DIAMETER 4.00"
WIDTH 1.00" BODY .50" THICK
HUB WIDTH 1.50" X .75" O.D.
GROOVE 45° AND CENTERED
CAST IRON
FN2 FIT
WASHER
INNER DIAMETER 5/8"
BRONZE
SUPPORT COLD
ROLLED STEEL
HEX HEAD NUT
1/2" - 13NC
HEX HEAD BOLT
.500–13UNC X 1.00
R .625
Ø 3.25
4.00
2.00
1.50
R .125
4X .500-13UNC
11.50
6.50
2.50
1.00
3.00
.75
BASE
COLD ROLLED STEEL
WHEEL SUPPORT

Figure 5.19 Wheel Support

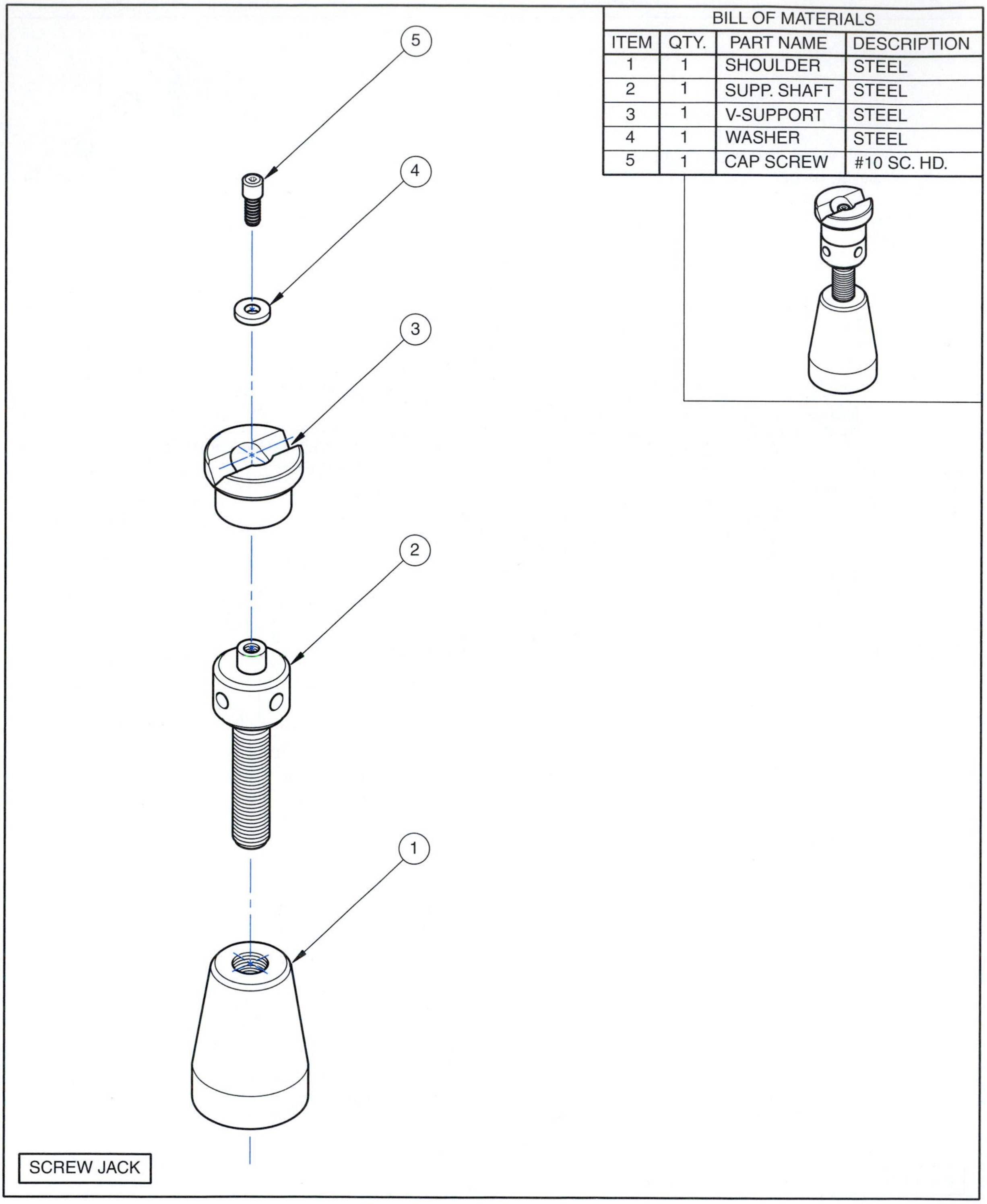

BILL OF MATERIALS			
ITEM	QTY.	PART NAME	DESCRIPTION
1	1	SHOULDER	STEEL
2	1	SUPP. SHAFT	STEEL
3	1	V-SUPPORT	STEEL
4	1	WASHER	STEEL
5	1	CAP SCREW	#10 SC. HD.

Figure 5.20 Screw Jack

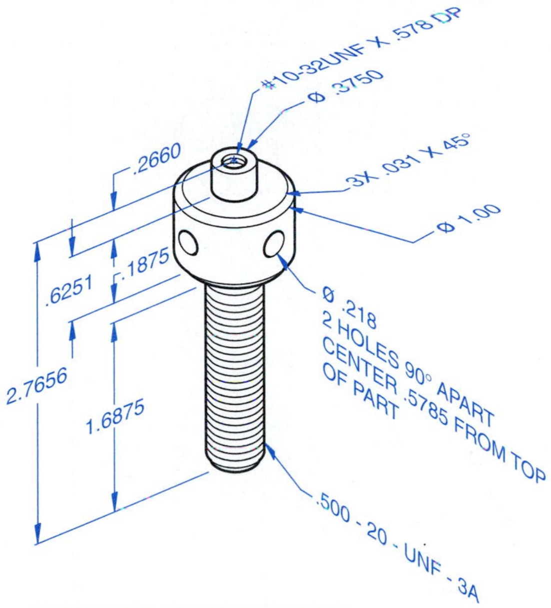

PART NAME: SUPPORT SHAFT

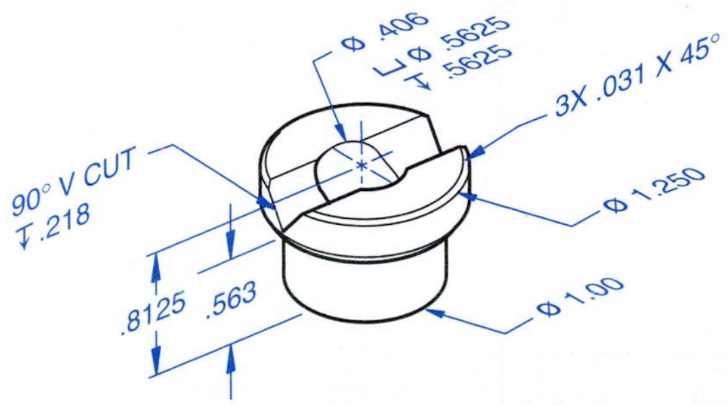

PART NAME: V-SUPPORT

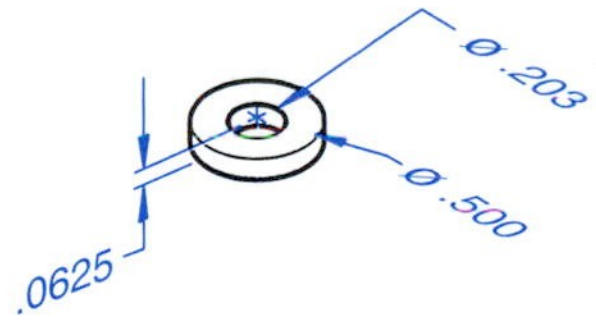

PART NAME: WASHER

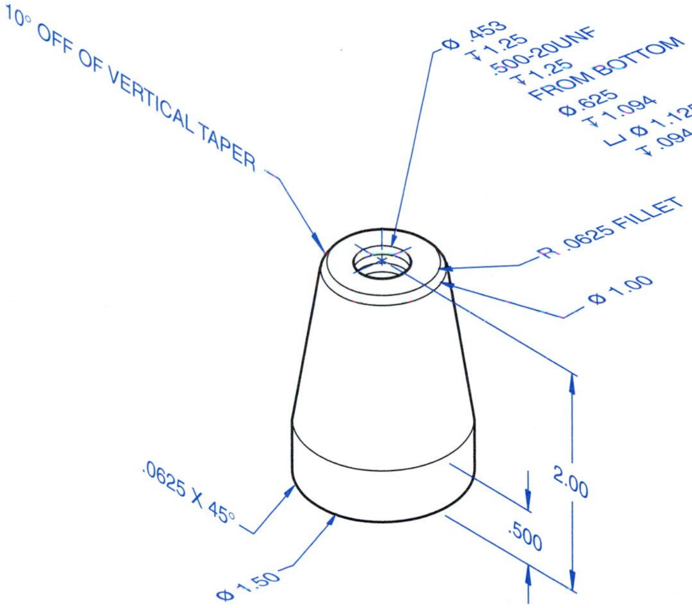

PART NAME: SHOULDER

Figure 5.20 Continued

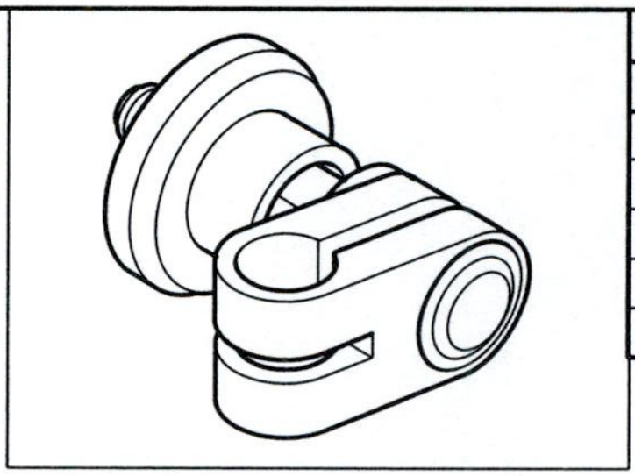

BILL OF MATERIALS			
ITEM	QTY.	PART NAME	DESCRIPTION
1	1	SHOULDER	STEEL
2	1	WASHER	STEEL
3	1	EXTENSION	STEEL
4	1	COUPLER	STEEL
5	1	SHAFT	STEEL

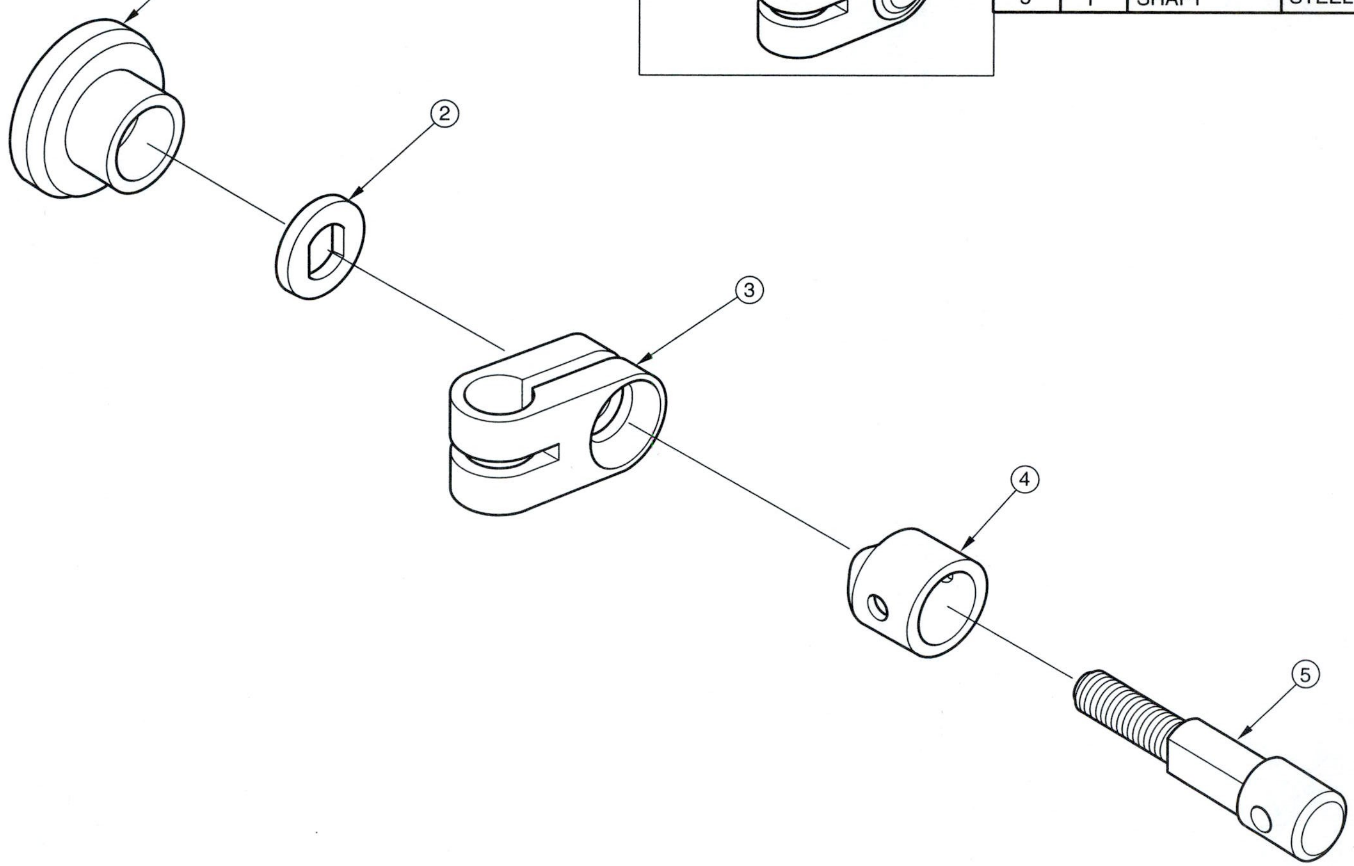

ADJUSTING ARM

Figure 5.21 Adjusting Arm

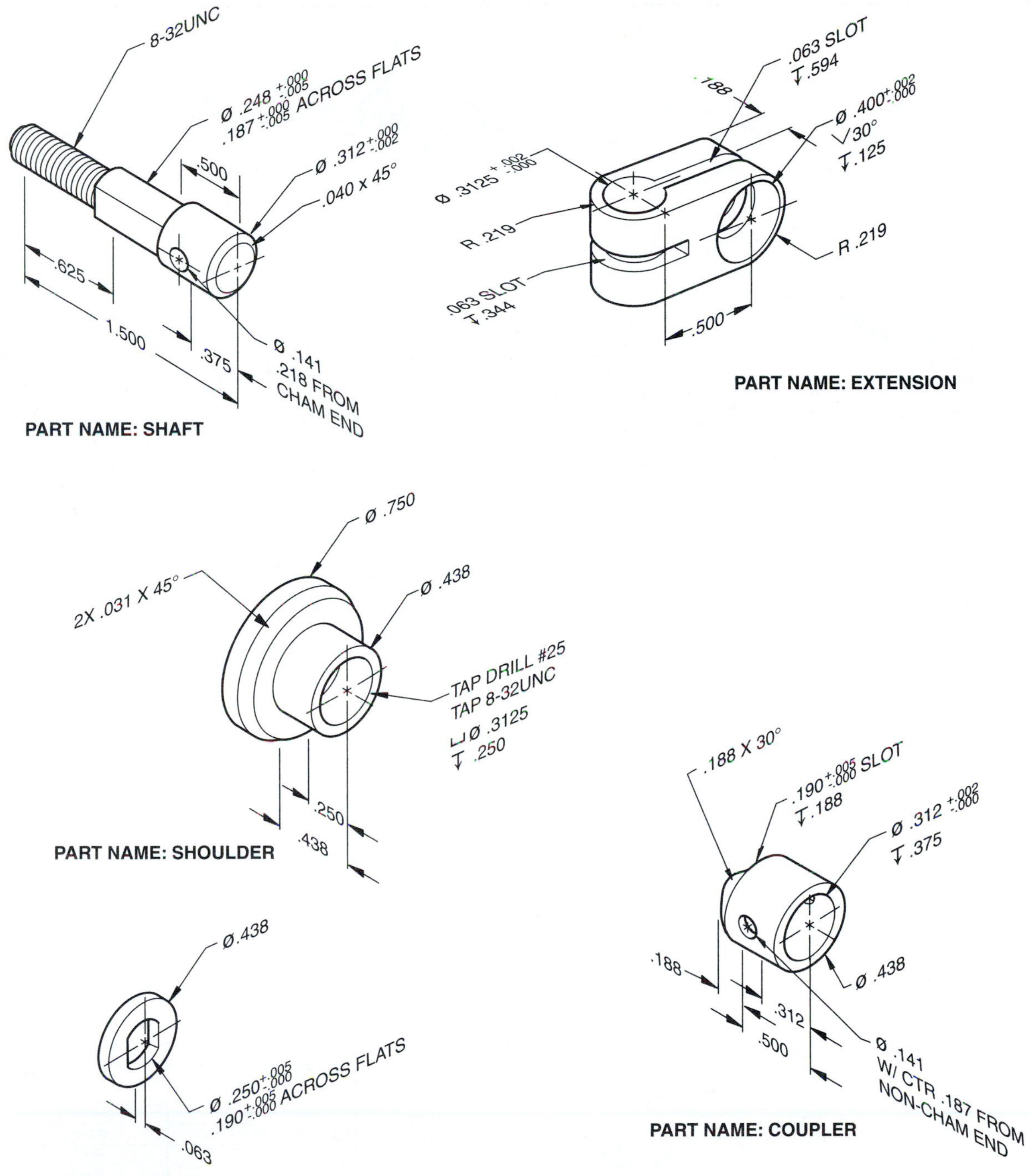

Figure 5.21 Continued

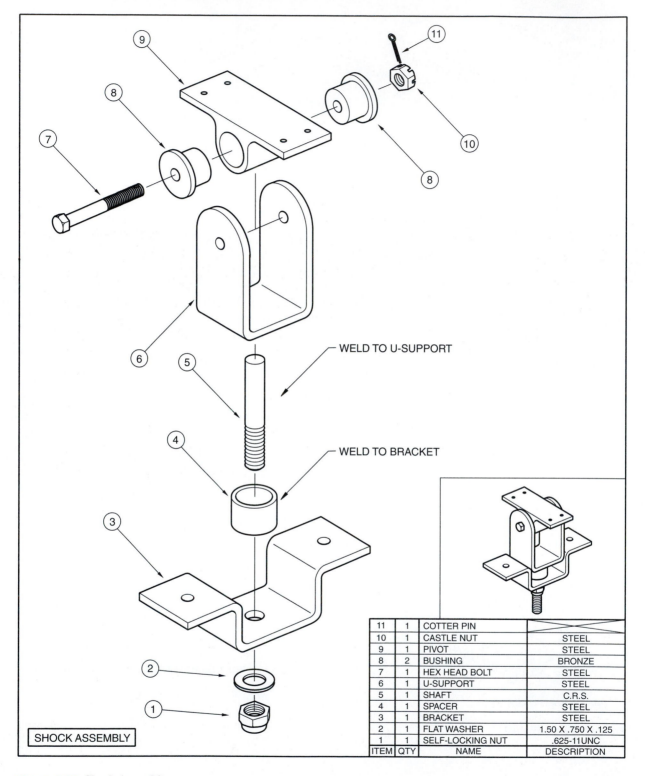

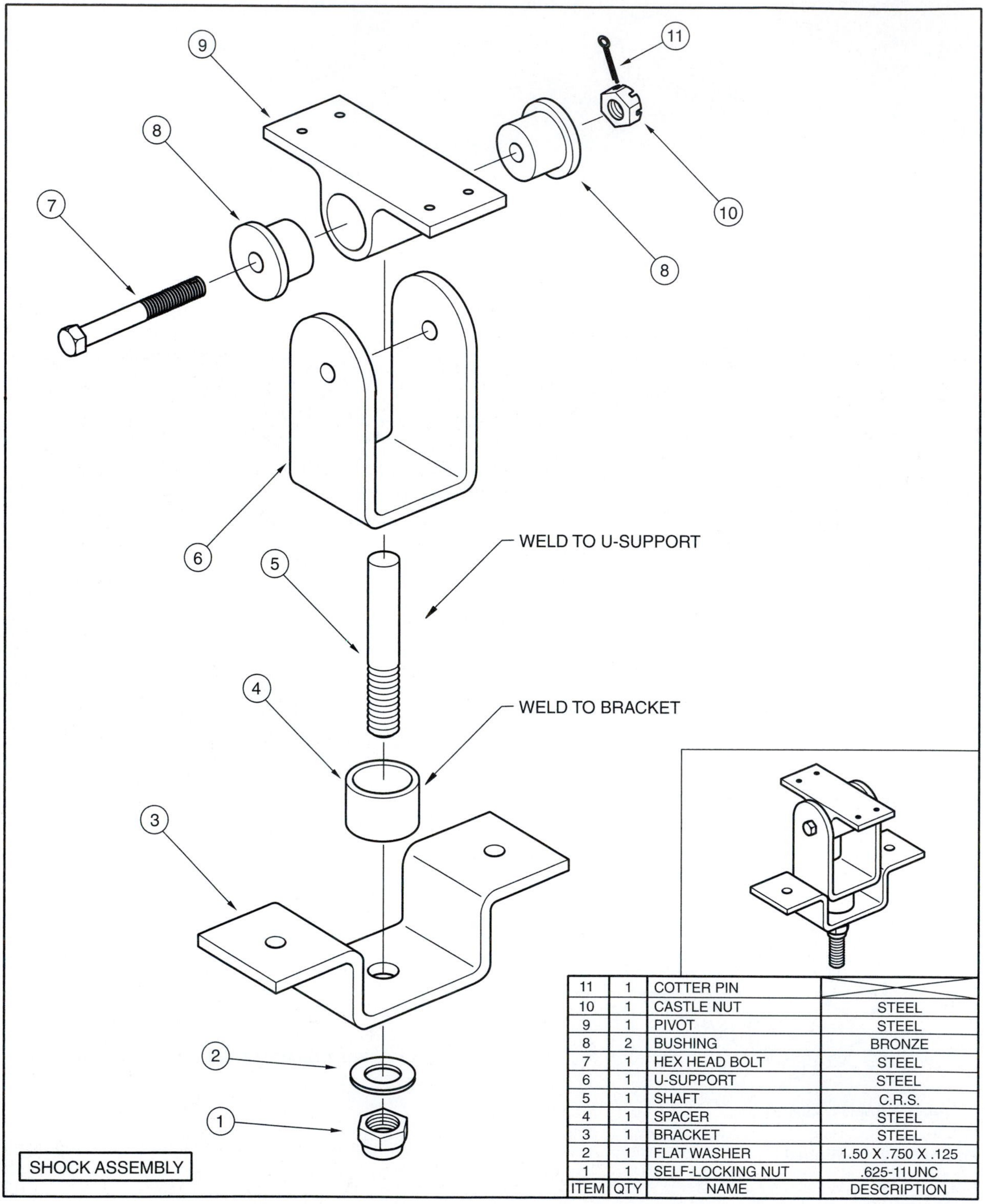

ITEM	QTY	NAME	DESCRIPTION
11	1	COTTER PIN	
10	1	CASTLE NUT	STEEL
9	1	PIVOT	STEEL
8	2	BUSHING	BRONZE
7	1	HEX HEAD BOLT	STEEL
6	1	U-SUPPORT	STEEL
5	1	SHAFT	C.R.S.
4	1	SPACER	STEEL
3	1	BRACKET	STEEL
2	1	FLAT WASHER	1.50 X .750 X .125
1	1	SELF-LOCKING NUT	.625-11UNC

Figure 5.22 Shock Assembly

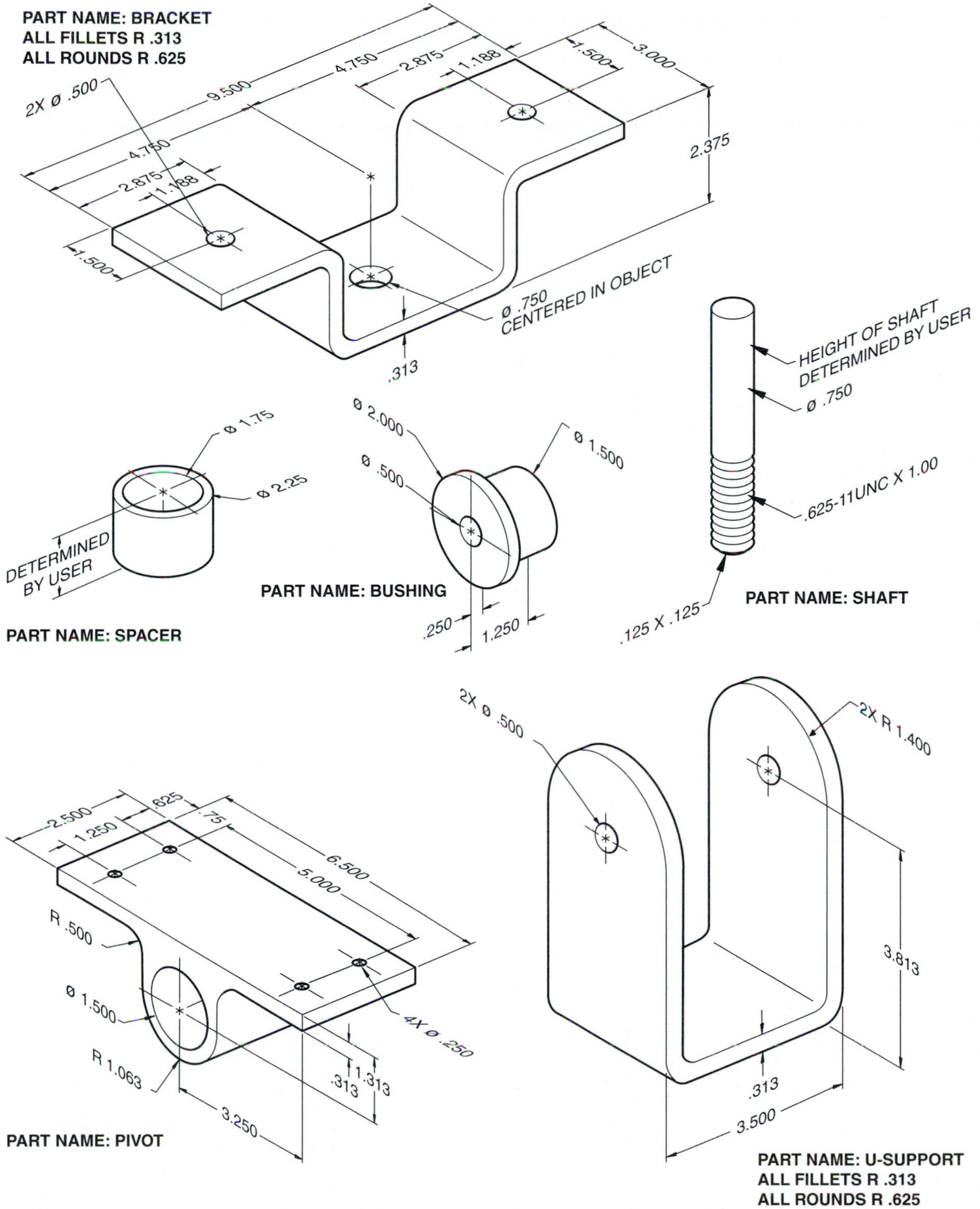
PART NAME: BRACKET
ALL FILLETS R .313
ALL ROUNDS R .625
2X ø .500
9.500
4.750
2.875
1.188
1.500
3.000
2.375
ø .750
CENTERED IN OBJECT
.313
PART NAME: SPACER
ø 1.75
ø 2.25
DETERMINED BY USER
PART NAME: BUSHING
ø 2.000
ø .500
ø 1.500
.250
1.250
PART NAME: SHAFT
HEIGHT OF SHAFT DETERMINED BY USER
ø .750
.625-11UNC X 1.00
.125 X .125
PART NAME: PIVOT
2.500
1.250
.625
.75
6.500
5.000
R .500
ø 1.500
R 1.063
4X ø .250
.313
1.313
3.250
PART NAME: U-SUPPORT
ALL FILLETS R .313
ALL ROUNDS R .625
2X ø .500
2X R 1.400
3.813
.313
3.500

Figure 5.22 Continued

Problems Refer to Figures 5.16 through 5.22

Sketch the necessary views, then add dimensions.

Sketch Number: ____________________

Name: ____________________

Div/Sec: ____________________

Date: ____________________

Problems Refer to Figures 5.16 through 5.22

Sketch the necessary views, then add dimensions.

Sketch Number: ____________________

Name: ____________________

Div/Sec: ____________________

Date: ____________________

Problems Refer to Figures 5.16 through 5.22

Sketch the necessary views, then add dimensions.

Sketch Number: ______________

Name: ______________

Div/Sec: ______________

Date: ______________

Problems Refer to Figures 5.16 through 5.22

Sketch the necessary views, then add dimensions.

Sketch Number: ____________

Name: ____________

Div/Sec: ____________

Date: ____________

Problems Refer to Figures 5.16 through 5.22

Sketch the necessary views, then add dimensions.

Sketch Number: ______________

Name: ______________

Div/Sec: ______________

Date: ______________

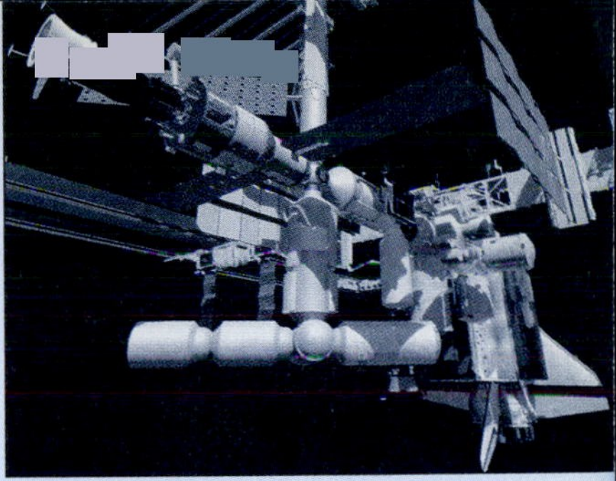

Chapter

6

Design and 3-D Modeling

OBJECTIVES

After completing this chapter, you will be able to:

1. Describe the engineering design process and the role graphics plays.
2. Describe concurrent engineering and design for manufacturability (DFM).
3. Describe the types of modeling systems used in 3-D modeling.
4. Explain the role 3-D modeling plays in the engineering design process.
5. List and describe the modeling techniques used in design.
6. Describe the rapid prototyping process.
7. List and describe the analysis techniques used in design.
8. Explain how graphing and visualization can be used in the design process.
9. Describe the total quality management (TQM) process.

6.1 ENGINEERING DESIGN

Design is the process of conceiving or inventing ideas mentally and communicating those ideas to others in a form that is easily understood. Most often the communications tool is graphics.

Engineering design is one of the processes normally associated with the entire business or enterprise, from receipt of the order or product idea, to maintenance of the product, and all stages in between (Figure 6.1). The design process requires input from such areas as customer needs, materials, capital, energy, time requirements, and human knowledge and skills.

6.1.1 Traditional Engineering Design

Traditional engineering design is a linear approach divided into a number of steps. For example, a six-step process

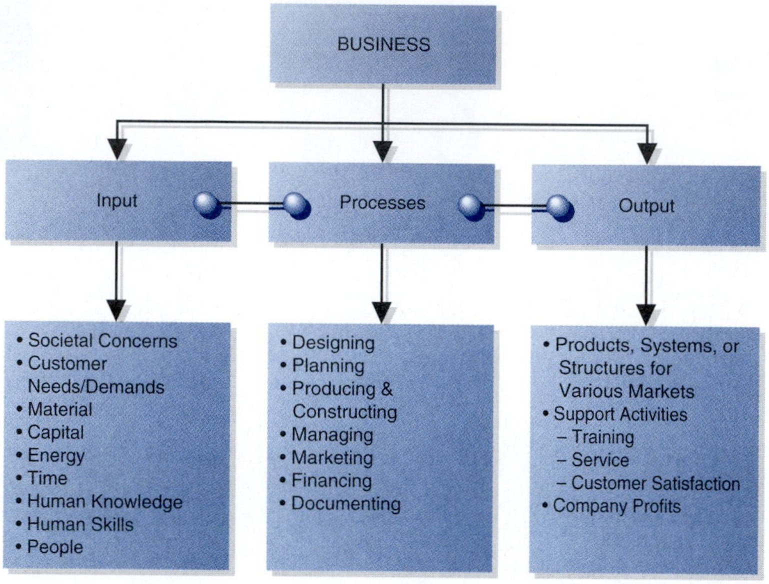

Figure 6.1 The Business Process

A manufacturing business or enterprise includes all the inputs, processes, and outputs necessary to produce a product or construct a structure. Designing is one of the major processes in such a business.

might be divided into problem identification, preliminary ideas, refinement, analysis, documentation, and implementation. The design process moves through each step in a sequential manner; however, if problems are encountered, the process may return to a previous step. This repetitive action is called **iteration** or looping. Many industries use the traditional engineering design process; however, a new process is developing that combines some features of the traditional process with a team approach that involves all segments of a business.

6.1.2 Concurrent Engineering Design

The **production process** executes the final results of the design process to produce a product or system. In the past, the creative design process was separated from the production process. With the advent of computer modeling, this separation is no longer necessary, and the modern engineering design approach brings both processes together.

Concurrent engineering is a nonlinear team approach to design that brings together the input, processes, and output elements necessary to produce a product. The people and processes are brought together at the very beginning, which is not normally done in the linear approach. The team consists of design and production engineers, technicians, marketing and finance personnel, planners, and managers, who work together to solve a problem and produce a product. Many companies are finding that concurrent engineering practices result in a better, higher-quality product, more satisfied customers, fewer manufacturing problems, and a shorter cycle time between design initiation and final production.

Figures 6.2 and 6.3 represent the concurrent approach to engineering design, based on 3-D modeling. The three intersecting circles represent the concurrent nature of this design approach. For example, in the ideation phase, design engineers interact with service technicians to ensure that the product will be easily serviceable by the consumer or technician. This type of interaction results in a better product for the consumer. The three intersecting circles also represent the three activities that are a major part of the concurrent engineering design process: ideation, refinement, and implementation. These three activities are further divided into smaller segments, as shown by the items surrounding the three circles.

The center area in Figure 6.3 represents the 3-D computer model and reflects the central importance of 3-D modeling and graphics knowledge in engineering design and production. With the use of a modeling approach, everyone on the team can have access to the current design through a computer terminal. This data sharing is critically important to the success of the design process.

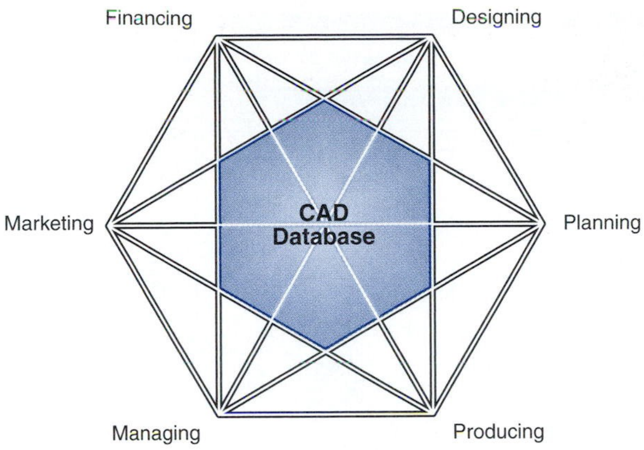

Figure 6.2 Sharing the CAD Database

The concurrent engineering model shows how every area in an enterprise is related, and the CAD database is the common thread of information connecting the areas.

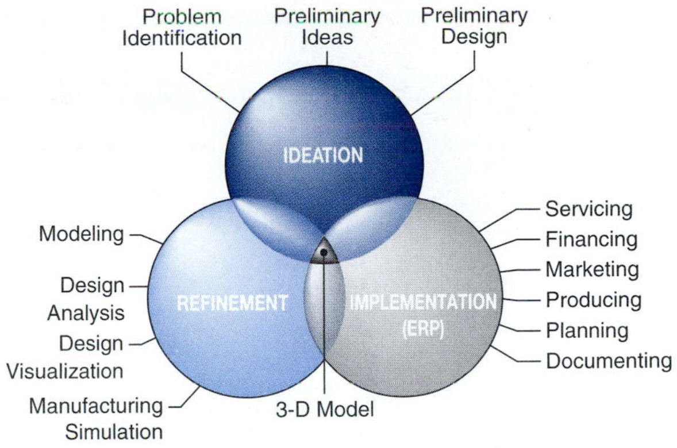

Figure 6.3 Concurrent Engineering Design

The engineering design process consists of three overlapping areas: ideation, refinement, and implementation, which all share the same 3-D CAD database.

Through the sharing of information, often in the form of a database, it is possible for all areas of the enterprise to work simultaneously on their particular needs as the product is being developed. For example, a preliminary 3-D model could be created by the design engineers early in the ideation phase. A mechanical engineer could use the same 3-D model to analyze its thermal properties. The information gained from this preliminary analysis could then be given to the design engineers, who could then make any necessary changes early in the ideation phase, minimizing costly changes later in the design process.

6.2 3-D MODELING

Traditionally, the means of communication in the engineering design process was through paper drawings done by hand. With the increasing availability of CAD tools, these 2-D technical drawings are being produced on computer. More recently, 3-D modeling software has become available on increasingly powerful PCs and increasingly inexpensive engineering workstations (Figure 6.4). Because 3-D modeling systems create models of the product being designed, such a system offers considerably more possibilities as to how it can be integrated into the design process than a 2-D CAD drawing does.

The following section offers an overview of the most common approaches for generating 3-D computer models. Because working with a 3-D model is so much different than creating a 2-D drawing, the ways in which these models are constructed and viewed are also discussed. As the various stages of the design process are presented, the ways in which 3-D modeling can play an integral role will be outlined.

6.2.1 Wireframe Modeling

The simplest 3-D modeler is a **wireframe modeler.** In this type of modeler, which is a natural outgrowth of 2-D CAD, two types of elements must be defined in the database: *edges* and *vertices* (Figure 6.5). For the tetrahedron in the figure, the vertex list contains the geometric information on the model. Each vertex is defined by an (X, Y, Z) coordinate, which anchors the model in space. The topology of the model is represented by the edge list. The edge list does not contain coordinate information. The location, orientation, and length of an edge must be derived indirectly, through calculations of the vertices at either end of the edge. For example, edge E1 consists of vertices V1 and V2. The coordinate locations of V1 (0,0,0) and V2 (1,0,0) indicate that E1 has a length of 1 and is oriented along the X axis.

6.2.2 Surface Modeling

Surface models define the surface features, as well as the edges, of objects.

Different types of spline curves are used to create surface patches with different modeling characteristics. For example, the advantage of Bezier surface patches is that they are easy-to-sculpt natural surfaces (Figure 6.6). The control points are an intuitive tool with which the user can work.

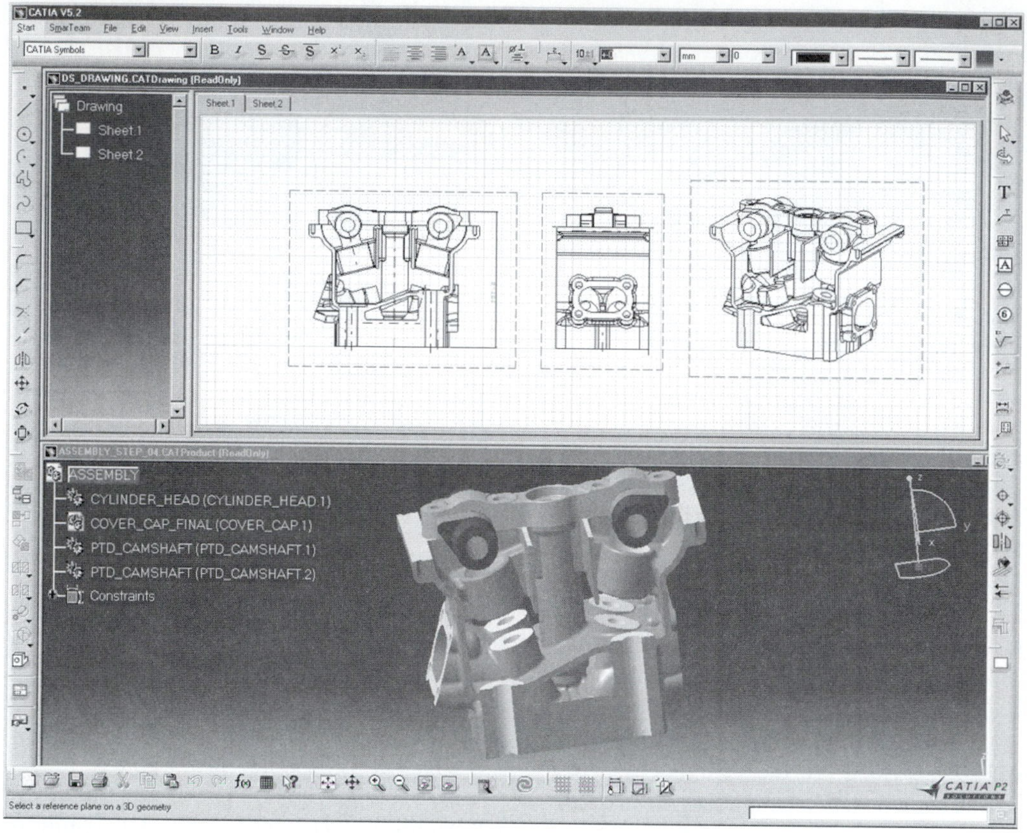

Figure 6.4 Production Drawing

PC workstations are popular for particularly demanding 3-D modeling work. Courtesy of Dassault Systems.

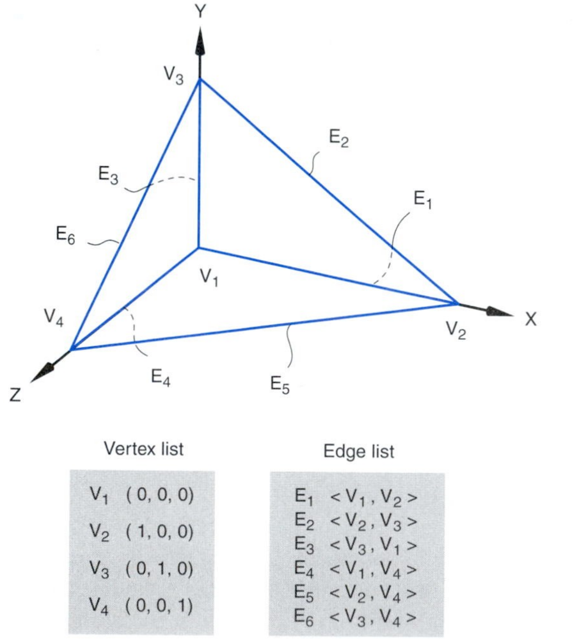
Y
X
Z
V_1
V_2
V_3
V_4
E_1
E_2
E_3
E_4
E_5
E_6
Vertex list
V_1 (0, 0, 0)
V_2 (1, 0, 0)
V_3 (0, 1, 0)
V_4 (0, 0, 1)
Edge list
E_1 < V_1, V_2 >
E_2 < V_2, V_3 >
E_3 < V_3, V_1 >
E_4 < V_1, V_4 >
E_5 < V_2, V_4 >
E_6 < V_3, V_4 >

Figure 6.5 The Vertex and Edge List of a Wireframe Model

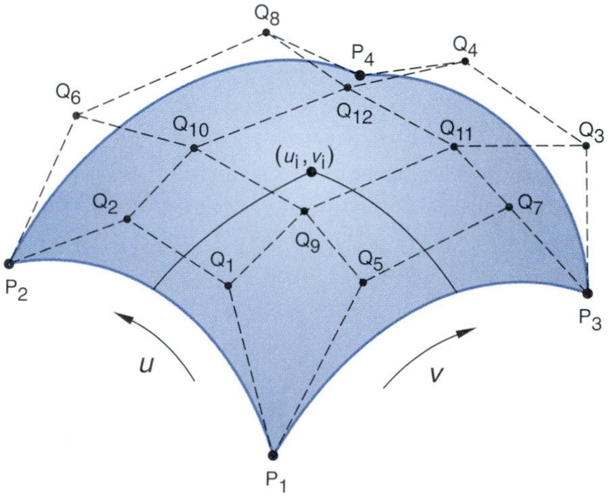

Figure 6.6 A Bezier Bicubic Surface Patch

The patch consists of four connected Bezier curves and 16 control points.

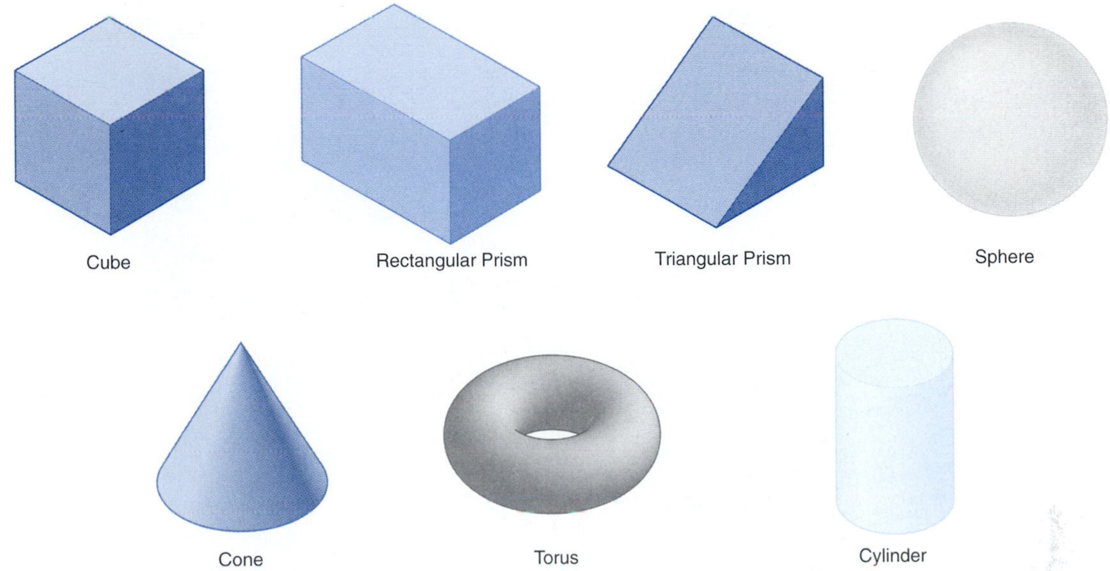

Figure 6.7 **Common Geometric Primitives**

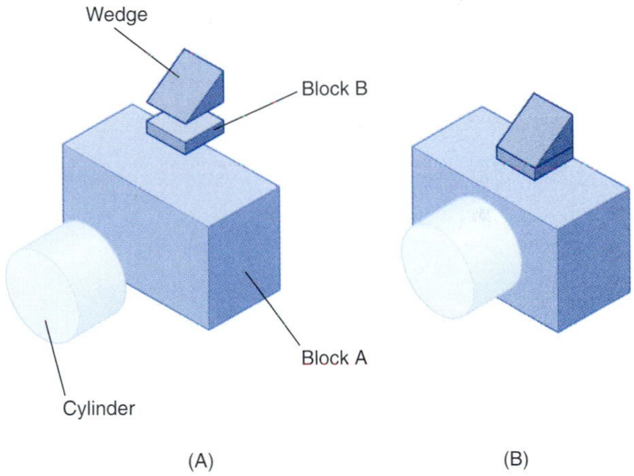

Figure 6.8 **A Camera Described with Geometric Primitives**
Additive modeling with geometric primitives allows a variety of objects to be represented.

In contrast, B-spline patches allow local control; moving one control point does not affect the whole surface. With B-splines, it is much easier to create surfaces through predefined points or curves. NURBS surfaces use *rational* B-splines, which include a *weighting* value at each point on the surface. The weighting value allows some points to have more influence over the shape of the curve than other points. This means that a wider variety of curved surfaces are possible than with regular B-splines. Because NURBS surfaces can also precisely describe conic surfaces, they are gaining popularity in many tasks previously handled by other types of 3-D modelers.

6.2.3 Solid Modeling

Through the use of parametric description, surface modelers can accurately describe the surface of an object. Often, however, information about the inside of an object, that is, its solidity, is also required. **Solid models** include volumetric information, that is, what is on the inside of the 3-D model, as well as information about the surface of an object. In this case, the surface of the model represents the boundary between the inside and outside of the object.

Primitive Modeling Many objects, including most mechanical parts, can be described mathematically using basic geometric forms. Modelers are designed to support a set of *geometric primitives,* such as cubes, right rectilinear prisms (i.e., blocks), right triangular prisms (i.e., wedges), spheres, cones, tori, and cylinders (Figure 6.7). Although most geometric primitives have unique topologies, some differ only in their geometry, like the cube and the right rectilinear prism.

Many modelers also allow primitives to be joined together to create more complex objects. The user mentally decomposes the object into a collection of geometric primitives and then constructs the model from these elements (Figure 6.8).

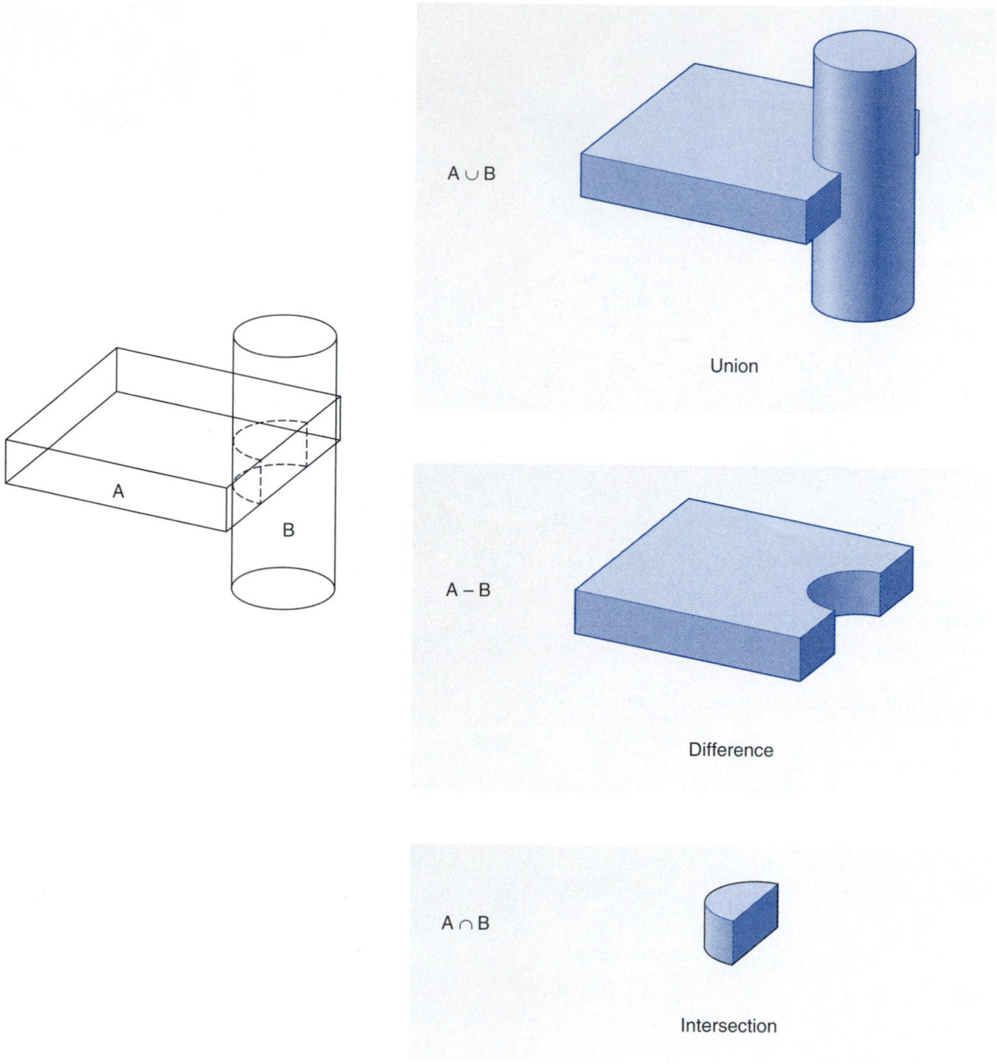

Figure 6.9 The Three Boolean Operations: Union, Difference, and Intersection
The three operations, using the same primitives in the same locations, create very different objects.

There are a number of positive aspects to using a limited set of primitives. First, limiting the allowed topologies reduces the risk of an *invalid* model. Typically, the only safeguards needed are limits or rules on parameter input. For example, the diameter of a cylinder could not be specified as zero. Second, the limited number of allowable shapes means a very concise and efficient database. Third, the limited range of primitives means a model that is unique and is easy to validate.

Constructive Solid Geometry (CSG) Modeling **Constructive solid geometry (CSG)** modeling is a powerful technique that allows flexibility in both the way primitives are defined and the way they are combined. The relationships between the primitives are defined with **Boolean operations.** There are three types of Boolean operations: **union** (∪), **difference** (–), and **intersection** (∩). Figure 6.9 shows how each of these operations can be used to create different forms. The critical area is the place where two objects overlap. This is where the differences between the Boolean operations are evident. The union operation is essentially additive, with the two primitives being combined. However, in the final form, the volume where the two primitives overlap is only represented once. Otherwise there would be twice as much material in the area of overlap, which is not possible in a real object. With a difference operation, the area of overlap is not represented at all. The final form resembles one of the original primitives with the area of overlap removed. With the intersection operation, *only* the area of overlap remains; the rest of the primitive volume is removed.

Boundary Representation (B-Rep) Modeling **Boundary representation (B-rep) modeling** and CSG modeling are the two most popular forms of solid modeling. With CSG modeling, surfaces are represented indirectly through half-spaces; with B-rep modeling, the surfaces, or *faces,* are themselves the basis for defining the solid. The face of a B-rep model is fundamentally different from a face in a wireframe modeler. Although both include a set of edges connected at vertices, a B-rep face explicitly represents an oriented surface. There are two sides to this surface: one is on the *inside* of the object (the solid side), and the other is on the *outside* of the object (the void side).

6.3 | CONSTRAINT-BASED MODELING

In a traditional 3-D modeling system, defining the size or shape of a geometric feature is typically done independently of any other feature. It is completely up to the user of the CAD system to make sure the resulting model reflects the real-world design constraints the final product will have to face. With a simple part consisting of three overall dimensions and one or two features (e.g., holes or slots), making sure these features are properly placed as the part is constructed is not a major task. On the other hand, with complex parts or large assemblies, the total number of features to track can be a major task. Traditionally with CAD systems, it has been up to the operator to ensure that when one part is modified, related features on other parts are also updated. **Constraint-based modeling** is a technique that can help the CAD operator manage the model modification process.

Although the use of solid modeling grew steadily during the 1980s, many of the productivity gains promised by CAD vendors were not being realized by companies. One of the reasons was that the process of creating a solid model was much more abstract than the process of designing real-world products.

In a constraint-based modeler, a part is created by describing the relationship of geometric elements with equations and logical relationships. For example, in a traditional

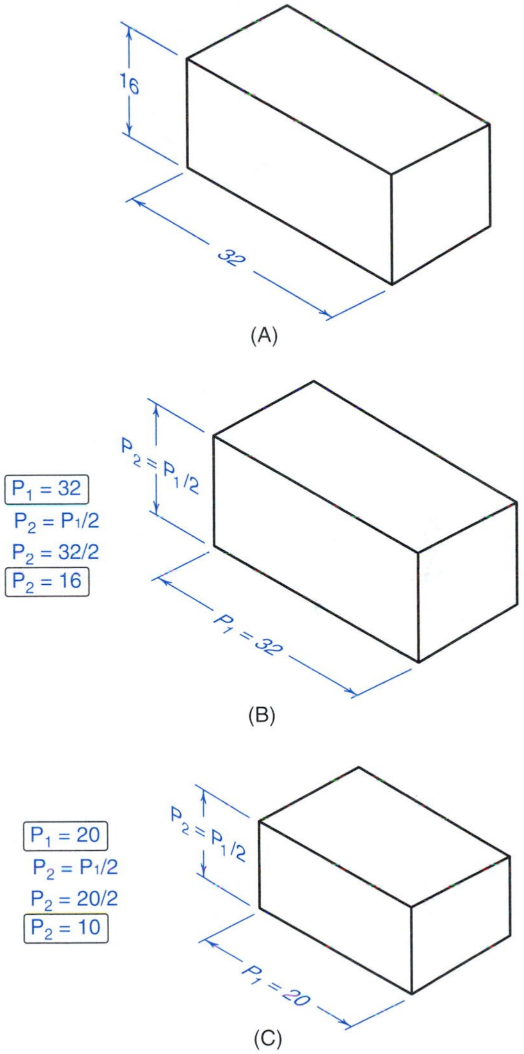

Figure 6.10 **Traditional and Parametric Part Definition**
(A) A traditional CAD modeler creates geometry of a specified size. (B) With a constraint-based modeler, geometry is defined through parameters represented as an equation. (C) Parameters can easily be altered to represent new values.

CAD modeler, if a plate is to have a length equal to 32 mm and a width of half the length, a rectangle 32 mm long and 16 mm wide is created (Figure 6.10A). In a constraint-based modeler, the geometric relationships are coded directly into the model with *parameters:* the length would be defined as $P_1 = 32$ mm and the width defined as $P_2 = P_1/2$ (Figure 6.10B). With the geometry defined with parameters, the equations are solved to derive dimensional values for each of the features. The power of this approach

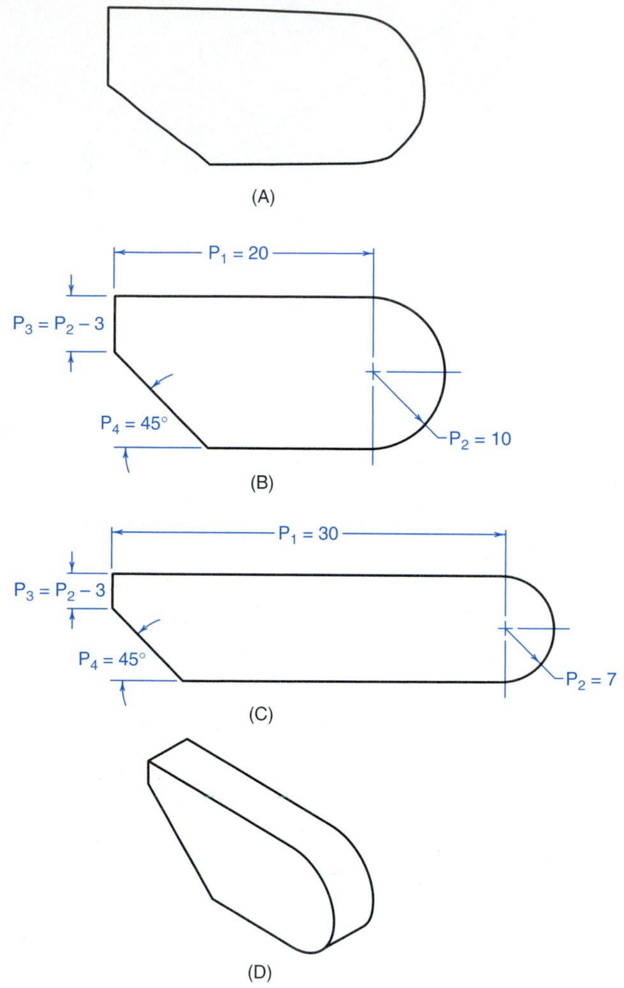

Figure 6.11 Creating a Constrained Model from a Sketch Profile

(A) A rough sketch defines the basic geometry of the 2-D profile. (B) Constraints are added to define the size and shape of the profile. (C) The size and shape of the profile can be altered at any time. (D) Once constrained, the profile can be extruded or swept into a 3-D part.

is seen when the model is modified. Instead of having to individually update all related dimensions, one dimension can be altered, and all dimensions linked through parameters automatically reflect the change. For example, if the length of the plate is reduced ($P_1 = 20$), the width automatically reduces by the appropriate amount to reflect the change in length (Figure 6.10C).

Like dimensions, parameters can be associated with geometric features such as lines, curves, and planes. Unlike dimensions, parameters do not have to represent a single static value. Parameters can be

- Assigned numeric values (e.g., $P_1 = 32$ mm).
- Related to other parameters through equations (e.g., $P_2 = P_1 + 20$).
- Related to other parameters through geometric relationships (e.g., P_3 is parallel to P_4).
- Varied based on logical relationships (e.g., IF $P_5 > 24$ mm, THEN $P_6 = 6$ mm, ELSE $P_6 = 4$ mm).

It is important to remember that parameters are always related back to geometric features. A parameter that represents a numeric value measures a distance or angle between two features (i.e., points or lines). Likewise, geometric relationships such as parallelism are made between features such as lines or planes.

In many constraint-based modelers, the modeling process begins by creating a 2-D *sketch profile* (Figure 6.11A). Unlike in a traditional 2-D or 3-D CAD package, the geometry does not have to be created with a high degree of accuracy; it only needs to represent the basic geometric features. The next step is to *constrain* the model by assigning enough parameters to fully define the size and shape of the 2-D profile (Figure 6.11B). Depending on the modeler, some of these parameters may be automatically assigned. For example, if the two horizontal elements are within a preset tolerance range (say, 5 degrees) of parallel, the two edges are constrained to be parallel. You might think of this as being like a snapping or grid system, but, in fact, it is much more powerful. As the part is resized, the sides will stay parallel no matter how far apart they are. Because all of the features are parametrically defined, one or more of the parametric values can be altered, and all related parameters will automatically update (Figure 6.11C). Once the sketch profile has been constrained and the parameters assigned appropriate values, a 3-D part can be created by an extrusion or sweep operation (Figure 6.11D).

In Figure 6.12A, two plates are created with holes and two different design intents. In the plate on the left, the hole was intended to be placed 10 mm from the left edge, while the plate on the right had the hole placed in the center. When the overall width is set to 20 mm, no difference is seen in the two models, but when the overall width is set to a different value, the difference in design intent is immediately seen (Figure 6.12B). With a traditional modeler, an engineer viewing the model in Figure 6.12A would not be able to tell what the design intent was and, therefore, how the hole should shift if the model was altered.

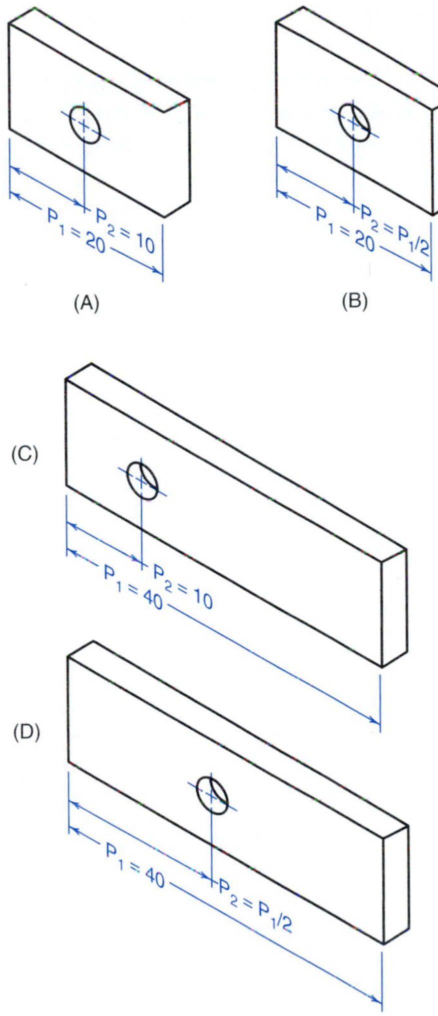

Figure 6.12 The Effect of Design Intent on Model Changes

(A) Though these two parts are initially the same, the parametric constraints represent two different design intents. The part on the left intends to have the hole fixed at an offset of 10 from the left edge whereas the part on the right intends to have the hole centered. (B) The differences in the design intent can be seen when the overall width of the part is altered. Only the hole on the part on the right shifted to the center.

6.4 | FEATURE-BASED MODELING

Another important advancement in 3-D solids modeling is the introduction of **feature-based modeling.** Like constraint-based modeling, feature-based modeling is an attempt to make modeling a more efficient process more in tune with how designers and engineers actually work. Feature-based modeling bundles commands together to automate the

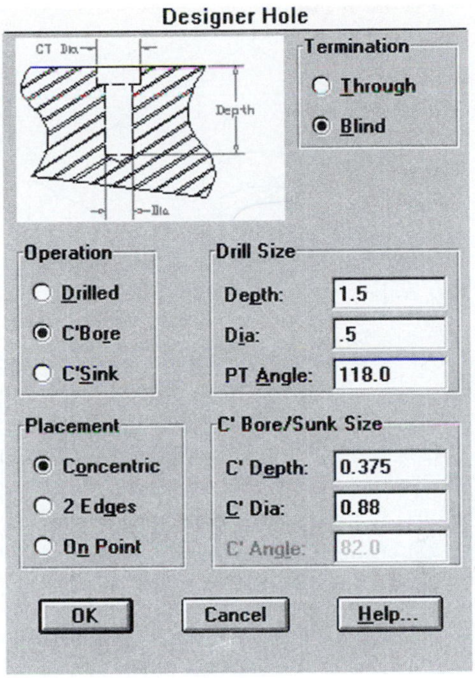

Figure 6.13 Example of a Dialog Box in a Feature-Based Modeler

Notice that all of the essential variables defining the feature are represented as inputs in the dialog box. Typically, many of these variables can also be defined as parameters linked to other features on the model.

process of creating and modifying features that represent common manufacturing operations. Usually implemented in modelers that also have constraint capabilities, feature-based modeling systems use special dialog boxes or other interface elements that allow users to input all of the variables needed to create a common manufactured feature (Figure 6.13).

Examples of manufactured features created through special feature-based dialog boxes include the following:

- Blind holes
- Counterbores and countersinks
- Slots
- Bosses

The hole dialog box shown in Figure 6.13 is a good example of automating the process of creating features in a model. The feature is broken down into its essential variables, with each variable represented by an input in the dialog box. Variables such as the hole's diameter have a value typed in, while the depth can be set to "through" by clicking a button or set to a finite value.

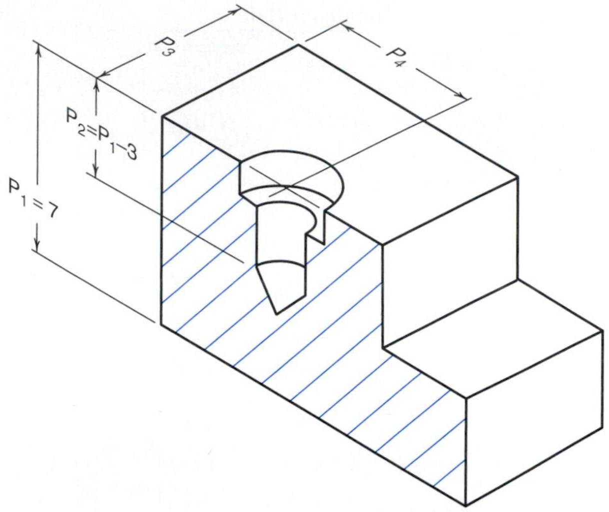

Figure 6.14 Locating a Feature on the Base Part
The blind counterbore is located on the base part relative to a face on the part. The overall depth of the counterbore is linked to the overall height of the part through a parametric equation.

The variables entered through the dialog box largely define the *shape* and *size*. Once these variables of the feature are defined, the *location* is defined. By convention, features usually don't constitute an entire part. For that reason, the feature is typically located somewhere on a *base part.* A feature such as a blind hole is located by indicating its orientation to a face and distance from two edges (Figure 6.14). In a constraint-based modeler, all of the variables of the feature—its shape, size, and location—are parametrically controlled and can be updated at any time. In addition, the parameters defining the feature can also be linked to other parameters defining the part. So, for example, the depth of a hole might be related to the overall thickness of the base part.

Figure 6.14

Earlier in the chapter, Boolean operations were introduced as the primary means of modifying a solids model. With feature-based modeling, one or more Boolean operations are manipulated using intelligent dialog boxes to create common features such as holes, slots, and bosses. Feature- and constraint-based techniques together give the user the ability to easily create and modify common manufactured features. When combined with parts libraries of common fasteners and other parts, assemblies using standard components and manufacturing techniques can be quickly modeled.

6.5 3-D MODELING AND THE DESIGN PROCESS

CAD was introduced into most businesses as an automated drafting tool and was then defined as computer-aided drafting. The introduction of 3-D modeling systems has transformed CAD into computer-aided *design.*

CAD 3-D modeling plays an important role in many newly emerging manufacturing techniques, including computer-aided manufacturing (CAM), computer-integrated manufacturing (CIM), concurrent engineering, and design for manufacturability (DFM). All of these manufacturing techniques are aimed at shortening the design cycle, minimizing material and labor expenditures, raising product quality, and lowering the cost of the final product. Central to these goals is better communications within a company. By sharing the 3-D database of the proposed product, more people can be working simultaneously on various aspects of the design problem. The graphics nature of the database has convinced many that 3-D modeling is a superior method of communicating many of the design intents.

Questions for Review

1. Describe the design process.
2. Define the three types of Boolean operations, and sketch examples of each one. Can you derive the same final object using different Boolean operations and/or primitives?
3. Describe feature-based modeling.
4. Describe constraint-based modeling.

Further Reading

Anand, V. *Computer Graphics and Geometric Modeling for Engineers.* New York: John Wiley and Sons, 1993.

Beakley, G. C., and H. W. Leach. *Introduction to Engineering Design Graphics.* New York: Macmillan, 1973.

Bolluyt, J., M. Stewart, and A. Oladipupo. *Modeling for Design Using SilverScreen.* Boston: PWS-Kent, 1993.

Chase, R. B., and N. J. Aquilano. *Production and Operations Management.* 6th ed. Homewood, IL: Richard D. Irwin, 1992.

3-D Modeling Projects

One of the best ways to learn engineering design graphics principles and to use 3-D modeling tools is to use a real object to simulate the process. When the author first learned CAD, he used this technique using a road bike. There are a number of inexpensive products that could be used for this exercise that can easily be purchased at stores, such as a disposable camera, mechanical pencil, or various tools. For this exercise you will be using a stapler like that shown in Figure 6A. The stapler in the figure is a Swingline Desktop Stapler #81800 that can be purchased at popular retail stores, such as Walmart, Kmart, and Staples. If you wish to avoid having curved surfaces, Swingline Desktop Stapler #74700 can be used. You will be assigned to teams to purchase a stapler, disassemble the item, measure the parts, create the design sketches, make simple design changes, model the parts, create working drawings for manufacturing, perform simple assembly analyses, and create rendered models of the assemblies.

To start on this assignment the stapler must be reverse engineered by disassembling the stapler and measuring all the parts using a caliper and a micrometer. As each part is measured, design sketches are made of each part with dimensions.

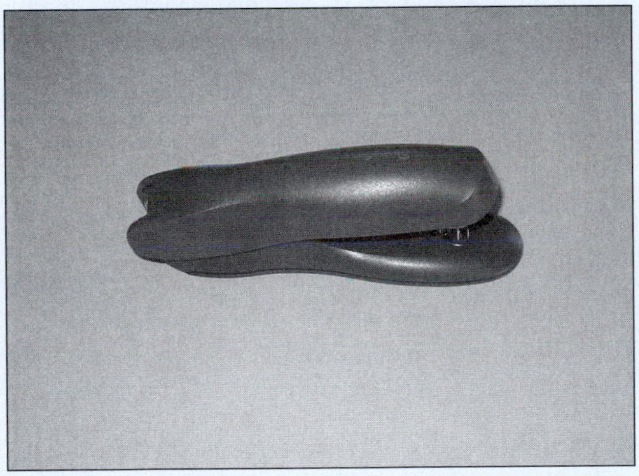

Figure 6A Desktop Stapler for Design Project

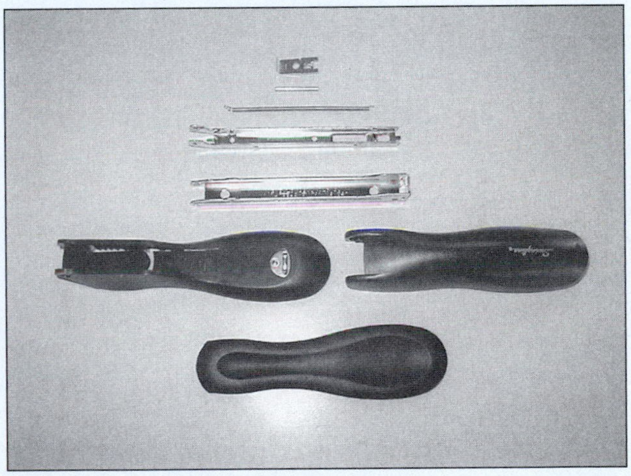

Figure 6B Parts to Be Sketched

Dieter, G. E. *Engineering Design: A Materials and Process Approach.* New York: McGraw-Hill, 1983.

Dreyfuss, H. *The Measure of Man, Human Factors in Design.* New York: Whitney Library of Design, 1967.

Foley, J. D., et al. *Computer Graphics: Principles and Practice.* Reading, MA: Addison-Wesley, 1992.

Guide for Patent Draftsmen (Selected Rules of Practice Relating to Patent Drawings). Washington, DC: U.S. Department of Commerce, Patent and Trademark Office, 1989.

Hunt, V. D. *Quality in America.* Homewood, IL: Business One Irwin, 1992.

LaCourse, D. E., ed. *Handbook of Solid Modeling.* New York: McGraw-Hill, 1995.

McCarthy, E. J., and W. E. Perreault, Jr. *Essentials of Marketing.* 5th ed. Homewood, IL: Richard D. Irwin, 1991.

Mortenson, M. *Geometric Modeling.* New York: John Wiley and Sons, 1985.

Oberg, E., F. D. Jones, H. L. Horton, and H. H. Ryffell. *Machinery's Handbook.* 24th ed. New York: Industrial Press, 1992.

Pillers, M. "Using AME for New Product Development." *CADENCE,* May 1993, pp. 45–48.

Taylor, D. L. *Computer-Aided Design.* Reading, MA: Addison-Wesley, 1992.

Zeid, I. *CAD/CAM Theory and Practice.* New York: McGraw-Hill, 1991.

Problems

Use the gridded sheets provided at the end of this section for the sketching problems.

6.1 In Figure 6.15, there are 12 objects swept, using 12 different profiles. Match the objects with the same profile used to create 3-D objects.

6.2 (Figures 6.16–6.22) Assign different Boolean operations to the seven assembled primitive parts, then do the following:

a. Sketch the resulting composite solids.

b. Use solid modeling software to create the primitives with the given dimensions, then verify your sketches by performing the same Boolean operations on the computer.

6.3 (Figures 6.23–6.24) Using the given information for feature-based modeling, do the following:

a. Using a scale assigned by your instructor, measure the profiles and workpiece. On isometric grid paper, sketch the resulting workpiece after the feature-based modeling is performed.

b. Do the same operations with CAD and compare the results with your sketch.

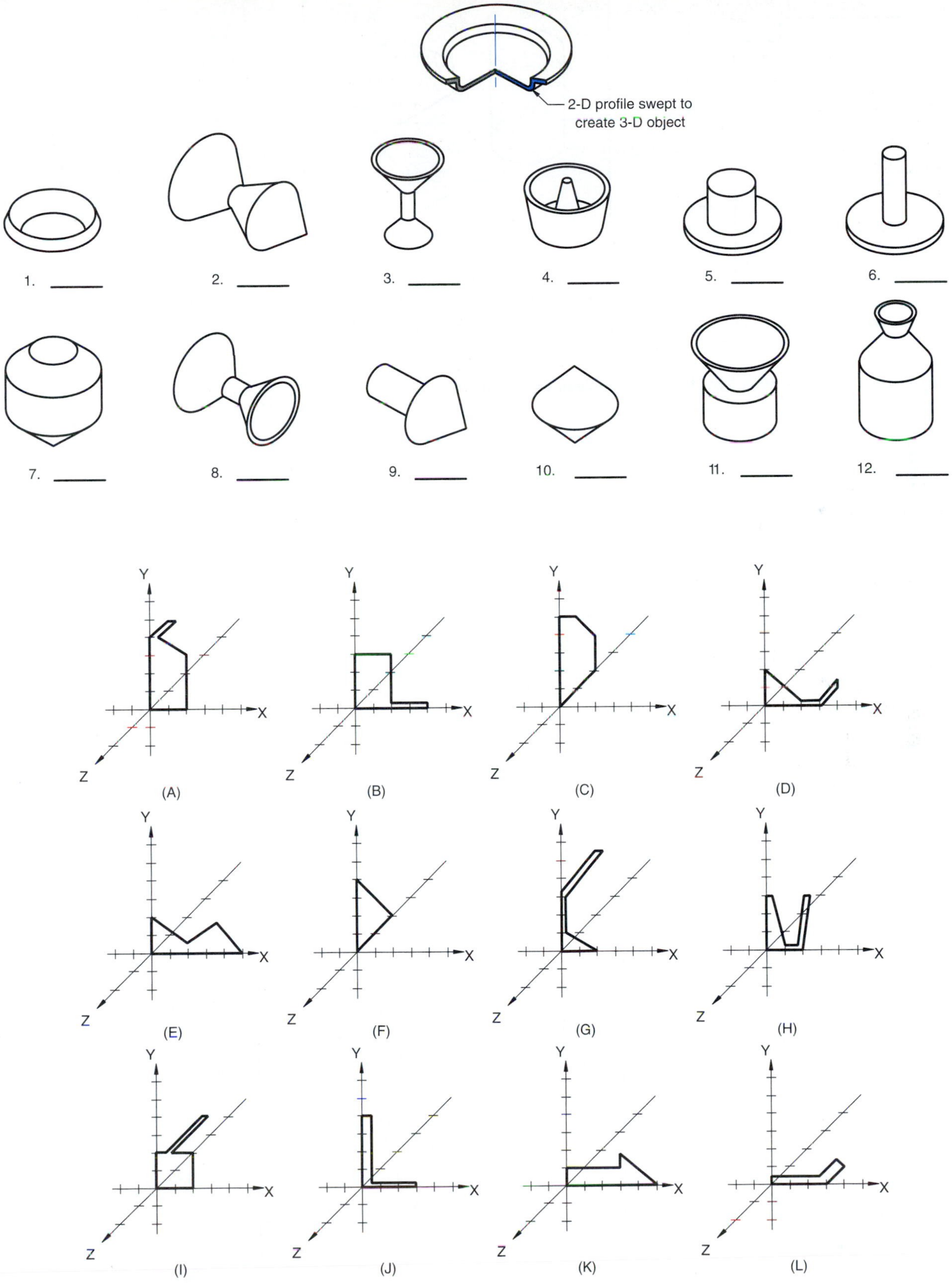

Figure 6.15 **Match 2-D Profiles to 3-D Objects**

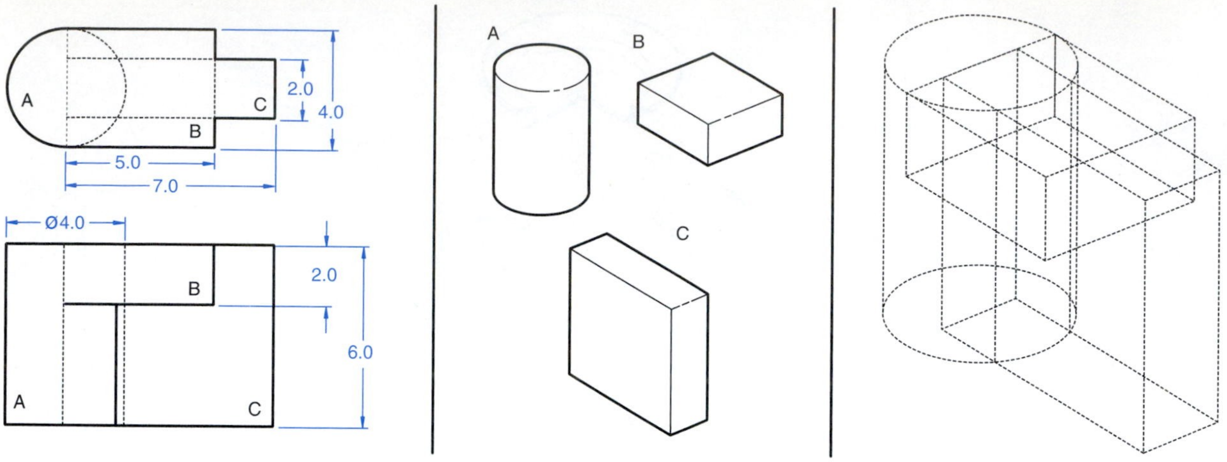

Figure 6.16 Assembled Primitive Parts for Boolean Operations

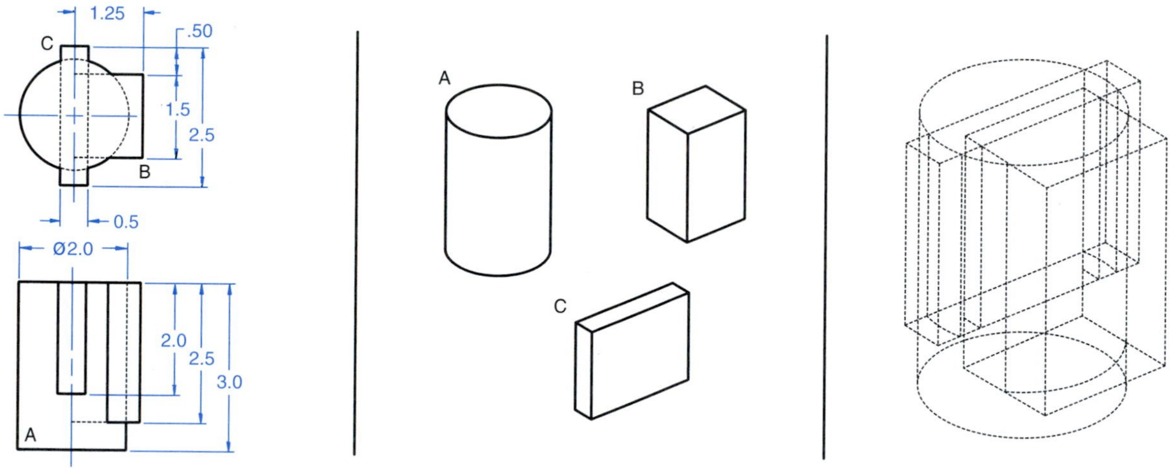

Figure 6.17 Assembled Primitive Parts for Boolean Operations

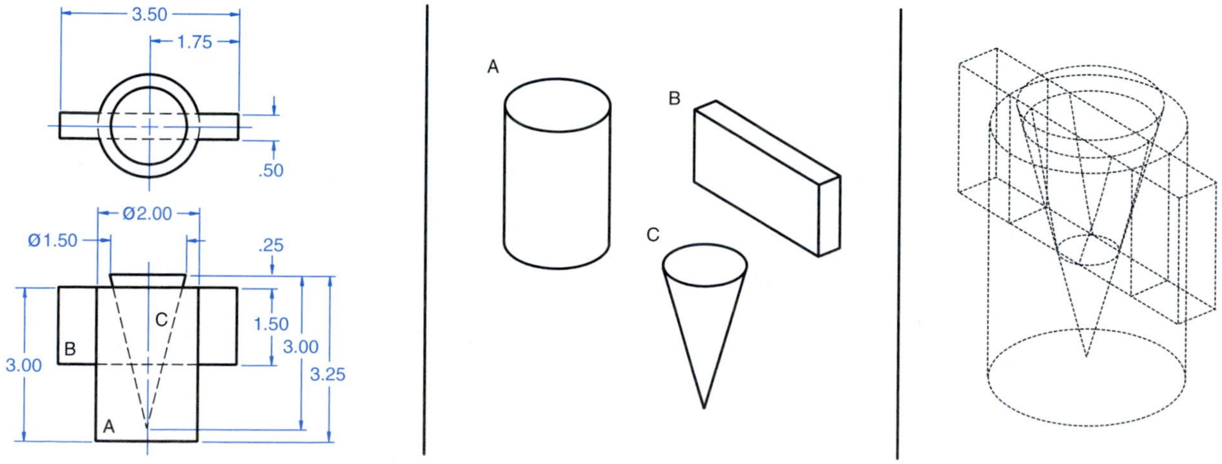

Figure 6.18 Assembled Primitive Parts for Boolean Operations

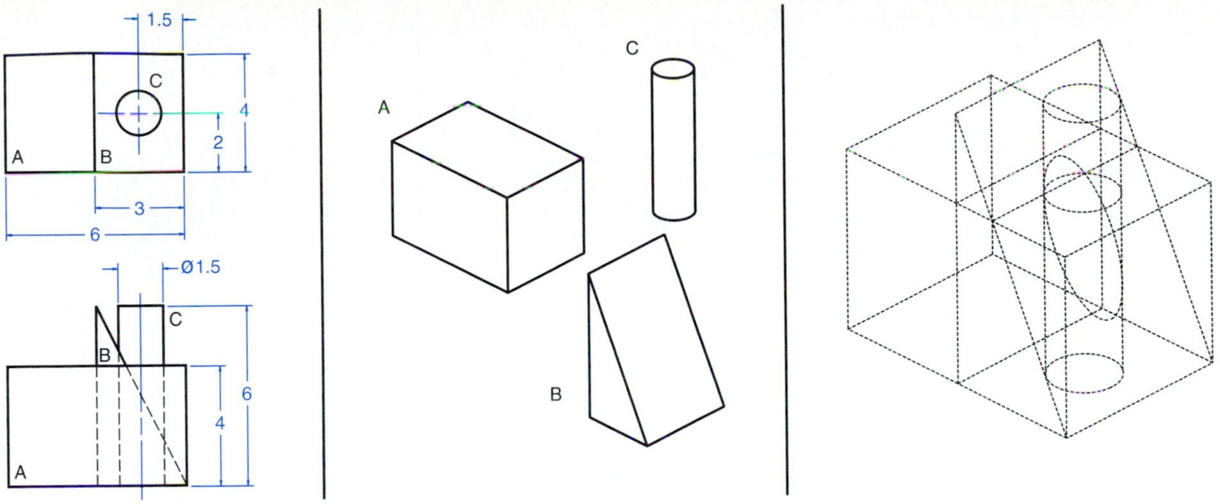

Figure 6.19 Assembled Primitive Parts for Boolean Operations

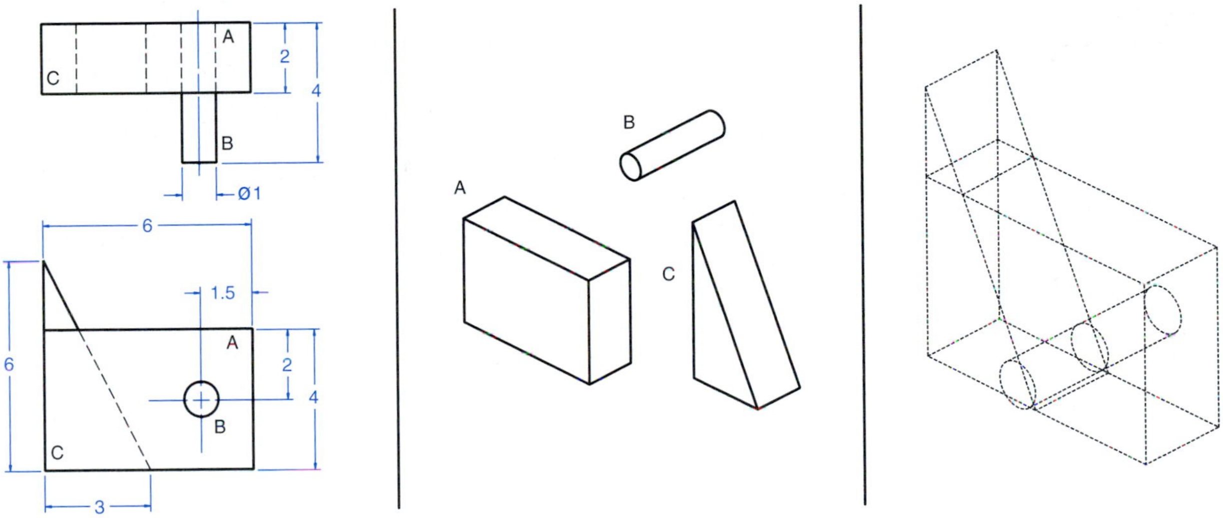

Figure 6.20 Assembled Primitive Parts for Boolean Operations

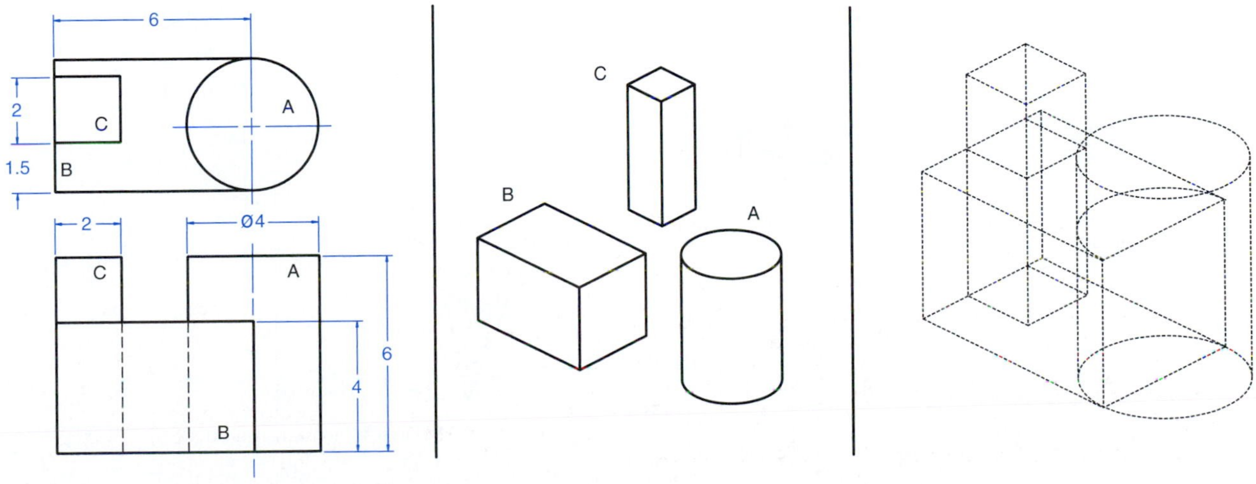

Figure 6.21 Assembled Primitive Parts for Boolean Operations

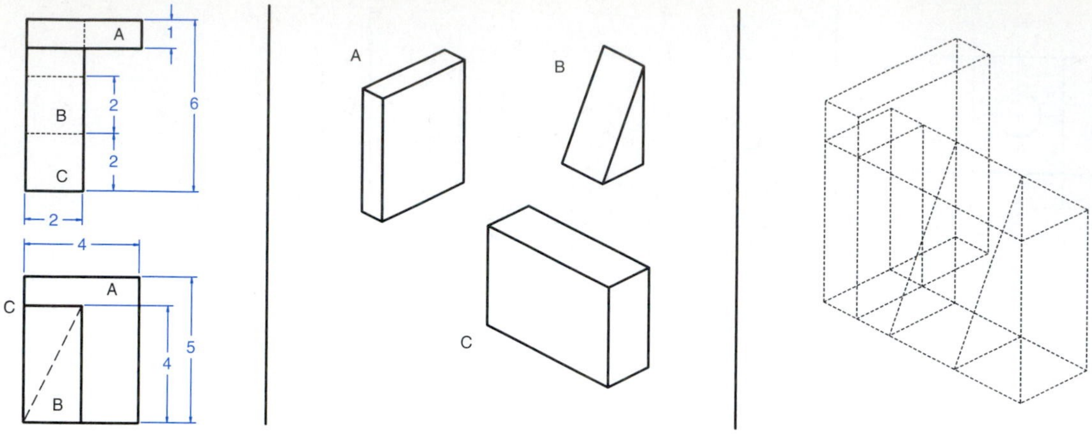

Figure 6.22 Assembled Primitive Parts for Boolean Operations

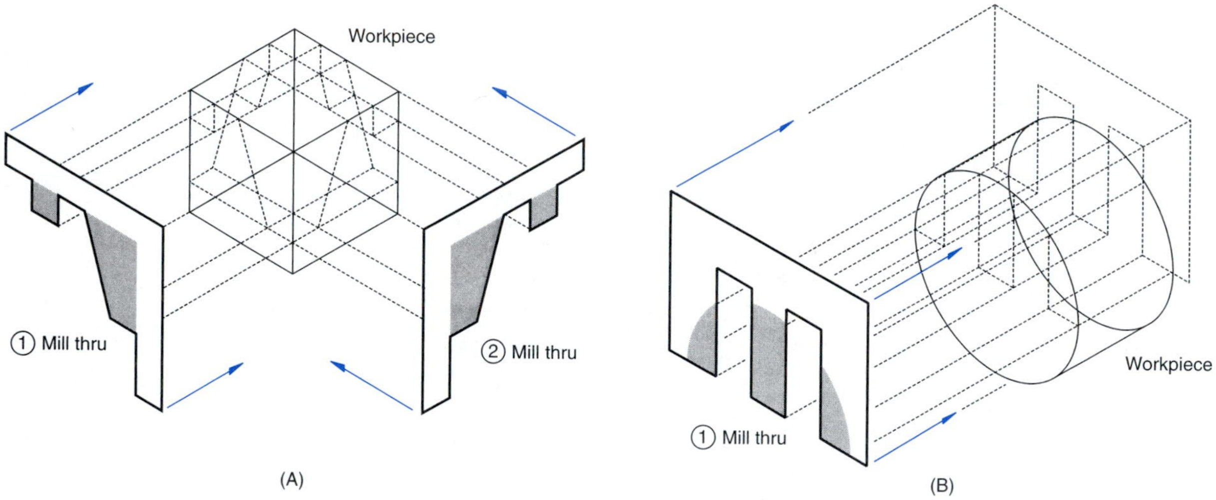

Figure 6.23 Feature-Based Modeling Information

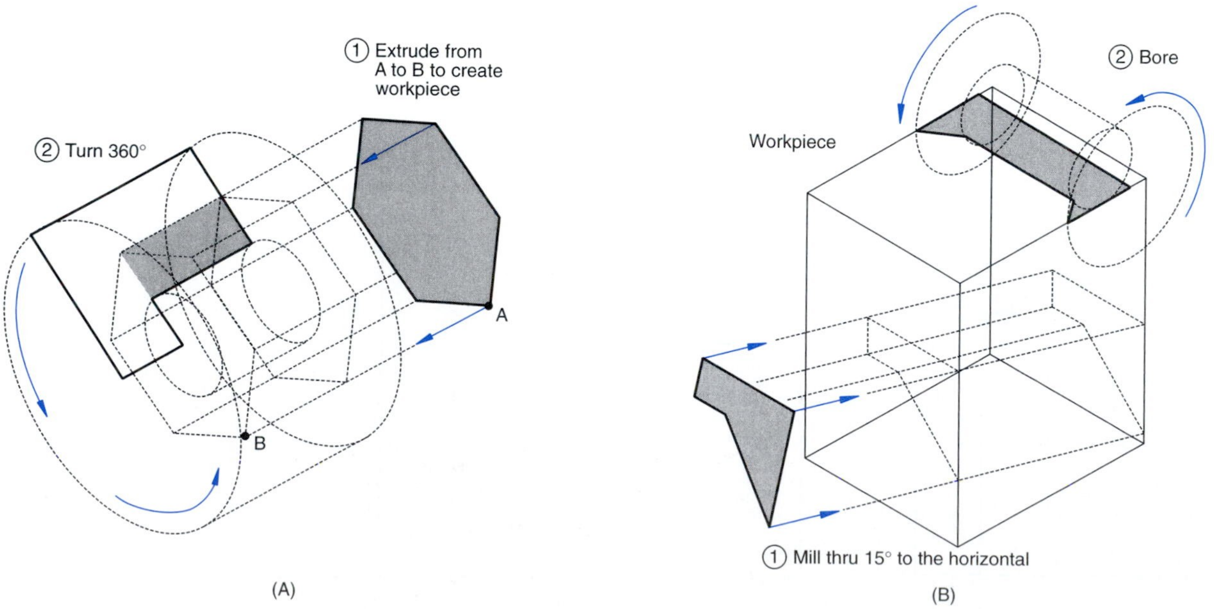

Figure 6.24 Assembled Primitive Parts for Boolean Operations

Problem 6.2

Refer to Figures 6.16 through 6.22. Assign different Boolean operations to the assembled primitive parts. Sketch the resulting composite solids.

Sketch Number: ____________________

Name: ____________________

Div/Sec: ____________________

Date: ____________________

Problem 6.2

Refer to Figures 6.16 through 6.22. Assign different Boolean operations to the assembled primitive parts. Sketch the resulting composite solids.

Sketch Number: ____________________

Name: ____________________

Div/Sec: ____________________

Date: ____________________

Problem 6.2

Refer to Figures 6.16 through 6.22. Assign different Boolean operations to the assembled primitive parts. Sketch the resulting composite solids.

Sketch Number: ____________________

Name: ____________________

Div/Sec: ____________________

Date: ____________________

Problem 6.2

Refer to Figures 6.16 through 6.22. Assign different Boolean operations to the assembled primitive parts. Sketch the resulting composite solids.

Sketch Number: ________________

Name: ________________

Div/Sec: ________________

Date: ________________

Problem 6.3

Refer to Figures 6.23 and 6.24. Using a scale assigned by your instructor, measure the profiles and workpiece, then sketch the resulting workpiece after the feature-based modeling is completed.

Sketch Number: ____________________

Name: ____________________

Div/Sec: ____________________

Date: ____________________

Problem 6.3

Refer to Figures 6.23 and 6.24. Using a scale assigned by your instructor, measure the profiles and workpiece, then sketch the resulting workpiece after the feature-based modeling is completed.

Sketch Number: ____________________

Name: ____________________

Div/Sec: ____________________

Date: ____________________

Supplement

Design Problems

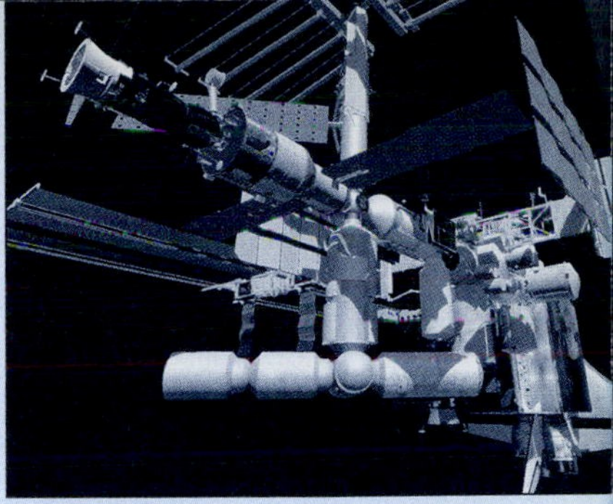

GENERAL INSTRUCTIONS

The following design problems are intended to challenge you to be creative, individually or in a group. Those problems designed specifically for a group are so labeled. The design problems are not meant to teach the design process as much as how to represent ideas graphically, using drawings and computer models. Any design problem labeled a **concept** means that all details of the solution are not necessary. For example, Problem 2 is a concept automobile highway bridge. Obviously, a highway bridge cannot be completely designed in 10, or even 16, weeks by students. However, the basic concept for a highway bridge can be developed and graphically represented in that time frame.

For each project, you must create the engineering drawings and models necessary to communicate your solution to others. The engineering drawings should include the following:

1. Initial design sketches.
2. Multiview drawings, with dimensions. (Concept designs only need critical dimensions and details, where necessary.)
3. Sectioned assembly drawings, with parts lists.
4. Pictorial drawings of the final design, with parts lists where appropriate.

Group projects use a team approach to solve the design problems. The team should consist of four to eight students, randomly selected. Each team should have a group leader to schedule meetings, assign tasks and deadlines, and make sure the team works together as a group to solve the problem.

PROBLEMS

1. **Concept solar-powered vehicle.** (Group) Design a solar-powered concept vehicle, for one passenger, that can travel up to 20 mph.
2. **Concept automobile highway bridge.** (Group) Design a bridge structure that will carry four lanes of traffic and cross a river. The bridge should be 125 feet across and 78 feet above the water in the center.
3. **Concept multipurpose outdoor stadium.** (Group) Design a multipurpose outdoor stadium for football, baseball, and soccer that will seat 54,000 people.
4. **Concept multipurpose indoor stadium.** (Group) Design a multipurpose indoor stadium for basketball, hockey, and volleyball that will seat 18,000 people.
5. **Concept Olympic-sized swimming pool.** (Group) Design an Olympic-sized swimming facility that will seat 3,500 people.
6. **Ergonomic three-button mouse.** Design an ergonomic computer mouse, with three buttons, that can be used by left- and right-handed people.
7. **Laptop computer case.** Design a hard-plastic carrying case for a notebook computer and mouse that weighs 7 pounds and is 2.25 inches by 8 inches by 11 inches.
8. **Computer disk storage unit.** Design a storage box to hold and organize 100 3½-inch computer disks.
9. **Computer keyboard tray.** Design a computer keyboard tray that will be fastened to a desk. The keyboard tray must be able to slide under the desk and accommodate a mouse.
10. **Reading lamp.** Design a reading lamp that can be attached to a desk and adjusted to various positions.
11. **Highway interchange.** (Group) Design an interchange between a four-lane highway and a six-lane interstate highway.
12. **Concept airport facility.** (Group) Design an airport facility for a small city to accommodate two airlines and jets no larger than a Boeing 727.
13. **Portable stadium seat.** Design a portable stadium seat, with a backrest that can be folded flat and with storage for a rain poncho.
14. **Cordless telephone.** Design a cordless telephone.
15. **Snow ski carrier.** Design a device that holds four pairs of downhill or cross-country skis and poles and attaches to an automobile.
16. **Computer desk.** Design a desk for a personal computer that uses a 15-inch monitor mounted below the glass top.
17. **Educational toy.** Design an educational toy or game for children ages 3 to 5.
18. **Beverage cooler.** Design a cooler that can accommodate six 12-ounce beverage cans or bottles.
19. **Chair storage device.** Design a portable storage system for 35 folding steel chairs.
20. **Concept railroad bridge.** (Group) Design a railroad bridge that spans a river 100 feet across. It should be 38 feet above the water at the center.
21. **Children's playground.** (Group) Design a neighborhood playground with tennis courts, a basketball court, and playground equipment.
22. **Football/track facility.** (Group) Design a high school football/track facility with concessions and seating for 2,500 people.
23. **Packaging for a computer monitor.** Design the packaging for a 15-inch or 17-inch computer monitor.
24. **Solar water heater.** (Group) Design a solar collector to heat water for a 2,500-square-foot house.
25. **Solar collector for heat.** (Group) Design a solar collector to heat a 1,200-square-foot house located in northern Michigan, at latitude 45 degrees.
26. **Stereo speaker cabinet.** Design compact, three-way stereo speakers for a 40-watt amplifier.
27. **Concept swept wing aircraft.** (Group) Design a swept wing combat aircraft.
28. **Concept commercial aircraft.** (Group) Design a high-speed 350-passenger commercial aircraft.
29. **Concept spacecraft.** (Group) Design a three-person space vehicle to be used by a space station for in-space satellite repair missions.
30. **Remote control.** Design a handheld infrared remote control for an entertainment center that has cable TV, a videocassette player/recorder, and a stereo with cassette tape, tuner, and audio CD player.
31. **Concept mountain bike.** (Group) Design a lightweight frame for a 26-inch mountain bike with a shock-absorbing front fork.
32. **Concept amusement park ride.** (Group) Design an amusement ride for a specific age group for a theme park.
33. **Weightlifting bench.** Design a weightlifting bench that uses a maximum of 600 pounds of free weights. The bench should include the following stations: bench press, dual action leg lift, and adjustable inclined bench.
34. **Stair-stepper exerciser.** Design an exercise stair-stepper with dual hydraulic cylinders that can be

adjusted for variable resistance. The unit must have a height adjustment for the handlebar and must fold flat for storage.

35. **Portable basketball hoop.** Design a portable basketball system with a water-filled base that provides 250 pounds of weight for stability. The pole should be adjustable every 6 inches between 7 feet and 10 feet, and the unit must have wheels for easy movement.

36. **Computer workstation desk.** Design a computer workstation desk for a PC-based CAD system. The workstation must accommodate a 17-inch monitor, desktop computer, keyboard, 18-inch digitizing tablet, and some storage.

37. **Concept sports car.** (Group) Design a sports car for two passengers.

38. **Concept electric car.** (Group) Design an electric car to be used by commuters in large metropolitan areas. The car should have a maximum speed of 60 mph and a range of 120 miles.

39. **Communications device.** Design a communications device that uses pictures, for use by a disabled child. For example, a drawing of a cup with the word "drink" could be used to communicate that the child is thirsty. The communications device must have a minimum of 25 words for basic living functions, such as eat, sleep, open, close, help, etc.

40. **Propane gas grill.** Design an outdoor propane gas grill with approximately 600 square inches of cooking area.

41. **Alphabet toy.** Design a toddler-age toy used to teach the alphabet.

42. **Parking facility.** Design a parking facility to accommodate 100 automobiles, including five handicapped spaces. Include a single entrance with a booth for collecting parking fees.

43. **Logsplitter.** Design a device to split hardwood logs for home fireplaces. The device should use hydraulics to split the logs.

44. **Aluminum can crusher.** Design a device that will automatically crush aluminum beverage cans for recycling. The device should have a feature that automatically loads, crushes, and discharges the cans.

45. **Sports car dashboard.** Design a sports car dashboard. Lay out all the standard features on the dashboard so that they are within reach or eyesight of the driver.

46. **Toothpaste dispenser.** Design a device that will automatically dispense toothpaste from a standard toothpaste tube. This device should be able to mount on a wall.

47. **Ball return.** Design a device that returns a basketball when a person is practicing free throws. This device must attach to a standard basketball rim, backboard, or pole.

48. **File cabinet wheels.** Design a wheeled device that attaches to a file cabinet. This device would make the file cabinet easier to move. The device should be adjustable so that it will fit on different-sized cabinets.

49. **TV/VCR remote control.** Design the controls for a TV/VCR remote control. Take into account the size of an average human finger and its reach. Think of the functions that would be needed and how they could be grouped on the controls. Consider what the overall size and shape should be and what kind of a display it should have, if any.

50. **Door latch/lock.** Design a door latching/locking system. Define the environment (office, home, industrial, etc.) and the kind of door with which it will be used. Take into consideration the characteristics and abilities of a wide range of people who might be using the door. Include such factors as height, strength, and physical handicap.

REVERSE ENGINEERING PROBLEMS

Reverse engineering is a process of taking existing products, evaluating and measuring them, and then creating the CAD database to reproduce them. The following problems can be used as reverse engineering projects. Use a micrometer, scale, and calipers to make measurements. Use manufacturers' catalogs to specify standard parts, such as bearings and fasteners. For each project, the following is required:

1. Disassemble, measure, and sketch each part.
2. Create 3-D models or engineering drawings of each nonstandard part, with dimensions.
3. Specify standard parts, using engineering catalogs.
4. Create an assembly drawing with parts list.
5. Create a written report that summarizes your project, lists the strengths and weaknesses of the product you reverse engineered, comments on the serviceability of the product, and recommends changes to the design, especially as it relates to DFM principles.

The products to be reverse engineered are:

1. ⅜-inch reversible electric hand drill
2. Staple gun
3. Electric kitchen hand mixer
4. Electric can opener
5. Electric hair dryer
6. Electric glue gun
7. Electric hand jigsaw
8. Dustbuster
9. Kant-twist clamp
10. Hold-down clamp
11. Drill press vise
12. Telephone
13. Computer mouse
14. Paper stapler
15. Paper feeder tray for a laser printer
16. Multiple pen holder for a plotter
17. Computer game joystick
18. Piston and connecting rod for an automobile engine

PROBLEM-SOLVING ACTIVITIES

The following problems can be used as individual or group activities. The activities involve the problem-solving process. Some of the activities also involve visualization ability. Most of the problems will require sketches to solve the problem and to communicate the solution.

1. Determine one solid object that will pass completely through the circular, triangular, and square holes (Figure 1). The object must pass through each hole one at a time and be tight enough that little or no light passes between it and the sides of the hole. Make an isometric sketch of your solution.
2. Sketch the missing right side view and an isometric view of the two given views in Figure 2. The solution cannot have warped surfaces.
3. Create a low-cost lightweight container that will prevent an egg from breaking when dropped from a third-story window onto concrete.
4. Create a method of filling a 2-liter plastic soda bottle resting on the ground, from a third-story window.
5. A mountain climber starting at sunrise takes a well-worn path from the base of a mountain to its top and completes the trip in one day. The mountain climber camps overnight on the mountaintop. At sunrise the climber descends the mountain along the same path. Even though the rates of ascent and descent are different, there is one point along the path which the mountain climber passes at the same time of the day. Prove that there is a single point along the path where this occurs, and make a sketch of your solution.
6. Build the longest cantilevered structure possible using 20 sticks of spaghetti and 24 inches of clear tape. The base of the cantilever must be taped to an 8-inch-square horizontal area. The cantilever must be constructed in 30 minutes or less. When you are finished, measure the length of the cantilever section from the point on the base nearest to the overhanging cantilever end to the end of the cantilevered section.

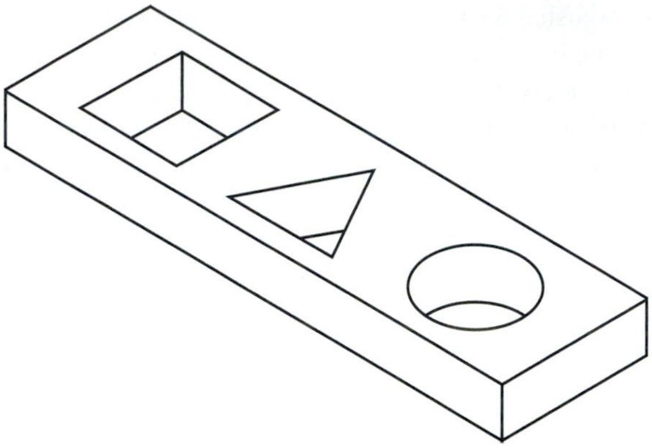

Figure 1 **Problem-Solving Activity 1**
Adapted from *Experiences in Visual Thinking,* R. H. McKim, 1972.

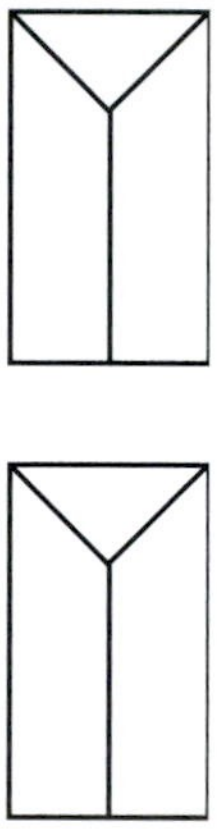

Figure 2 **Problem-Solving Activity 2**

Problems 1 and 2 are adapted from *Experiences in Visual Thinking,* R. H. McKim, 1972.

WHERE TO FIND CASE STUDIES IN ENGINEERING

This is a collection of information on engineering cases. These are accounts of real engineering projects that are written for use in engineering education. The accounts are not highly technical and are quite readable by those with the appropriate interest. Many engineering case studies, histories, and problems have already been developed and can be used to design your own cases for little or no cost. Here are some sources:

http://www.civeng.carleton.ca/ECL/

National Center for Case Study Teaching in Science
http://ublib.buffalo.edu/libraries/projects/cases/ubcase.html

Engineering Ethics Case Studies
http://ethics.tamu.edu/
http://onlineethics.org/cases/index.html

Other sources of cases:

H. Scott Fogler, *Strategies for Creative Problem Solving*. Englewood Cliffs, NJ: Prentice-Hall, 1994.

Henry Petroski, *Design Paradigms: Case Histories of Error and Judgment in Engineering*. New York: Cambridge University Press, 1994.

Design Case Study
The Motorola V600 Mobile Phone

The proliferation of wireless communication products has brought with it a desire to integrate various services into one device. Increased efforts to miniaturize and combine capabilities have afforded great opportunities for designers to solve the problems of high function and small format. As wireless products migrate from novelty gadgets for early adopters to mainstream must-haves, designers, engineers, and product planners must concentrate on providing intuitive simplicity and positive product emotion for the end user.

This product, enabled by several breakthroughs in wireless technology, began with three major industrial design goals in mind:

- Generate a universal appeal to attract first-time wireless users by restating the traditional "rectangular box" product in a less threatening, more intuitive package. Convey an image of consumer approachability while maintaining the client's heritage of reliable, rugged products.
- Create a desirable physical embodiment for a communication product that reinvents Motorola's long-standing equity in wireless communications by integrating paging, cellular, two-way, and data technologies.
- Develop a design compatible with Motorola's currently accepted manufacturing methods, environmental requirements, and cost goals.

Generate a Universal Appeal

Initial analysis of the "mess" involved multidisciplinary sessions where mindmapping and other creative problem-solving methods were used to identify the problem and create a profile of the user, the product, and the features. Marketing, engineering, human factors, and design team members generated and synthesized results into distinct descriptions of what was to become the V600. Early inputs included such diverse "wish list" aspects as the desire to incorporate a hands-free speakerphone, battery life goals, and desired product emotion targets. A clear desire to reach beyond Motorola's traditional commercial user base and connect with white-collar professionals and consumers emerged as a key theme in this early exploration. Establishing a method of capturing and evaluating preresearch data is essential to prevent the loss of early breakthrough ideas.

Effective form factor research in the development of the V600 was important in light of the desire to generate a universal appeal and enhance the user experience. Three hundred end users in various locations were interviewed and presented with nine different form factor solutions to gauge preference in such areas as wearability, fit to the face while in use, folding scheme, and general appeal. Forms ranging from crisp to soft and incorporating different folding schemes were generated by industrial designers using solid modeling software and stereolithography tools. This testing yielded data on the form factor direction to take. The "clamshell," or top-hinged format, emerged as the most preferred, affording users a high degree of portability while also offering a degree of privacy while in use.

Additional observation during research proved to be a great benefit to designers when, in addition to relying on written research reports, they were able to personally observe and participate in one-on-one interviews to discern "unspoken" opinions and comments from users. Research of this type often reveals a lot of peripheral information that is outside the scope of the interviewer's script and doesn't get recorded. Solutions to these unarticulated needs can be those which bring the most satisfaction to the end user.

An observing designer can learn much from the intensity and nuance with which a user discounts a certain feature, embraces another, or casually suggests a third. "Reading between the lines" was especially helpful as many users voiced a need to access certain display information at all times on their communication device. This need conflicted with a high percentage of users who also desired a hinged "clamshell" door to protect and cover the device and provide a face-fitting form when opened. Combining these two seemingly conflicting requirements resulted in the idea of creating a window in the clamshell door which affords protection but also allows the user to access many functions of the product in its closed state. A pair of side mounted soft keys actuate various functions by pushing through to corresponding keys on the main body of the phone. The paradox of providing a simplified device with only a few visible keys yet allowing access to the display information was solved by listening to what users really wanted and forming a solution on the fly in the field, not in the isolated confines of the design studio.

Create a Desirable Communication Product

The "clamshell" solution is one Motorola designers had looked at since 1991 for various telephone-only products, and it seemed to have good potential to convey the dual personality needed for this application.

It was agreed early on that the form of the V600 must present a variety of complex technologies in a user-friendly embodiment. Early design goals called for the shape of this product to invite the user to hold it. A subtle tapering of the form from top to bottom allows the product to fit well in the hand. This borrows from the heritage of Motorola's iconic handheld mobile microphones. The appearance is intended to give the product a character which hints at the underlying

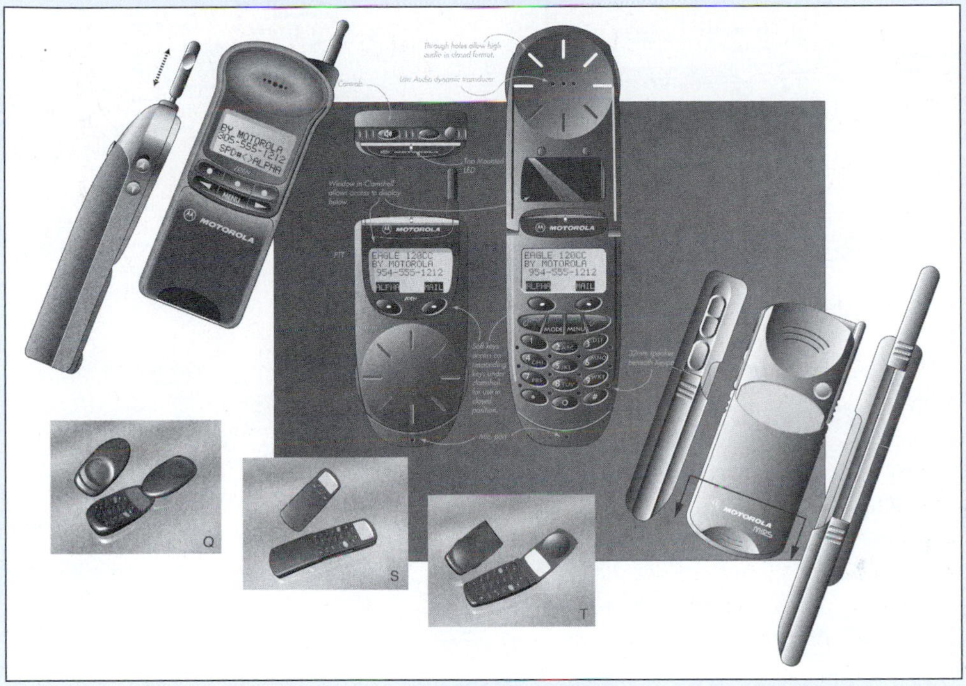

Montage
Several concepts were evaluated as the development of the V600 progressed.
(Courtesy of Motorola.)

features without looking overly complex. The domed feature on the front surface adds visual softness while forming the reverse side of the ear cup.

Meet Motorola's Product Goals

Important in any high-volume product development program are goals regarding cost, time, and manufacturability. Once the initial industrial design solution was agreed upon, a "vision rendering" was archived by Motorola's division general manager and referred to as a master drawing throughout the program to assure that the team stayed on course.

The handheld nature of this product demanded that it be modeled in 3-D and evaluated as quickly as possible. Designers generated the external form of the product in its entirety using the same solids modeling software as the engineering development groups. Extensive use of Motorola's internal rapid prototyping center enabled the creation of numerous wax deposition and stereolithography models to analyze concepts and gain additional user feedback in a greatly reduced time frame.

This common tool platform made for a seamless transfer of data with the design intent remaining intact. It also allowed designers to participate in various inevitable last-minute

Flip Style Design
The final design decision was to create a flip style or clamshell design for the phone.
(Courtesy of Motorola.)

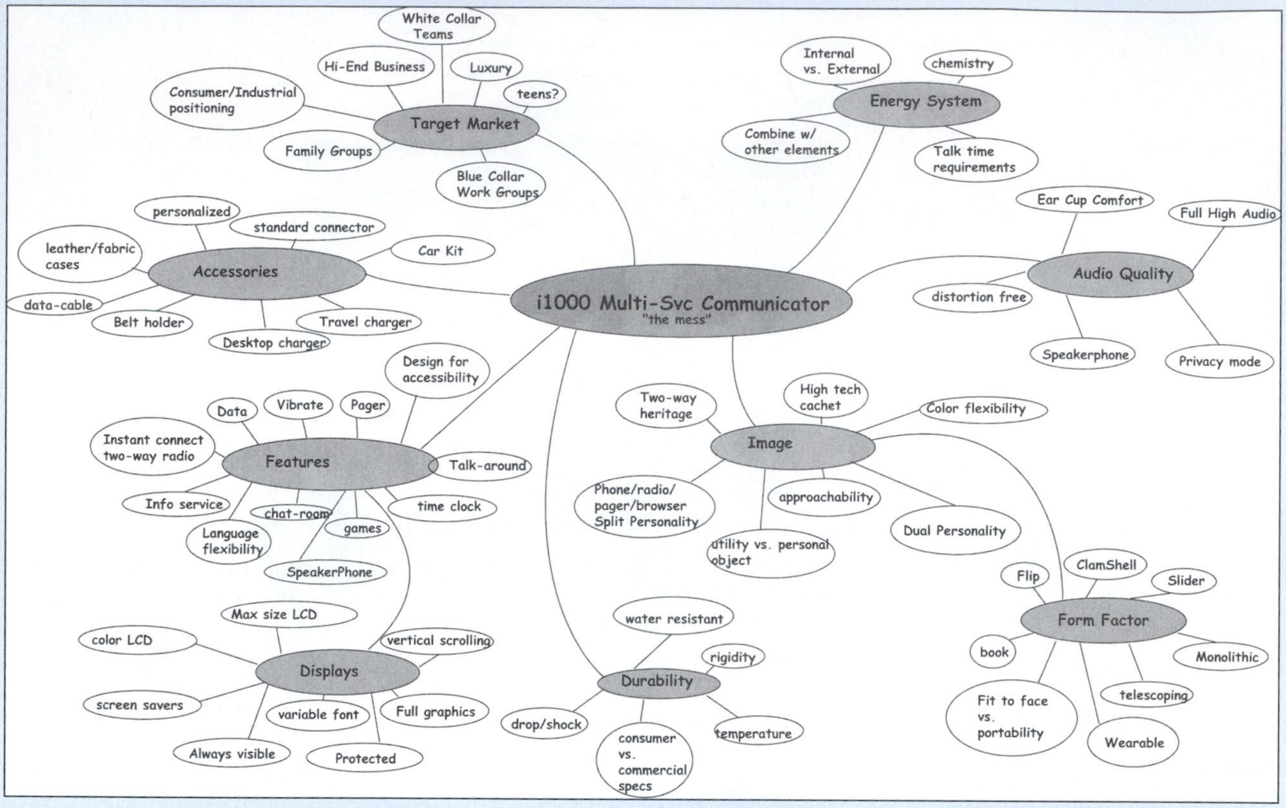

Mind Map

Predesign "mind maps" incorporated diverse requirements and "wish list" elements.
(Courtesy of Scott Richards: Motorola.)

modifications directly on the engineering databases before tooling was initiated. The iterative nature of design is facilitated by the quick results available from common databases and the existence of a common 3-D "language."

Motorola's V600 Final Design Features

The new Motorola V600 is the ideal heir of the series Motorola promoted with their models V60i and V66i. From these models it inherits its characteristic design. The Motorola V600 is characterized by its double display, internal and external. The external one is blue, two lines, 96 × 32 pixels and reversed. It displays the time or the caller ID and can be personalized with the lit-up profiles of callers. The internal one is graphic, color (65,536), TFT, and 176 × 220 pixels. The V600 also integrates a VGA digital camera, which makes it possible to take photos and share them with a few simple moves. The wide color display of the new line is ideal for viewing photos, creating personalized screensavers or adding images to the visual ID of the caller. Not to mention, at any moment, flipping through

The V600 compact and very function style.

your own virtual photo album, seeing friends and reliving special occasions. The V600 also has a large internal memory for saving your photos or those you download.

As far as connectivity is concerned, the Motorola V600 is a cellular phone that is Quad Band (GSM 850/900/1800/1900 MHz) and GPRS Class 10 (2 up/4 down). It can function on any GSM network in any part of the world. The integrated browser is dual mode, for mobile Internet access with WAP 2.0 protocol and xHTML. The V600 also supports WAP push (it can receive SMS messages by direct WAP connection), EMS technology version 5.0 and MMS (Multimedia Messaging Service). There is also support for video playback MPEG4 and H.263. Another strength of the Motorola V600 is its versatility: it is possible to connect this terminal to other compatible devices and PCs via Bluetooth and bus CE (USB/serial cable). Among the accessories sold with the V600, there is also a special Bluetooth headphone with an LCD display.

Aside from the ability to take and save photos thanks to the integrated digital camera and MMS support, the V600 offers numerous other multimedia functions. Among these is the ability to compose, save, or receive polyphonic ring tones, which can be played through the polyphonic speakers (22 Khz, 24 tones). Thanks to this function, it is possible to use the V600 in speaker mode. It is possible to download, via WAP, screensavers, wallpaper, ring tones, photos, and games in GPRS connection mode. The Motorola V600 also supports themes, meaning ring tones, images, animation, and audio downloaded in a single file. The telephone supports the standard J2ME, which allows you to personalize your cellular phone even more by downloading Java games and applications. In particular, the V600 arrives with a few new preloaded games, including the popular MotoGP.

Among the other functions:

- E-mail: the ability to send and receive e-mail
- Lexicus software: makes creating texts easier, automatically recognizing the word that you are inserting, and offering eventual options
- Voice commands: the ability to associate names with numbers in the phone book, using voice commands
- VibraCall function
- Calculator/clock/calendar
- PIM function: the ability to manage an agenda with alarms and contacts; can be synchronized with a PC (with TrueSync software)

The V600 has a metal structure with interchangeable covers, which allows you to personalize your telephone with a vast range of colorful front and back covers. The dimensions of the Motorola V600 are 47.3 × 87.5 × 22.5 mm, in 95 grams of weight and 75 cc of volume. The autonomy of the standard LiIon battery of 750 mAh is quite high, given the presence of the color screen: up to 6.5 hours in conversation and up to 285 hours in standby.

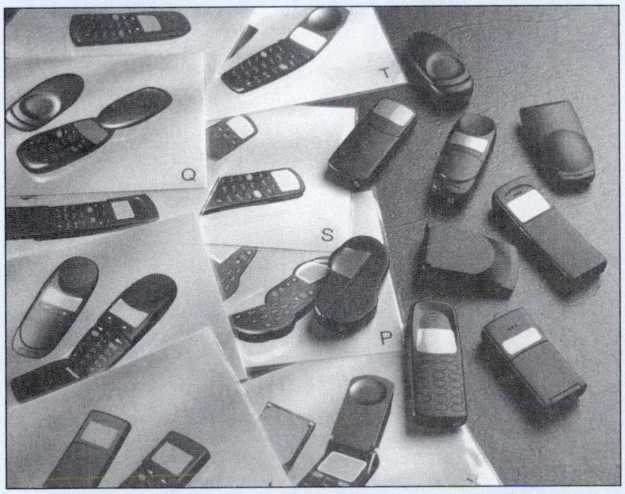

Form Test

Various form factors were tested with individual users to gauge preferences in such areas as wearability, fit to face while in use, folding scheme, and general appeal.

(© John Mazey.)

The Motorola V600 at a Glance

- Enhance integrated camera.
- Vivid color display, capable of supporting up to 65,000 colors.
- Bluetooth™ wireless technology allows you to connect wirelessly to compatible accessories and devices.
- Multimedia Messaging (MMS) capabilities allow you to send a photo along with a personalized voice message.
- Quad-band GSM technology gives you the ability to stay in touch worldwide.
- Customizable incoming ring tune and color notification lights.
- Polyphonic ring tones for rich, melodic sound.
- J2ME™ for downloadable games and productivity applications.
- Interchangeable metal covers.
- WAP browser with GPRS.
- Fast, "always-on" Internet connection.

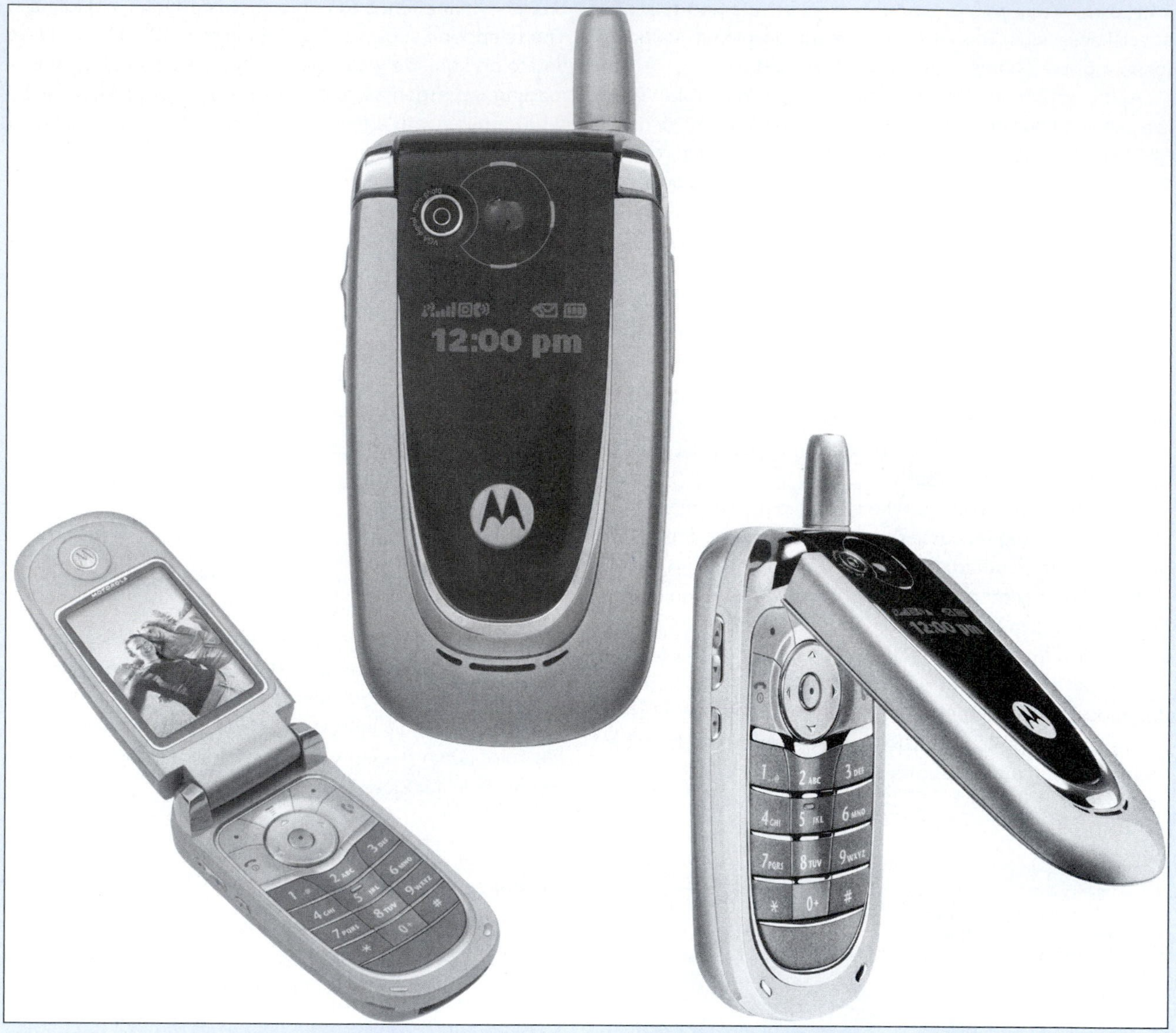

Use "Constructive Conflict" as an Advantage

While industrial design played a key role in the excitement of this product, it is important to note that it succeeds only when combined with similar success stories in the areas of electrical, software, materials, acoustics, manufacturing, and mechanical engineering. The ability of these teams to work together is central to the success of increasingly complex products. This does not imply that these disciplines see eye to eye and agree on every aspect of the process. To the contrary, the "constructive conflict" that results from so many complex requirements often nets the most successful products.

Those who will succeed in the future will be those who successfully integrate outstanding technologies with a usable human interface and compelling design. These factors combine to create the intangible positive emotion which transcends mere form, material, or features and makes a visceral connection with the user.

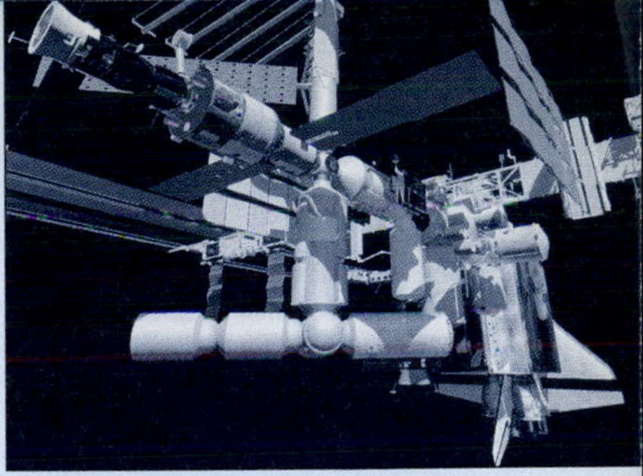

Additional Problems and Worksheets

Problems

7.1 Create a single-line drawing of the process flow diagram shown in Figure 7.1.

7.2 Create a single-line drawing of the mechanical flow diagram shown in Figure 7.2.

7.3 (Figure 7.3) Redraw the block diagram sketch as a complete production drawing.

7.4 (Figure 7.4) Redraw the cabling sketch as a complete production drawing.

7.5 Use a CAD software program to make a complete set of production drawings for the electronic circuit shown in Figure 7.5.

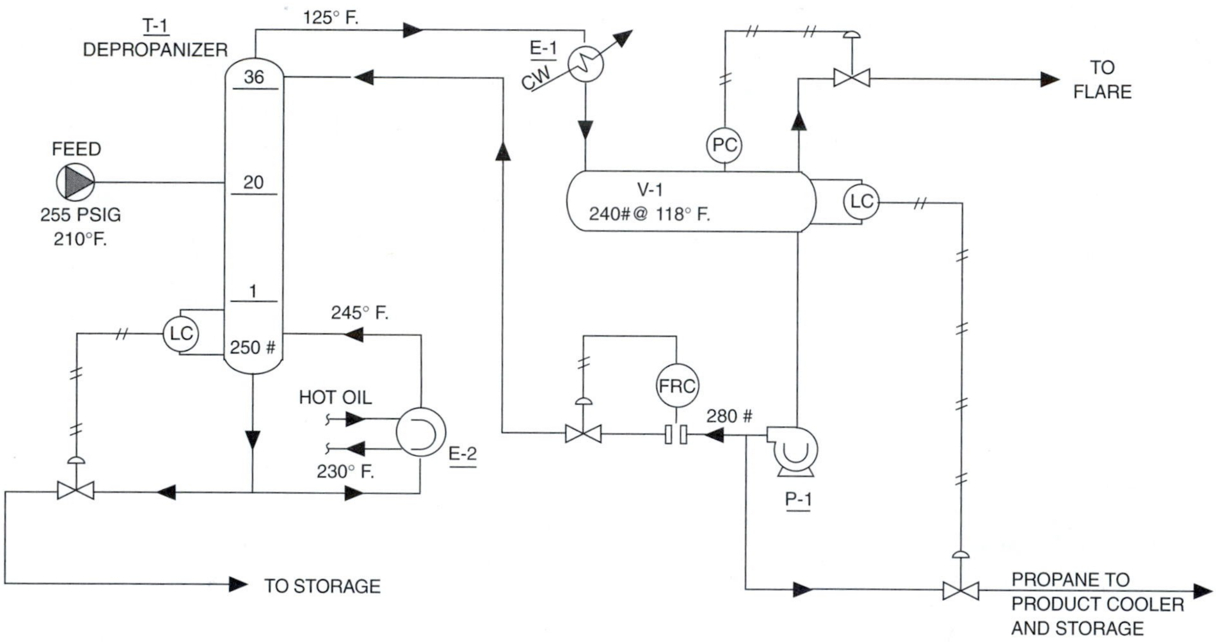

Figure 7.1 **Process Flow Diagram**

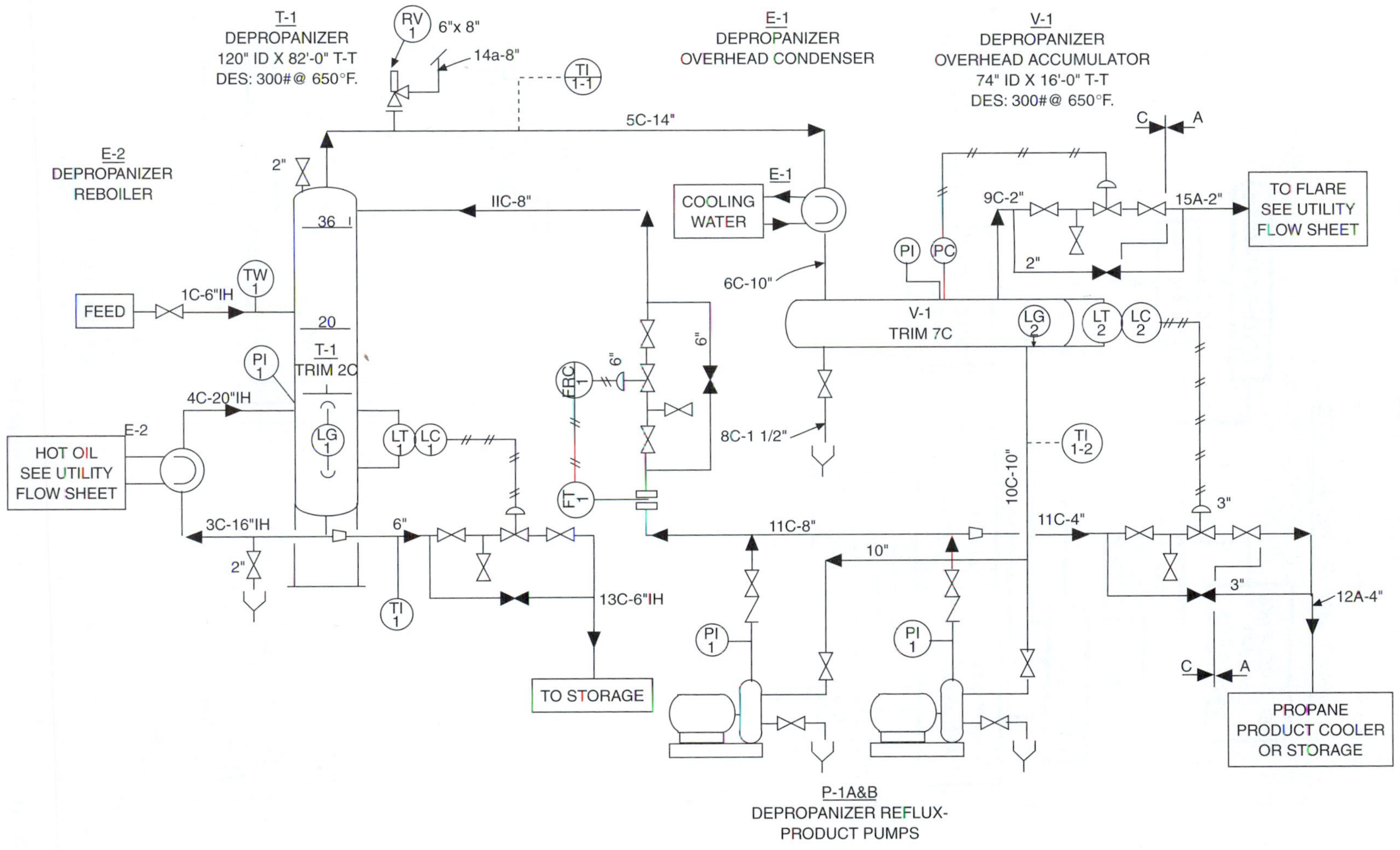

Figure 7.2 **Mechanical Flow Diagram**

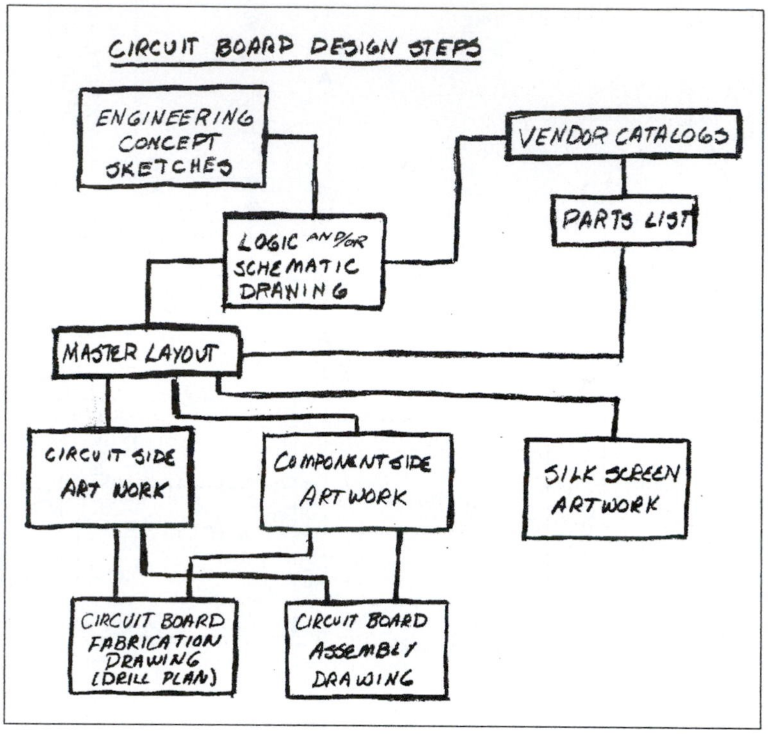

Figure 7.3 Block Diagram Sketch

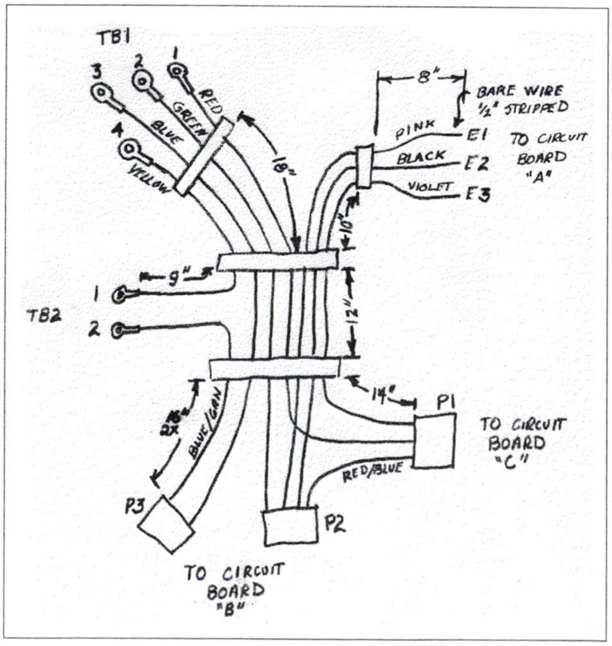

Figure 7.4 Cabling Sketch

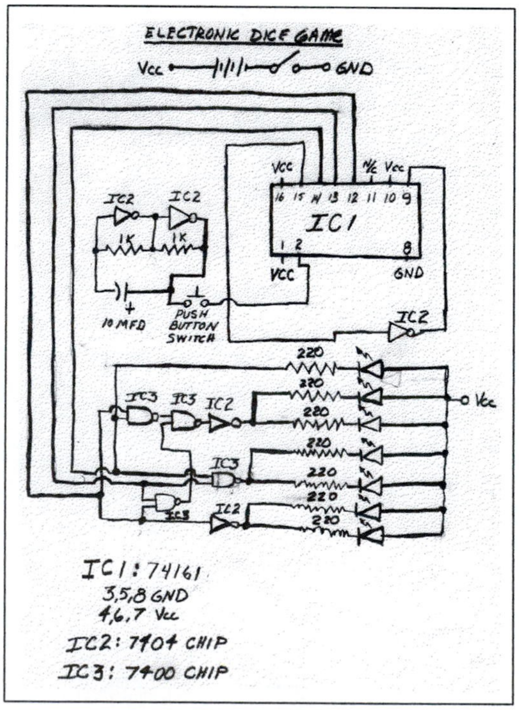

Figure 7.5 Electronic Circuit

7.6 (Figure 7.6) Draw the highway interchange, using the given dimensions.

7.7 Model the production facility shown in Figure 7.7. Before beginning the modeling process on the computer, divide the facility into a series of logical parts, giving each one a name. You may find it helpful to create sketches of the individual parts.

7.8 (Figure 7.8) Many power plants are installing large hyperbolic natural-draft cooling towers similar to that shown. They are thin, lightweight concrete, often no more than 4″ thick. An open grid of beams at the bottom supports the shell and permits the entry of air. Draw or sketch the cooling tower.

7.9 (Figure 7.9) An industrial building has a roof supported by split-ring connector timber trusses as shown. (The 4″×6″ struts all lie centered on the truss center lines.) Draw or sketch the roof truss.

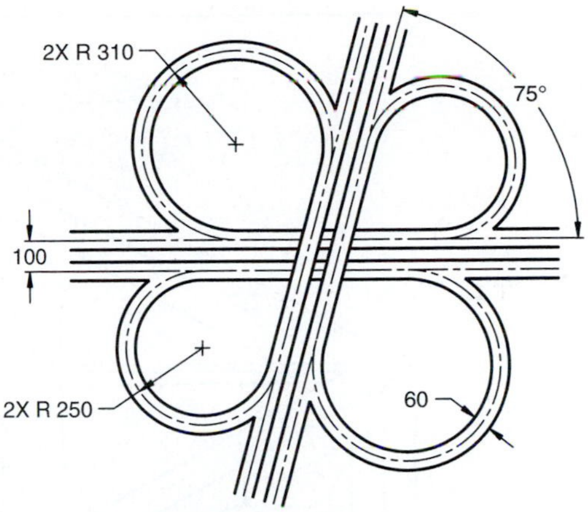

Figure 7.6 **Highway Interchange**

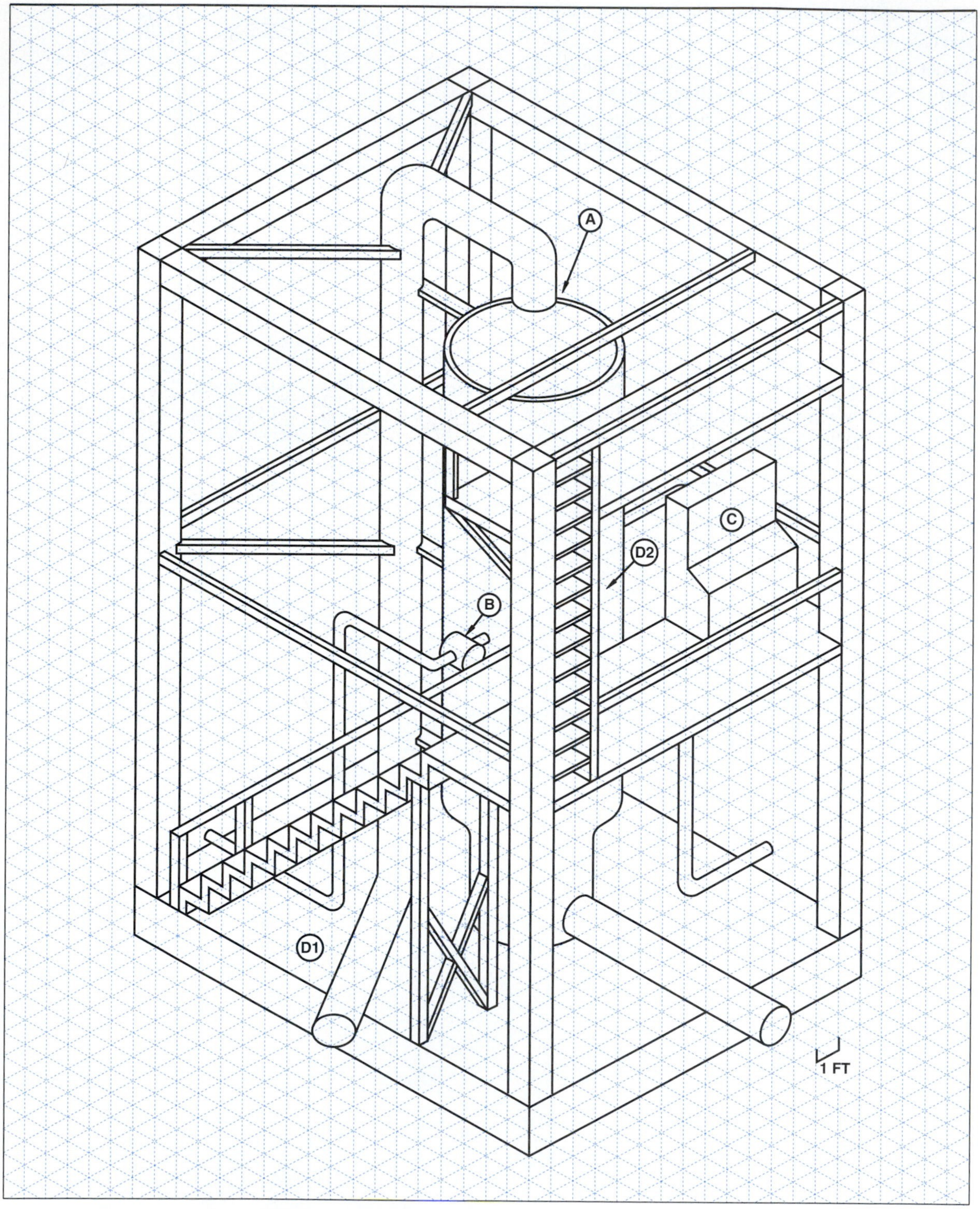

Figure 7.7 Create a 3-D Model of This Production Facility

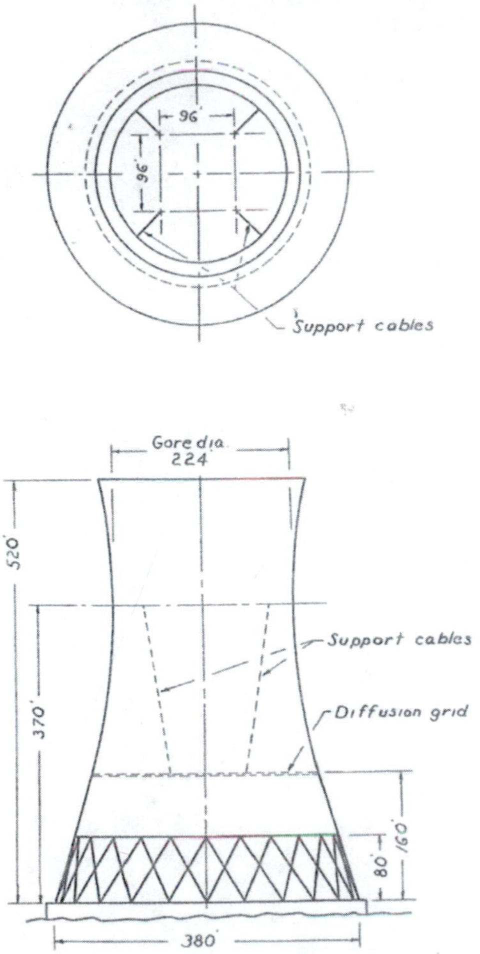

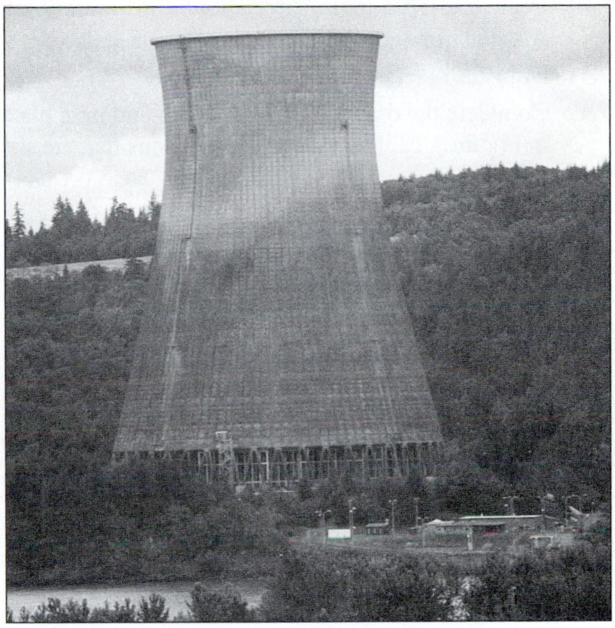

Figure 7.8 Hyperbolic Cooling Tower

(Courtesy of the Marley Company, Kansas City, Mo.)

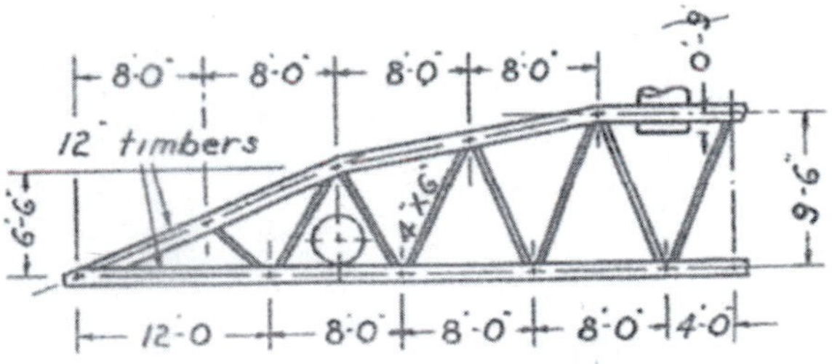

Figure 7.9 Timber Truss

7.10 Draw the three logic gates shown in Figure 7.10. Assume symmetry about a central horizontal axis.

7.11 Draw the open traverse shown in Figure 7.11.

7.12 Complete the drawing of the column and base plate in Figure 7.12. Estimate all dimensions that are not given.

7.13 Complete the drawing of the retaining wall shown in Figure 7.13.

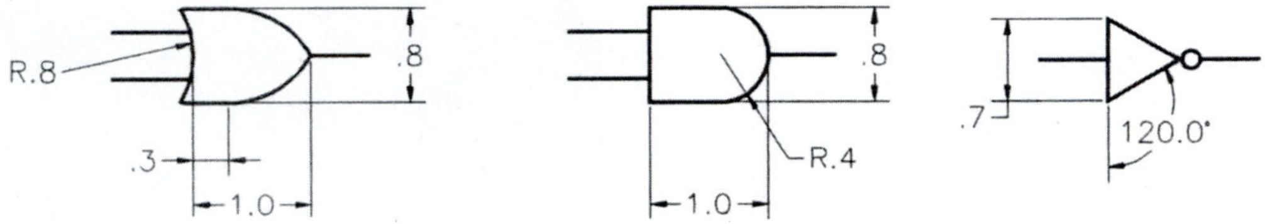

Figure 7.10 Logic Gates

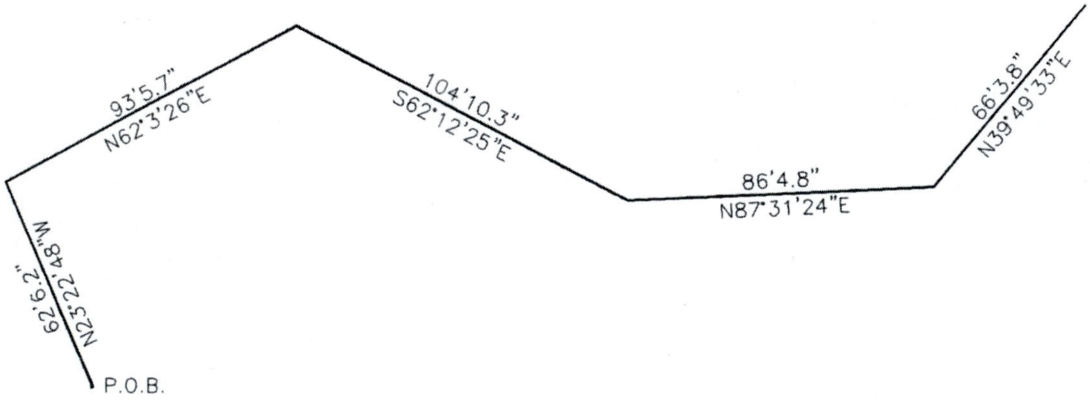

Figure 7.11 Open Traverse

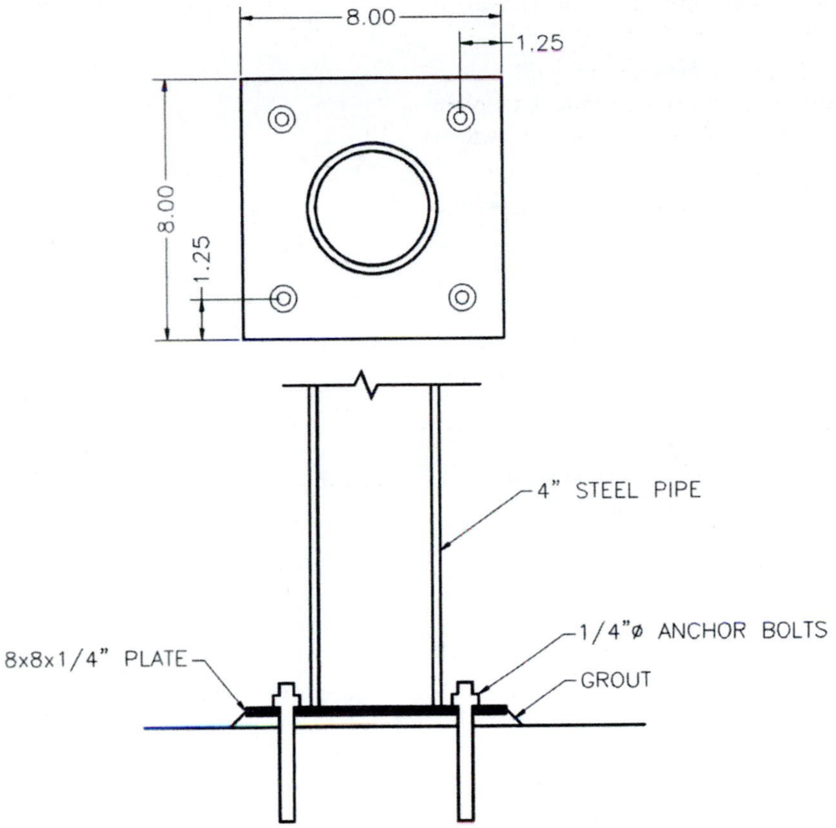

Figure 7.12 Steel Connection Detail

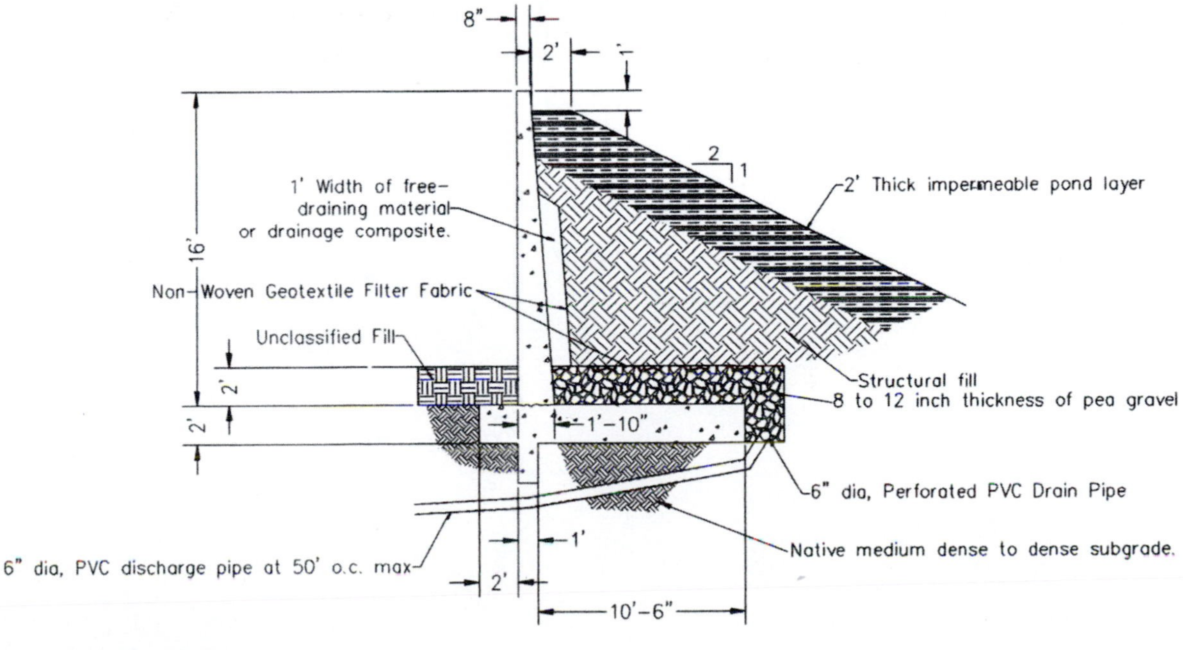

Figure 7.13 Retaining Wall Section

7.14 Make a drawing of the piping layout shown in Figure 7.14.

7.15 The slide gate shown in Figure 7.15. will hold a ½″ steel plate to give access to a tank. Complete the oblique and orthographic drawings shown in Figure 7.15.

7.16 The isometric drawing in Figure 7.16 shows the method for securing 1½″ steel grating to a 2 × 2 angle bracket. Complete the drawing as shown, estimating any dimensions that are not given.

7.17 An engineer wishes to locate a house on a plot of land with a 15° angle from due north. Draw the house per the dimensions shown in Figure 7.17 and then *rotate* it to place on the plot plan.

7.18 Complete the drawing of PCB1 shown in Figure 7.18. Change the sizes of the holes to create PCB2.

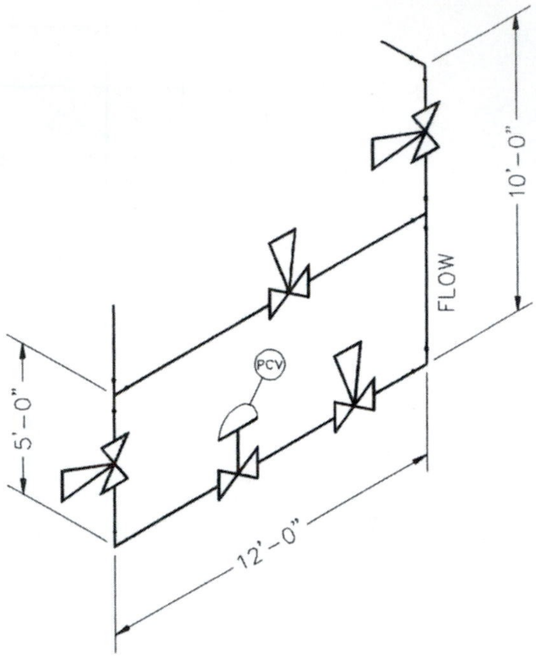

Figure 7.14 Piping Isometric

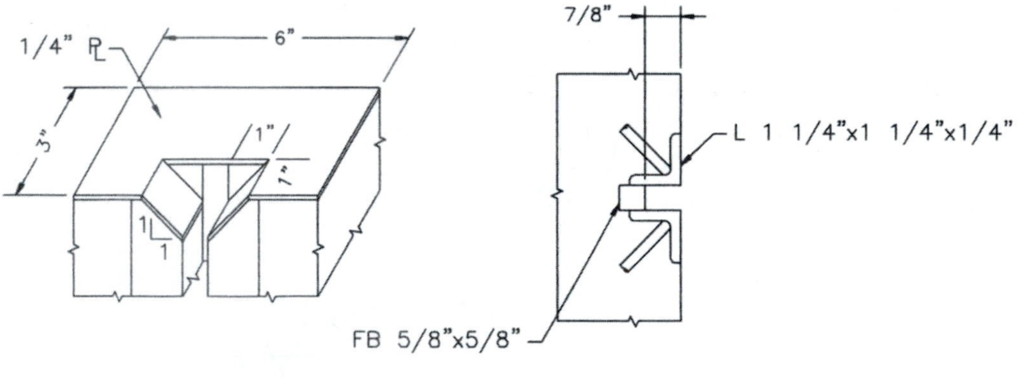

Figure 7.15 Slide Gate

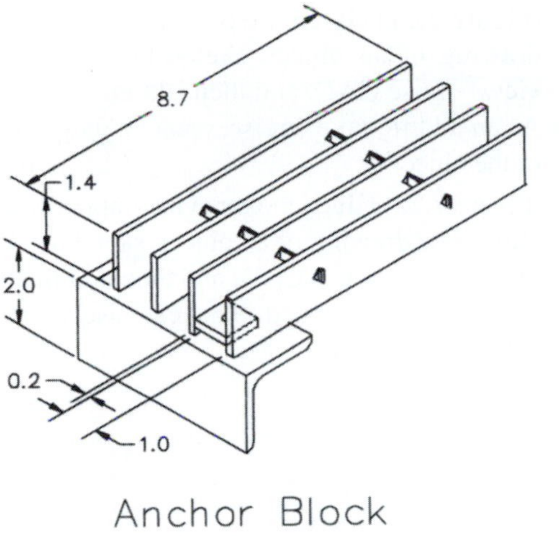

Figure 7.16 Anchor Block

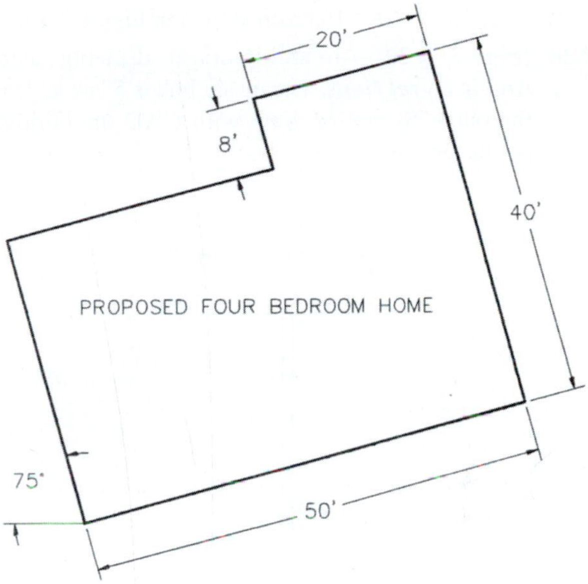

Figure 7.17 Plot Plan

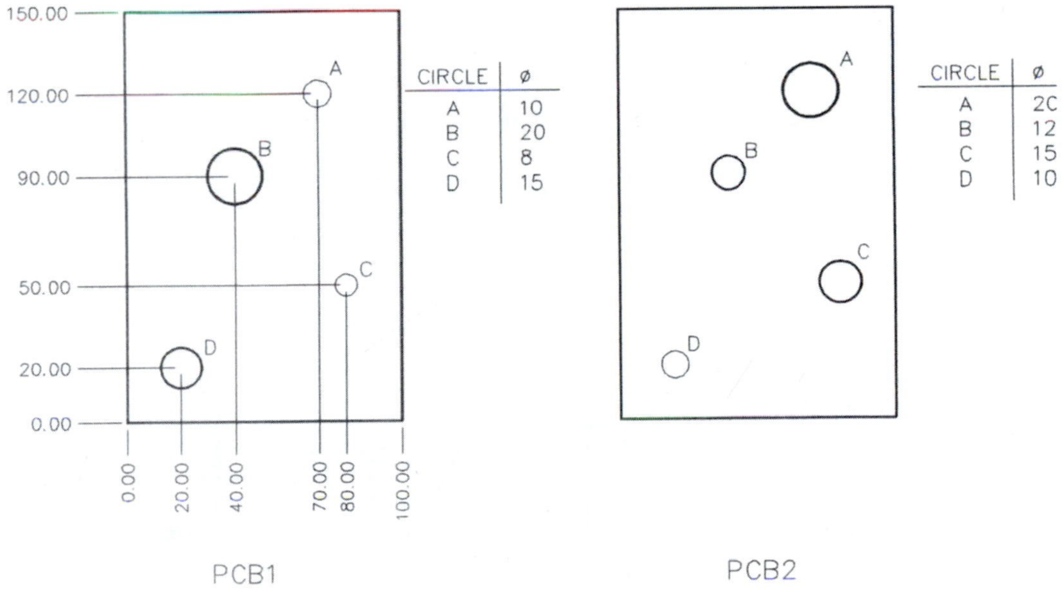

Figure 7.18 Printed Circuit Boards

7.19 Draw the W10 × 15 beam shown in Figure 7.19.

7.20 (Figure 7.20) An architectural drawing using ***Architectural Units.*** The inside line is 5″ away from the outer. Sketch or draw with CAD the building perimeter plan.

7.21 (Figure 7.21) Given the two views of a multiview drawing of an object, sketch or draw the given views or use CAD, and then add the missing view. As an additional exercise, create a pictorial sketch of the object.

7.22 (Figure 7.22) Given the three incomplete views of a multiview drawing of an object, sketch or draw the given views or use CAD, and then add the missing line or lines. As an additional exercise, create a pictorial sketch of the object.

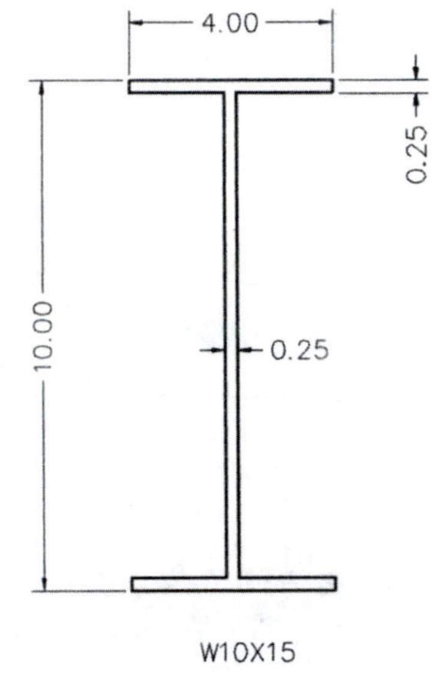

Figure 7.19 I-Beams

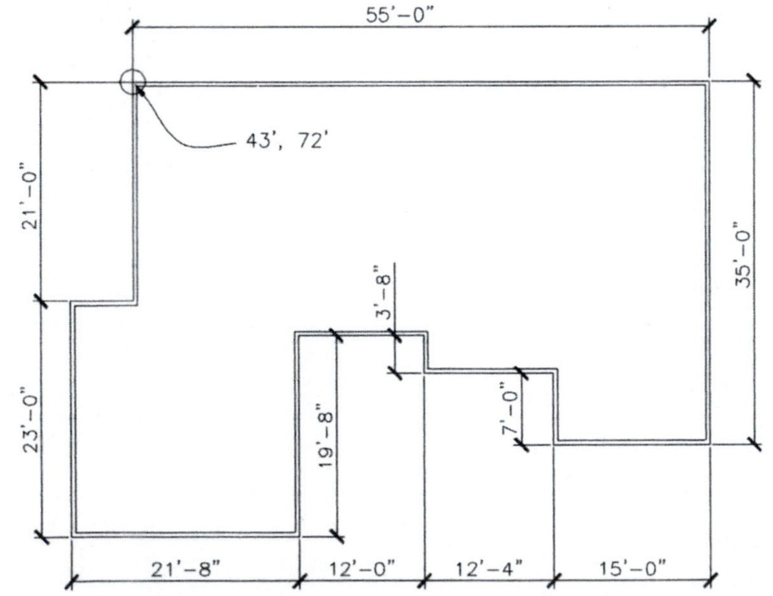

Figure 7.20 Floor Plan

(1)

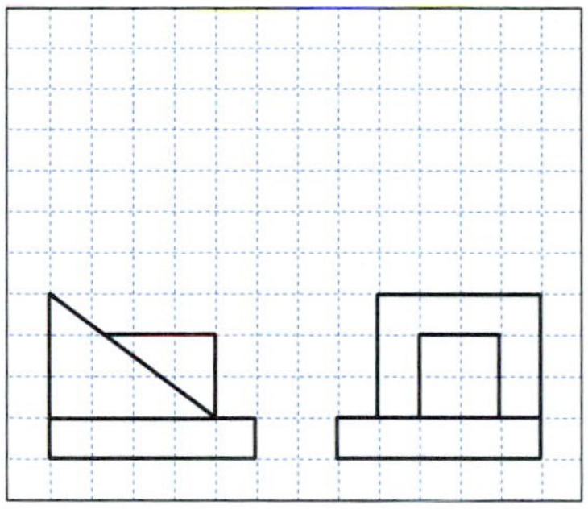

(2)

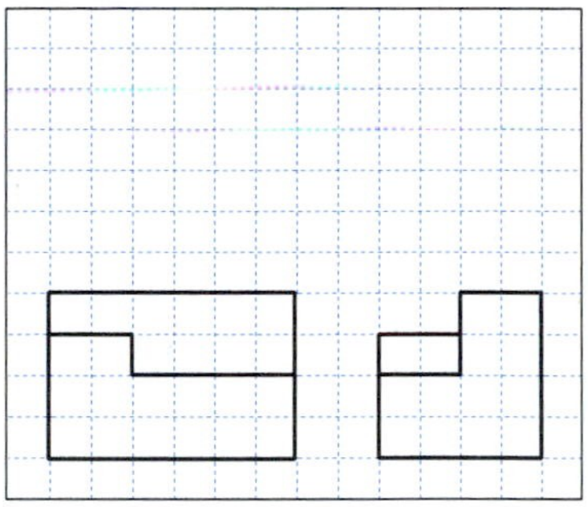

(3)

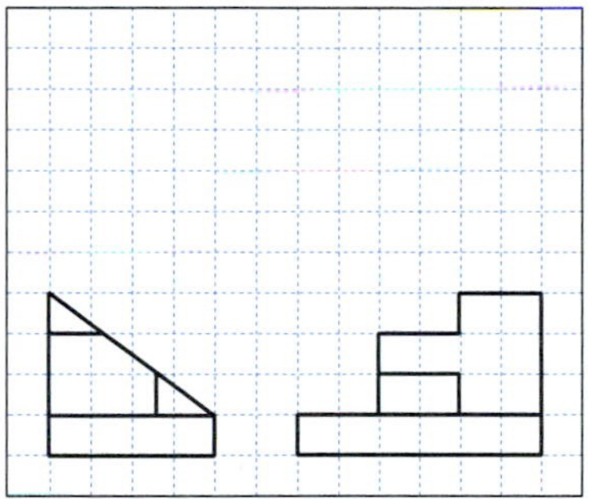

(4)

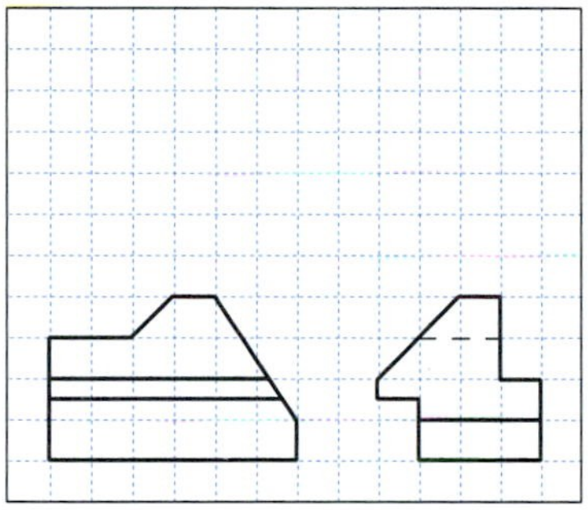

(5)

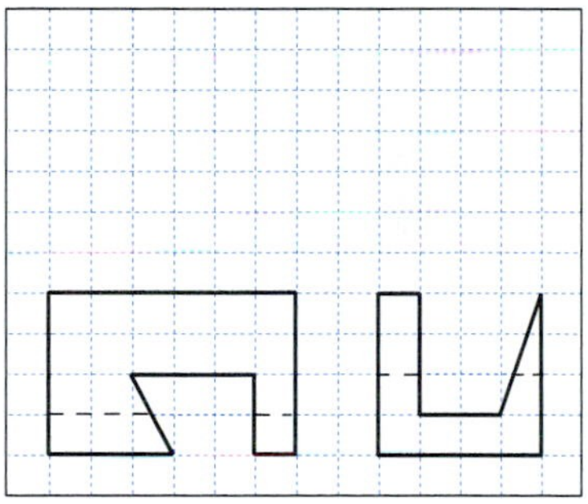

(6)

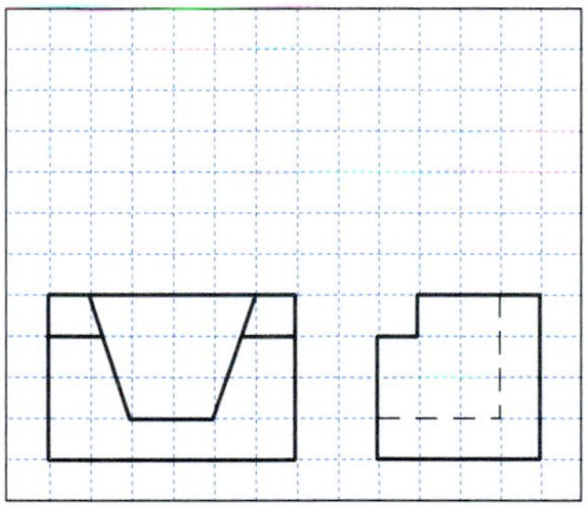

(7)

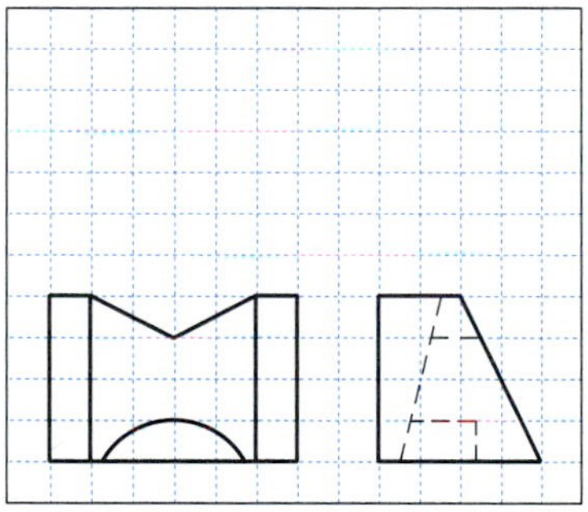

(8)

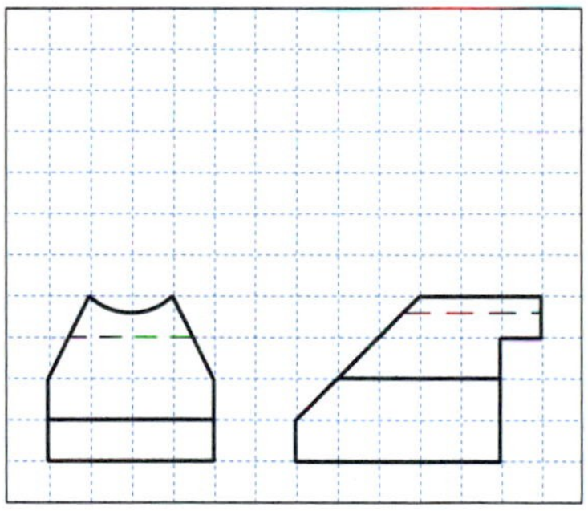

(9)

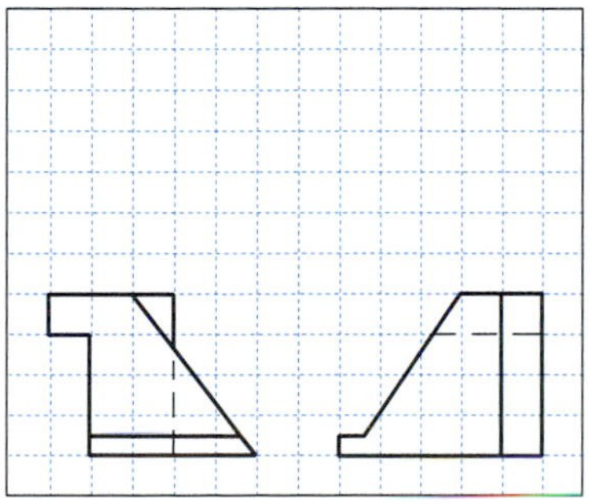

(10)

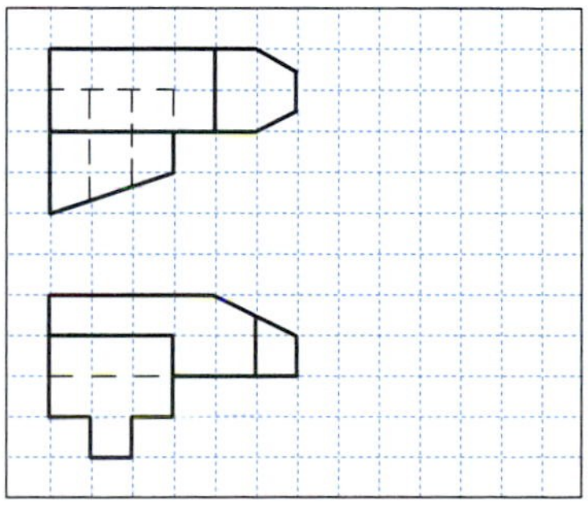

(11)

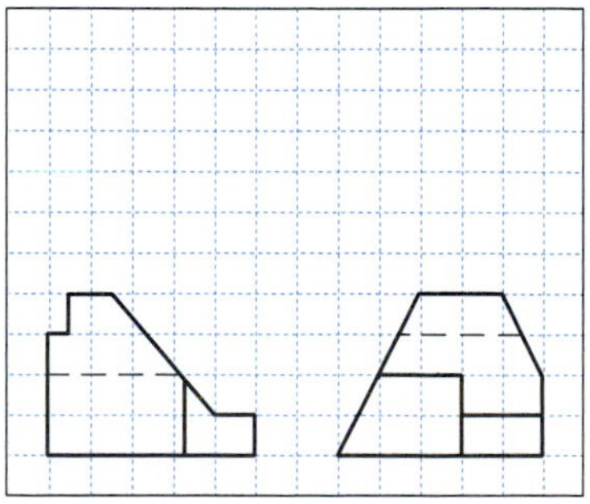

(12)

Figure 7.21 **Two-View Drawings of Several Objects, for Problem 7.21**

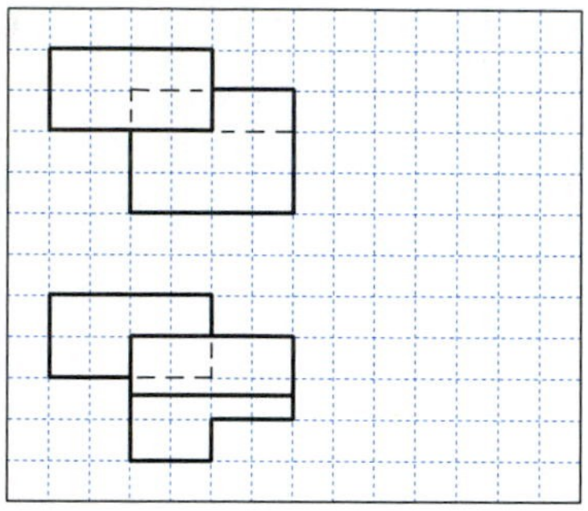
(13)

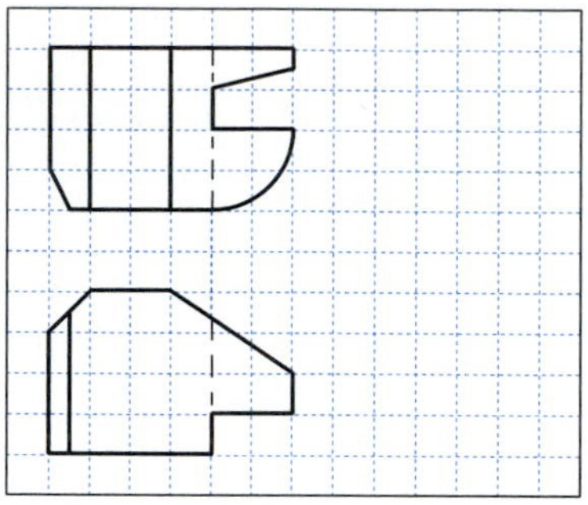
(14)

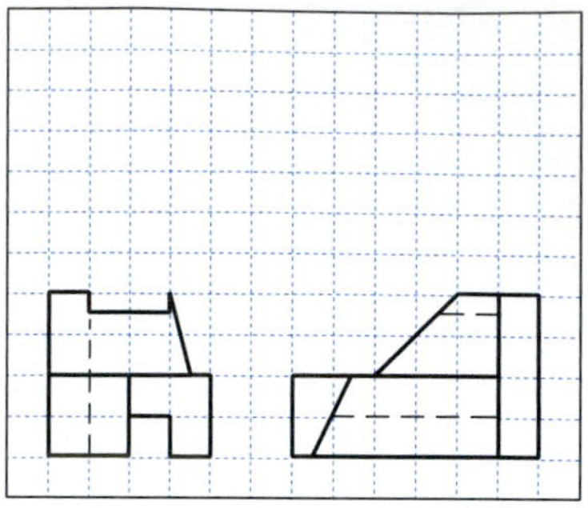
(15)

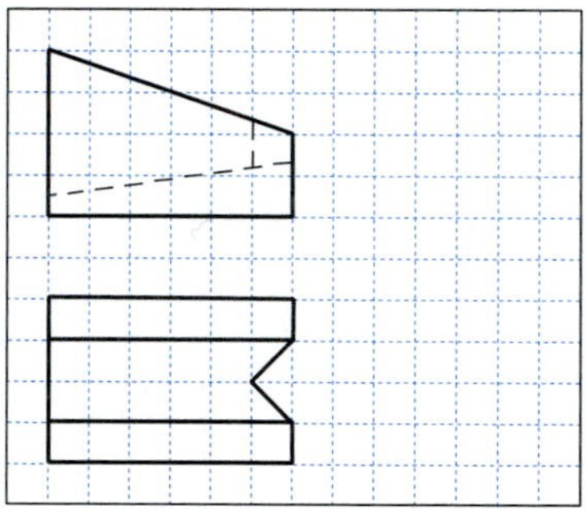
(16)

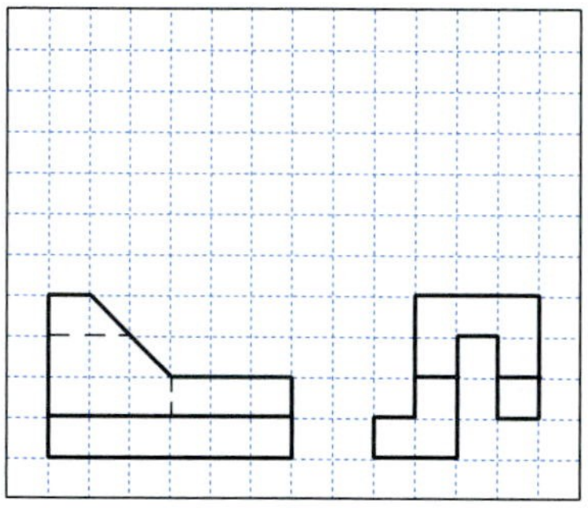
(17)

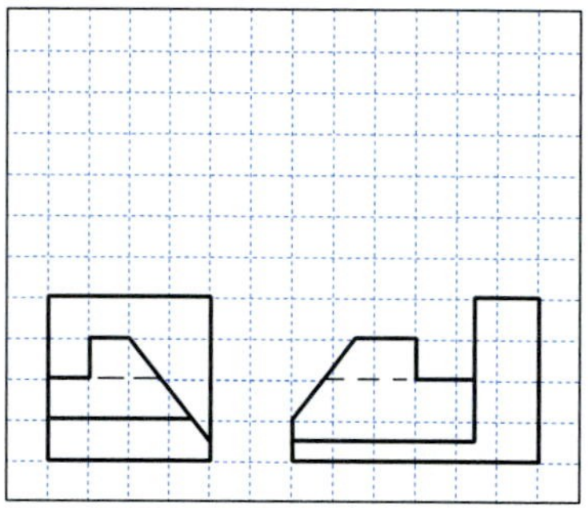
(18)

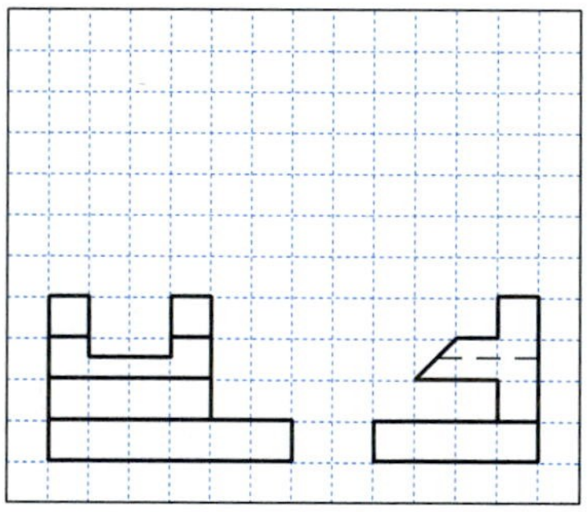
(19)

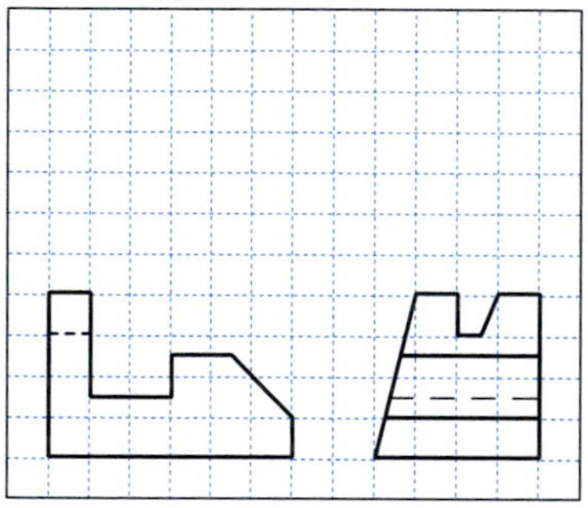
(20)

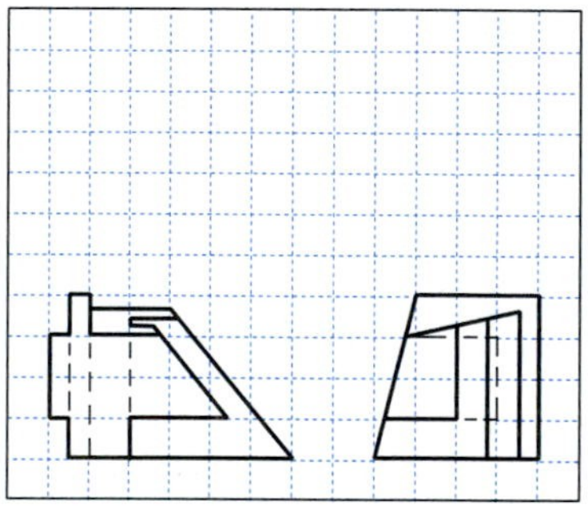
(21)

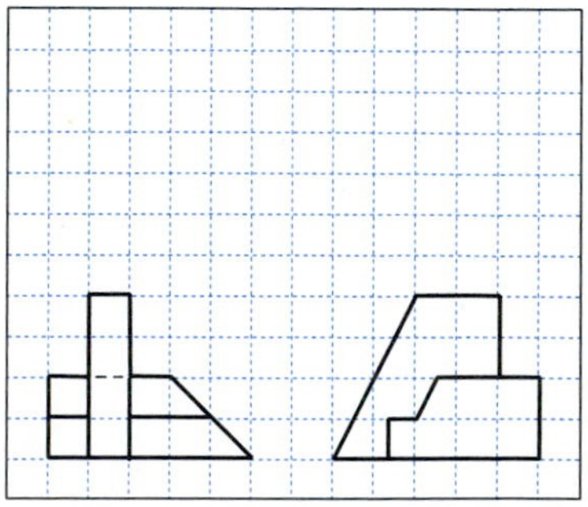
(22)

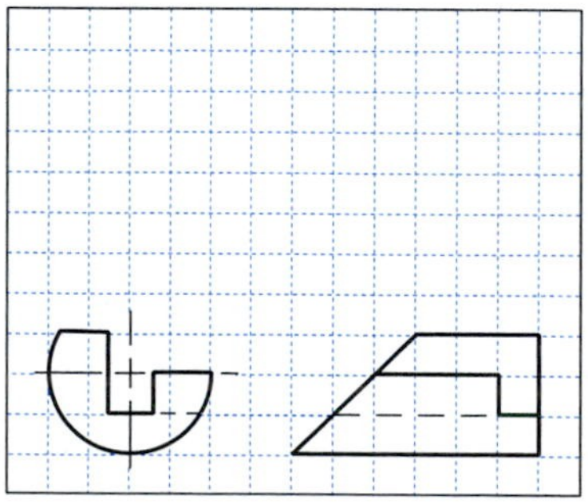
(23)

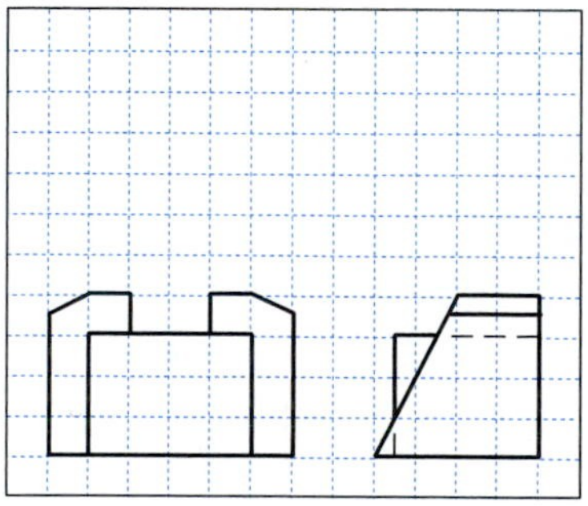
(24)

Figure 7.21 Continued

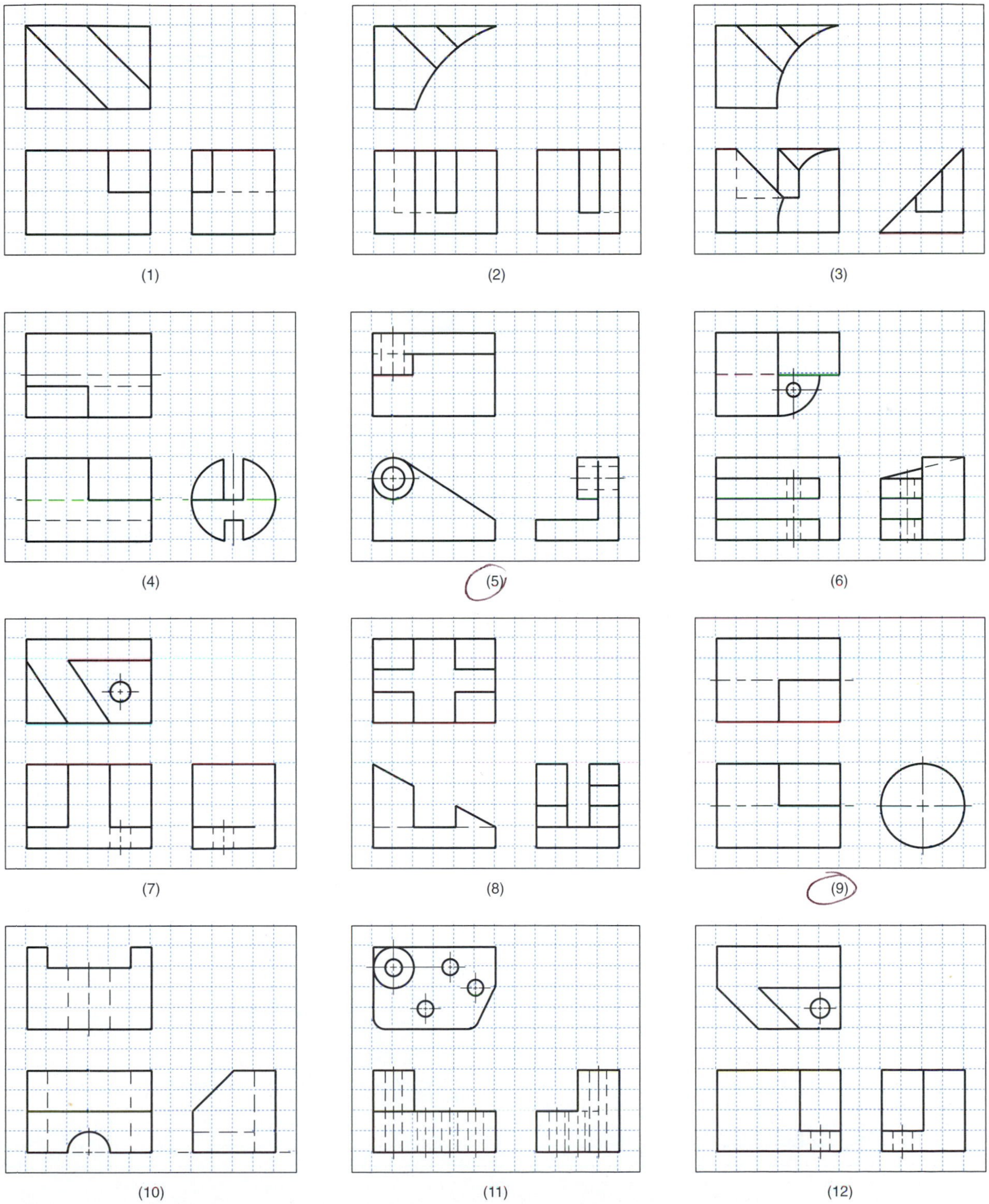

Figure 7.22 **Three Incomplete Views of a Multiview Drawing of an Object, for Problem 7.22**

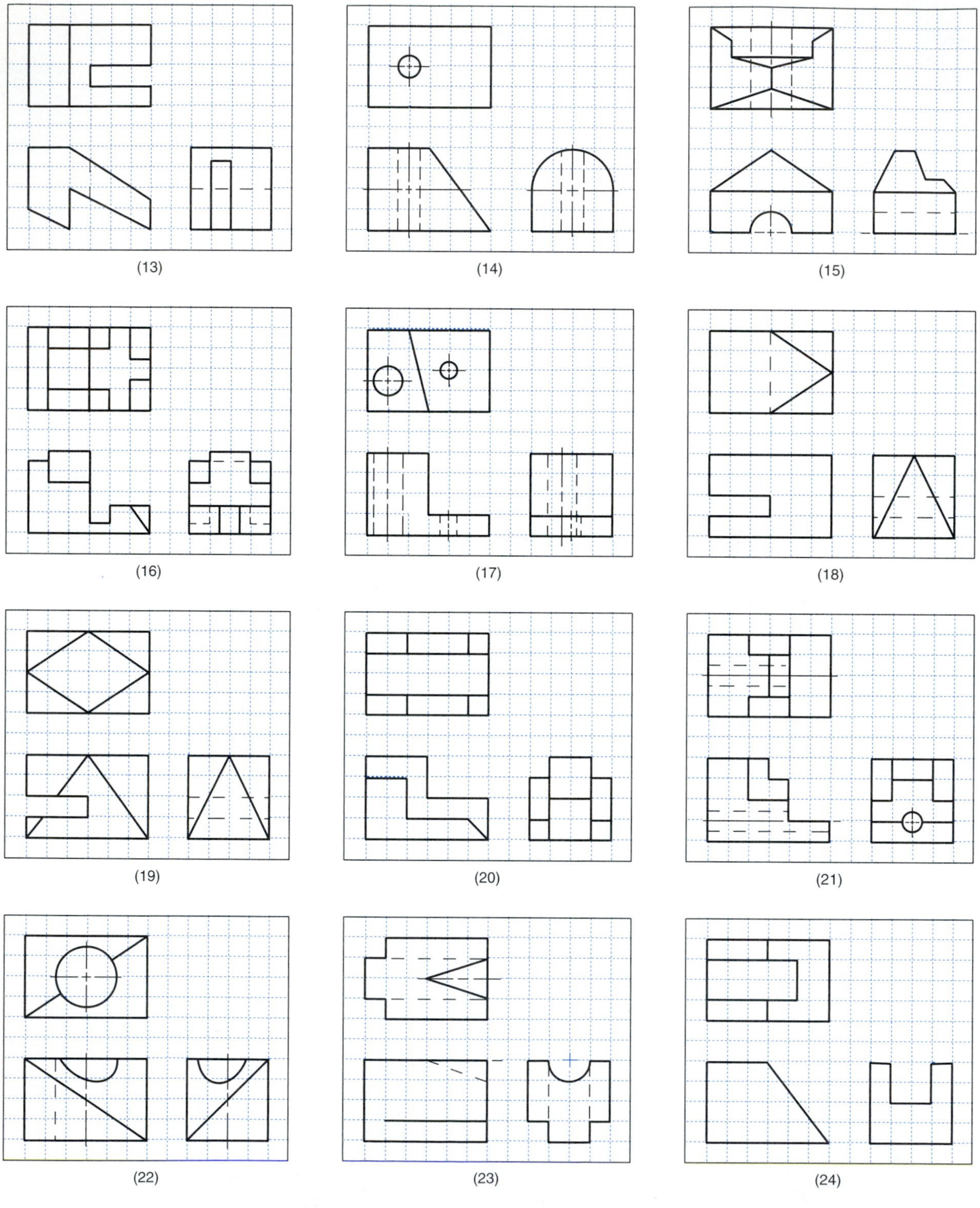

Figure 7.22 Continued

Orthographic Sketch Paper

Sketch Number: ____________________

Name: ____________________

Div/Sec: ____________________

Date: ____________________

Orthographic Sketch Paper

Sketch Number: ____________________

Name: ____________________

Div/Sec: ____________________

Date: ____________________

Orthographic Sketch Paper

Sketch Number: ______

Name: ______

Div/Sec: ______

Date: ______

Orthographic Sketch Paper

Sketch Number: ____________________

Name: ____________________

Div/Sec: ____________________

Date: ____________________

Orthographic Sketch Paper

Sketch Number: ______________________

Name: ______________________

Div/Sec: ______________________

Date: ______________________

Isometric Sketch Paper

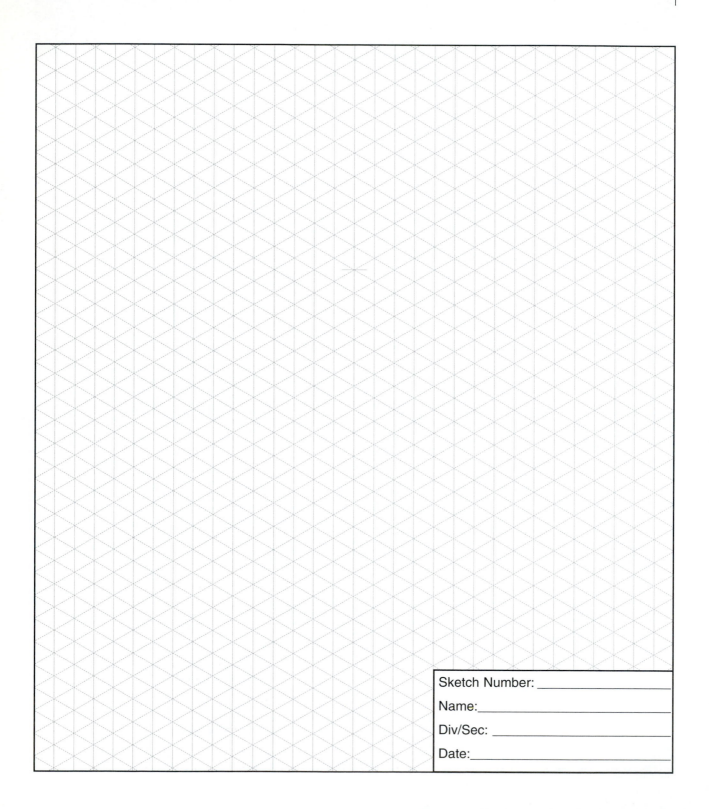

Sketch Number: ____________________

Name: ____________________

Div/Sec: ____________________

Date: ____________________

Isometric Sketch Paper

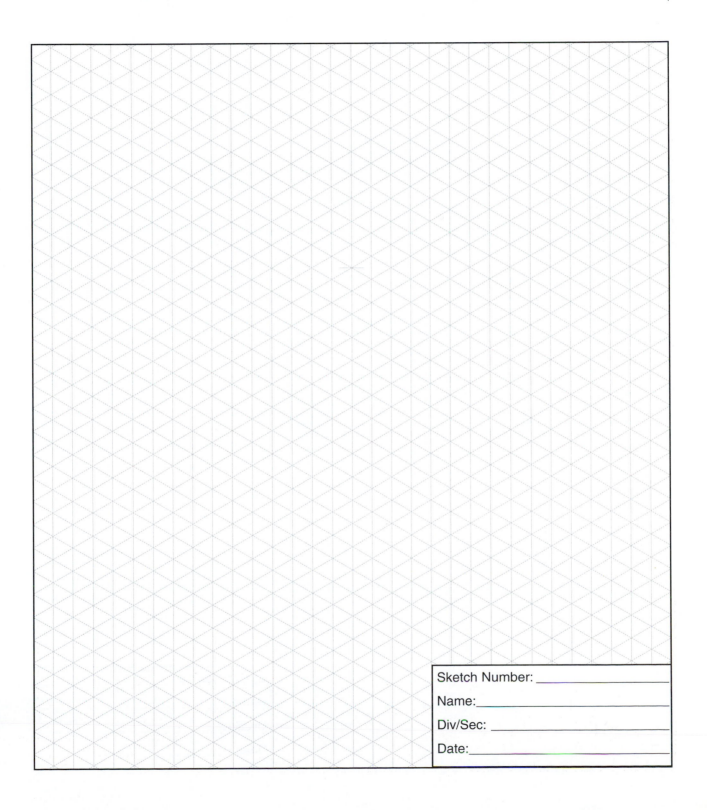

Sketch Number: ____________________

Name: ____________________________

Div/Sec: __________________________

Date: _____________________________

Isometric Sketch Paper

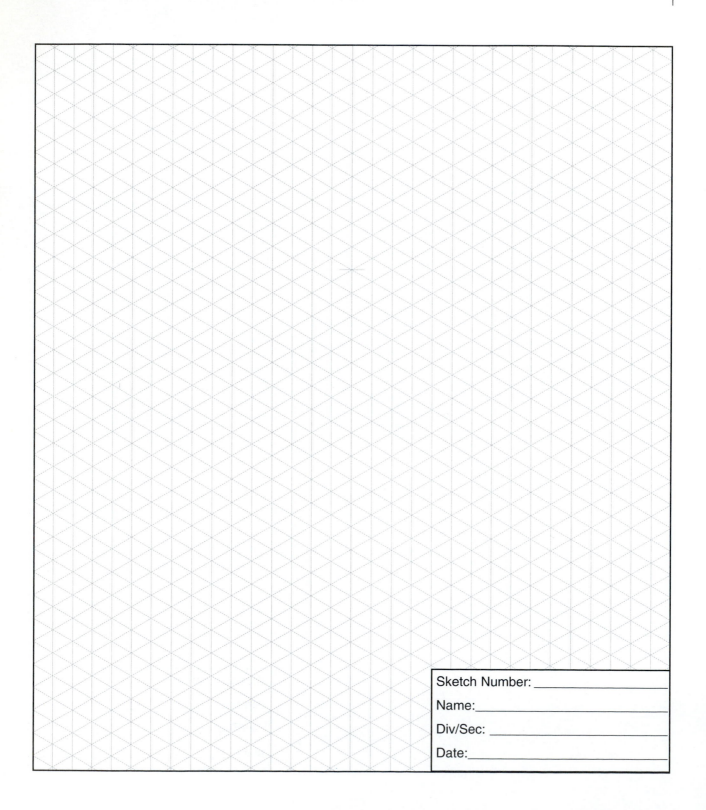

Sketch Number: ____________________

Name: ____________________

Div/Sec: ____________________

Date: ____________________

Isometric Sketch Paper

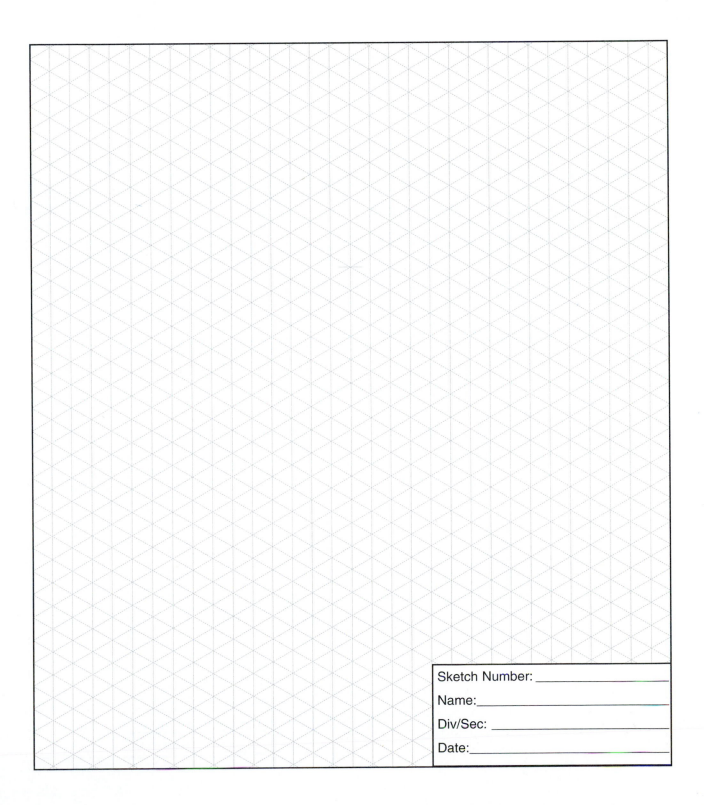

Sketch Number: ____________________

Name: ____________________

Div/Sec: ____________________

Date: ____________________

Isometric Sketch Paper

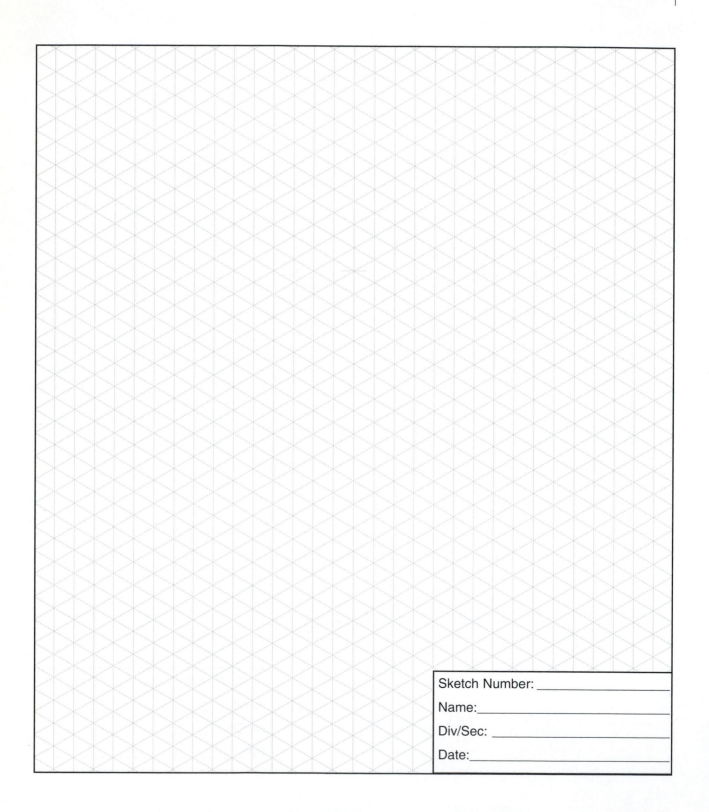

Sketch Number: ____________________

Name: ____________________

Div/Sec: ____________________

Date: ____________________

Problem Worksheet

Sketch Number: ____________________

Name: ____________________

Div/Sec: ____________________

Date: ____________________

Problem Worksheet

Sketch Number: ____________________

Name: ____________________

Div/Sec: ____________________

Date: ____________________

Problem Worksheet

Sketch Number: ____________________

Name: ____________________

Div/Sec: ____________________

Date: ____________________

Problem Worksheet

Sketch Number: ____________________

Name: ____________________

Div/Sec: ____________________

Date: ____________________

Index

D

E

F

G

H

I

L

M

T

U

V

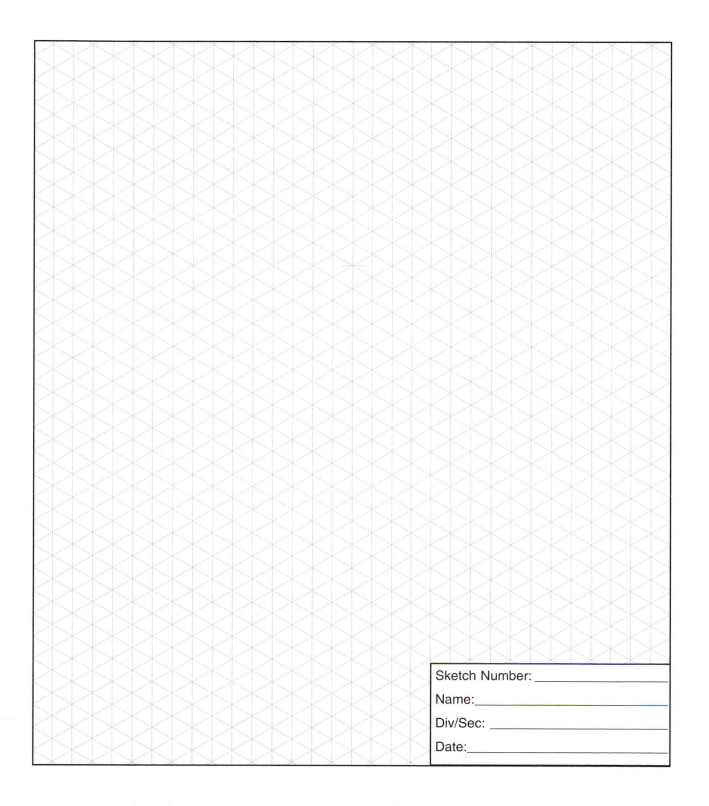

Sketch Number: ____________________

Name: ____________________

Div/Sec: ____________________

Date: ____________________

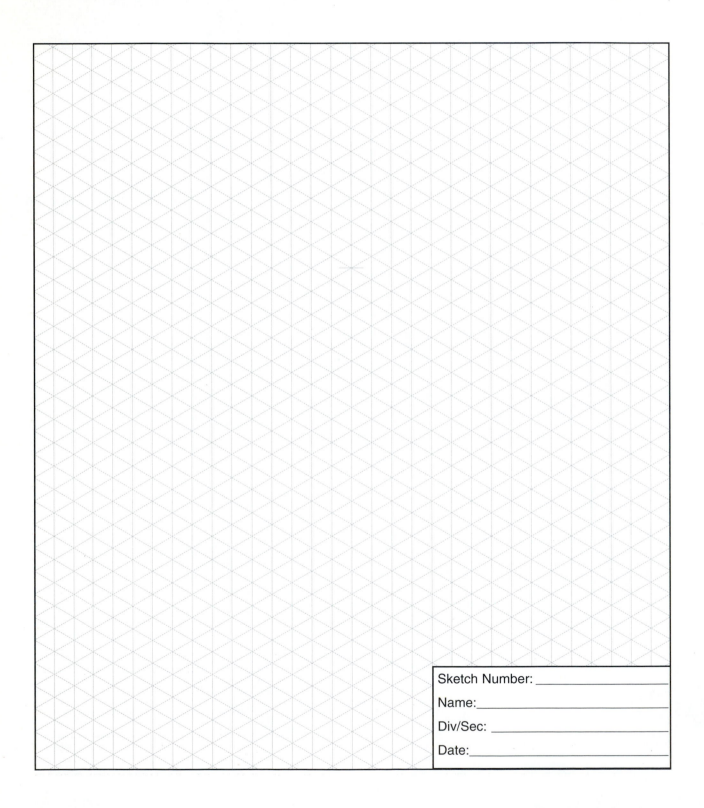

Sketch Number: ____________________

Name: ____________________

Div/Sec: ____________________

Date: ____________________

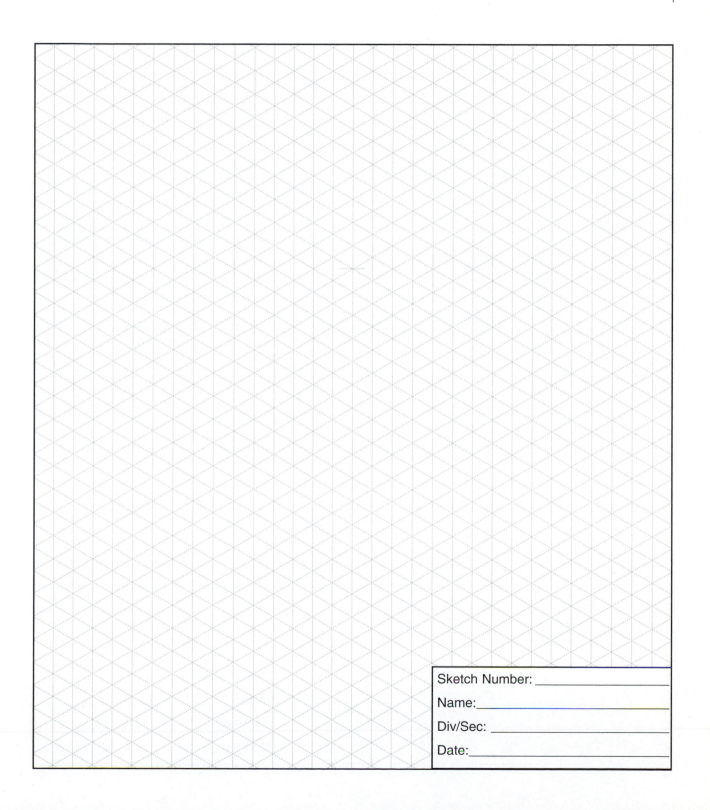

Sketch Number: ____________________

Name: ____________________

Div/Sec: ____________________

Date: ____________________

Isometric Sketch Paper

Sketch Number: ____________________

Name: ______________________________

Div/Sec: ___________________________

Date: ______________________________

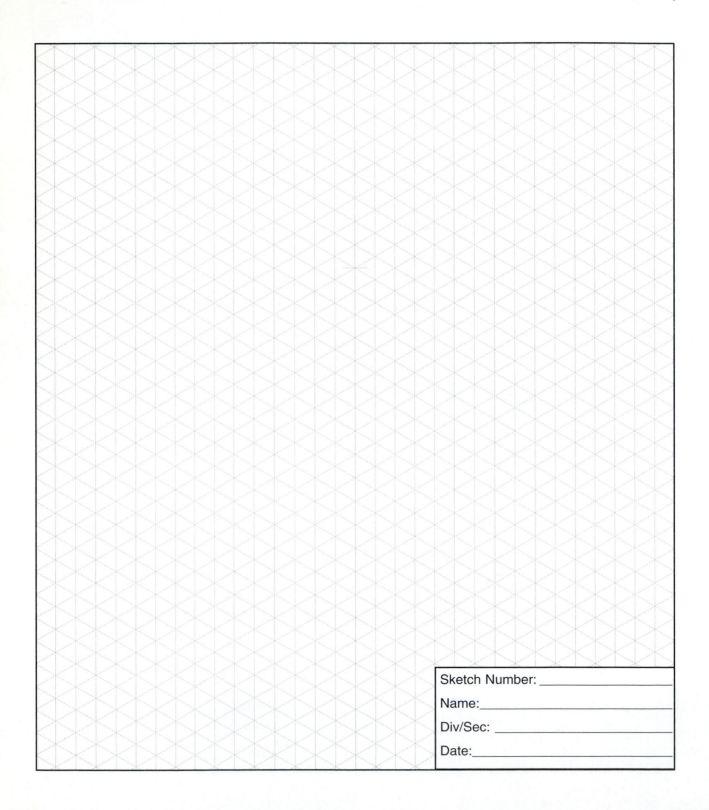

Isometric Sketch Paper

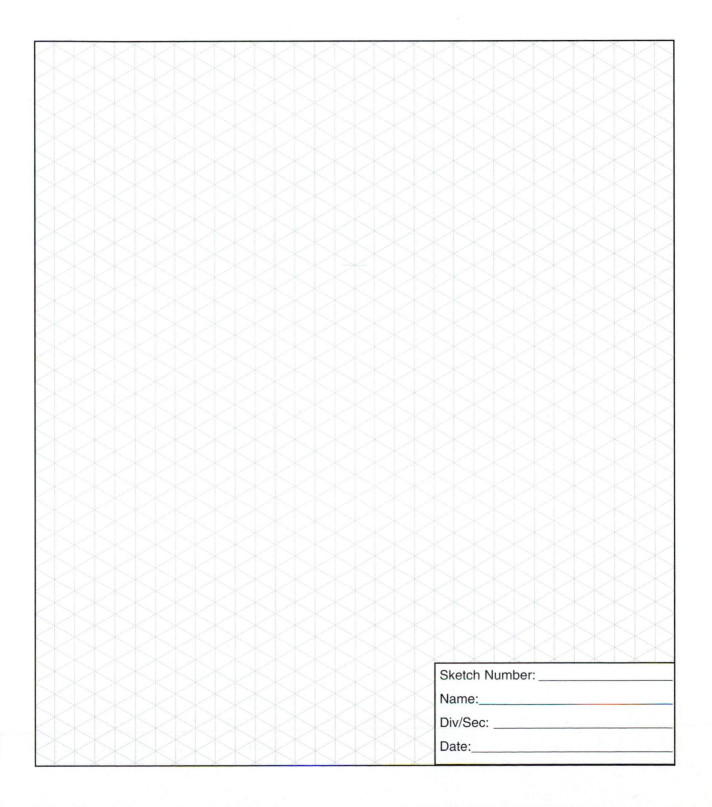

Sketch Number: ____________________

Name: ____________________

Div/Sec: ____________________

Date: ____________________

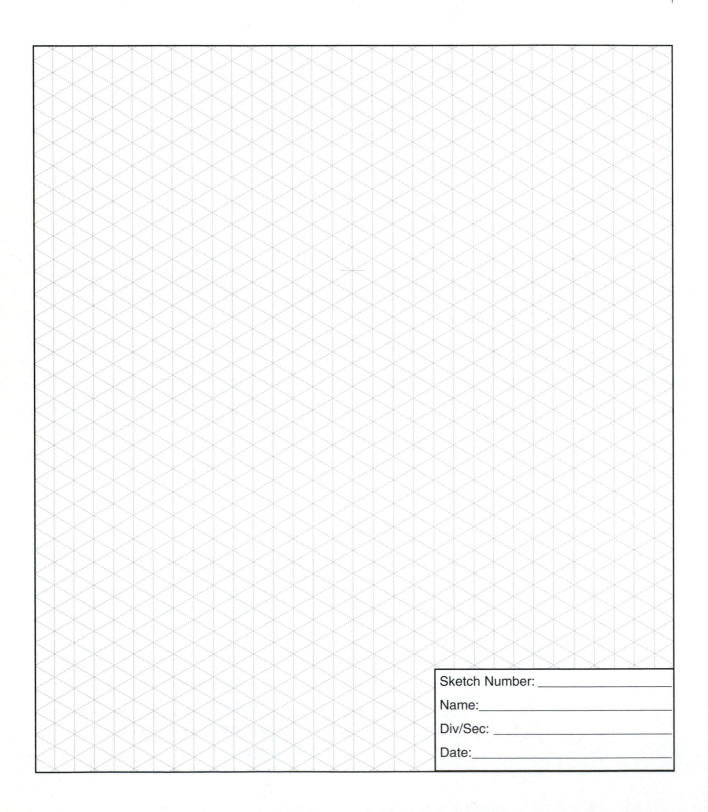

Sketch Number: ______________________

Name: ______________________

Div/Sec: ______________________

Date: ______________________

Isometric Sketch Paper

Sketch Number: ______________________

Name: ______________________

Div/Sec: ______________________

Date: ______________________

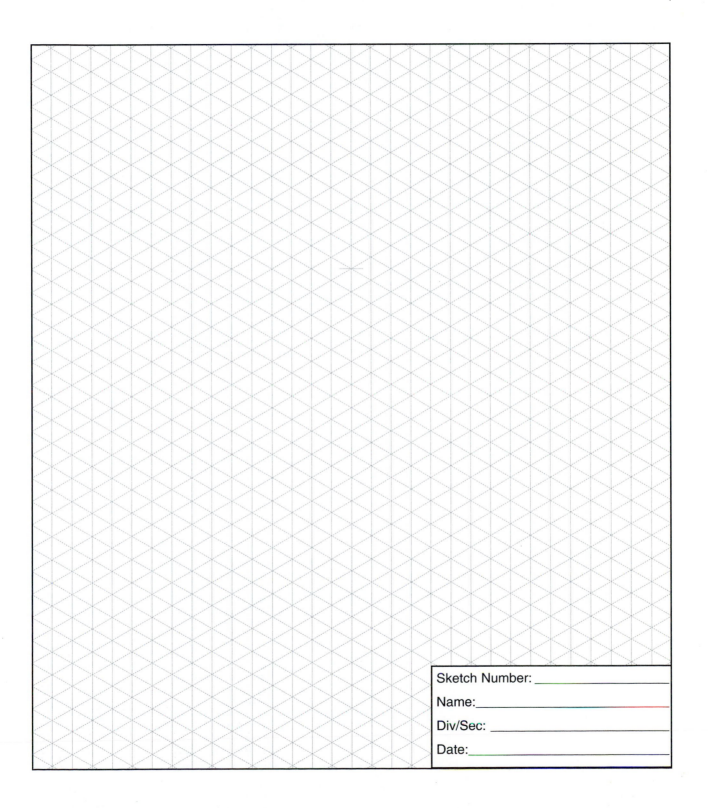

Isometric Sketch Paper

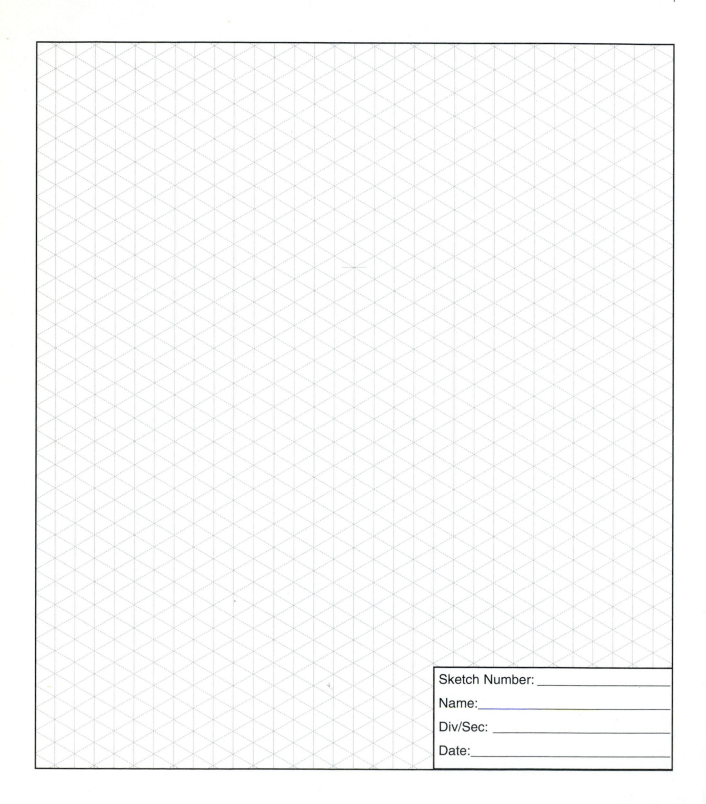

Sketch Number: ____________________

Name: ____________________

Div/Sec: ____________________

Date: ____________________